A. S. *Basra, PhD*
L. S. *Randhawa, PhD*
Editors

Quality Improvement in Field Crops

Quality Improvement in Field Crops has been co-published simultaneously as *Journal of Crop Production*, Volume 5, Numbers 1/2 (#9/10) 2002.

Pre-publication
REVIEWS,
COMMENTARIES,
EVALUATIONS . . .

"This book provides an exposition of the various factors that define seed quality in major crop plants, especially the cereal grains and the grain legumes, but also reviewed are the refinement of sugar beet roots, assessment of cotton fiber quality, and attributes of forage quality for animal feeds. Each chapter contains a wealth of historical data and a quick look at how modern molecular-based methods may be adopted to improve quality. The book emphasizes that quality is variously defined depending on the crop, ranging from how seed storage proteins of wheat define breadmaking quality of wheat flour to the long list of native antinutrition or toxin compounds found in beans and other leguminous crop plants.

This book will be a valuable reference for biotechnologists who are interested in the critical properties of plants that must be maintained to make them valuable food, feed, and fiber crops while also giving insights to defining targets for improvement by gene transformation. The book emphasizes genetics and breeding techniques for crop quality improvement, but the role of the crop production environment as a major determinant is recognized and discussed for some of the the the crops."

Calvin O. Qualset, PhD
Director
Genetics Resources Conservation Program
Professor Emeritus
University of California, Davis

Quality Improvement
in Field Crops

Quality Improvement in Field Crops has been co-published simultaneously as *Journal of Crop Production*, Volume 5, Numbers 1/2 (#9/10) 2002.

The *Journal of Crop Production* Monographic "Separates"

Below is a list of "separates," which in serials librarianship means a special issue simultaneously published as a special journal issue or double-issue *and* as a "separate" hardbound monograph. (This is a format which we also call a "DocuSerial.")

"Separates" are published because specialized libraries or professionals may wish to purchase a specific thematic issue by itself in a format which can be separately cataloged and shelved, as opposed to purchasing the journal on an on-going basis. Faculty members may also more easily consider a "separate" for classroom adoption.

"Separates" are carefully classified separately with the major book jobbers so that the journal tie-in can be noted on new book order slips to avoid duplicate purchasing.

You may wish to visit Haworth's website at . . .

http://www.HaworthPress.com

. . . to search our online catalog for complete tables of contents of these separates and related publications.

You may also call 1-800-HAWORTH (outside US/Canada: 607-722-5857), or Fax 1-800-895-0582 (outside US/Canada: 607-771-0012), or e-mail at:

getinfo@haworthpressinc.com

Quality Improvement in Field Crops, edited by A. S. Basra, PhD, and L. S. Randhawa, PhD (Vol. 5, No. 1/2 #9/10, 2002). *Examines ways to increase nutritional quality as well as volume in field crops.*

Allelopathy in Agroecosystems, edited by Ravinder K. Kohli, PhD, Harminder Pal Singh, PhD, and Daizy R. Batish, PhD (Vol. 4, No. 2 #8, 2001). *Explains how the natural biochemical interactions among plants and microbes can be used as an environmentally safe method of weed and pest management.*

The Rice-Wheat Cropping System of South Asia: Efficient Production Management, edited by Palit K. Kataki, PhD (Vol. 4, No. 1 #7, 2001). *This book critically analyzes and discusses production issues for the rice-wheat cropping system of South Asia, focusing on the questions of soil depletion, pest control, and irrigation. It compiles information gathered from research institutions, government organizations, and farmer surveys to analyze the condition of this regional system, suggest policy changes, and predict directions for future growth.*

The Rice-Wheat Cropping System of South Asia: Trends, Constraints, Productivity and Policy, edited by Palit K. Kataki, PhD (Vol. 3, No. 2 #6, 2001). *This book critically analyzes and discusses available options for all aspects of the rice-wheat cropping system of South Asia, addressing the question, "Are the sustainability and productivity of this system in a state of decline/stagnation?" This volume compiles information gathered from research institutions, government organizations, and farmer surveys to analyze the impact of this regional system.*

Nature Farming and Microbial Applications, edited by Hui-lian Xu, PhD, James F. Parr, PhD, and Hiroshi Umemura, PhD (Vol. 3, No. 1 #5, 2000). *"Of great interest to agriculture specialists, plant physiologists, microbiologists, and entomologists as well as soil scientists and evnironmentalists. . . . very original and innovative data on organic farming." (Dr. André Gosselin, Professor, Department of Phytology, Center for Research in Horticulture, Université Laval, Quebec, Canada)*

Water Use in Crop Production, edited by M.B. Kirkham, BA, MS, PhD (Vol. 2, No. 2 #4, 1999). *Provides scientists and graduate students with an understanding of the advancements in the understanding of water use in crop production around the world. You will discover that by utilizing good management, such as avoiding excessive deep percolation or reducing runoff by increased infiltration, that even under dryland or irrigated conditions you can achieve improved use of water for greater crop production. Through this informative book, you will discover how to make the most efficient use of water for crops to help feed the earth's expanding population.*

Expanding the Context of Weed Management, edited by Douglas D. Buhler, PhD (Vol. 2, No. 1 #3, 1999). *Presents innovative approaches to weeds and weed management.*

Nutrient Use in Crop Production, edited by Zdenko Rengel, PhD (Vol. 1, No. 2 #2, 1998). *"Raises immensely important issues and makes sensible suggestions about where research and agricultural extension work needs to be focused." (Professor David Clarkson, Department of Agricultural Sciences, AFRC Institute Arable Crops Research, University of Bristol, United Kingdom)*

Crop Sciences: Recent Advances, Amarjit S. Basra, PhD (Vol. 1, No. 1 #1, 1997). *Presents relevant research findings and practical guidance to help improve crop yield and stability, product quality, and environmental sustainability.*

Quality Improvement in Field Crops

A. S. Basra, PhD
L. S. Randhawa, PhD
Editors

Quality Improvement in Field Crops has been co-published simultaneously as *Journal of Crop Production*, Volume 5, Numbers 1/2 (#9/10) 2002.

CRC Press
Taylor & Francis Group
Boca Raton London New York

CRC Press is an imprint of the
Taylor & Francis Group, an **informa** business

Published by

Food Products Press®, 10 Alice Street, Binghamton, NY 13904-1580 USA

Food Products Press® is an imprint of The Haworth Press, Inc., 10 Alice Street, Binghamton, NY 13904-1580 USA.

Quality Improvement in Field Crops has been co-published simultaneously as *Journal of Crop Production*, Volume 5, Numbers 1/2 (#9/10) 2002.

The development, preparation, and publication of this work has been undertaken with great care. However, the publisher, employees, editors, and agents of The Haworth Press and all imprints of The Haworth Press, Inc., including The Haworth Medical Press® and Pharmaceutical Products Press®, are not responsible for any errors contained herein or for consequences that may ensue from use of materials or information contained in this work. Opinions expressed by the author(s) are not necessarily those of The Haworth Press, Inc. With regard to case studies, identities and circumstances of individuals discussed herein have been changed to protect confidentiality. Any resemblance to actual persons, living or dead, is entirely coincidental.

Cover design by Thomas J. Mayshock Jr.

Library of Congress Cataloging-in-Publication Data

Quality improvement in field crops / A.S. Basra, L.S. Randhawa, editors.
     p. cm.
     "Improvement in field crops has been co-published simultaneously as Journal of crop production, volume 5, numbers 1/2 (#9/10) 2002."
     Includes bibliographical references.
     ISBN 1-56022-100-3 (hard : alk. paper)–ISBN 1-56022-101-1 (pbk : alk. paper)
     1. Field crops. 2. Crop improvement. I. Basra, Amarjit S. II. Randhawa, L. S.
SB185.Q35 2002
631.5'23–dc21

2002005988

Indexing, Abstracting & Website/Internet Coverage

This section provides you with a list of major indexing & abstracting services. That is to say, each service began covering this periodical during the year noted in the right column. Most Websites which are listed below have indicated that they will either post, disseminate, compile, archive, cite or alert their own Website users with research-based content from this work. (This list is as current as the copyright date of this publication.)

Abstracting, Website/Indexing Coverage . . . . Year When Coverage Began

- *AGRICOLA Database <www.natl.usda.gov/ag98>* 1998

- *BIOBASE (Current Awareness in Biological Science)*
  <URL: http://www.elsevier.nl> 1998

- *BUBL Information Service, an Internet-Based Information*
  Service for the UK higher education community
  <URL:http://bubl.ac.uk/> 2000

- *Cambridge Scientific Abstracts (Water Resources*
  Abstracts/Agricultural & Environmental Biotechnology
  Abstracts) <www.csa.com> 2001

- *Chemical Abstracts Services <www.cas.org>* 1998

- *CNPIEC Reference Guide: Chinese National Directory*
  of Foreign Periodicals 1998

- *Derwent Crop Production File* 1998

- *Environment Abstracts. Available in print–CD-ROM–*
  on Magnetic Tape. For more information check:
  <www.cispubs.com> 1998

(continued)

Special Bibliographic Notes related to special journal issues (separates) and indexing/abstracting:

- indexing/abstracting services in this list will also cover material in any "separate" that is co-published simultaneously with Haworth's special thematic journal issue or DocuSerial. Indexing/abstracting usually covers material at the article/chapter level.
- monographic co-editions are intended for either non-subscribers or libraries which intend to purchase a second copy for their circulating collections.
- monographic co-editions are reported to all jobbers/wholesalers/approval plans. The source journal is listed as the "series" to assist the prevention of duplicate purchasing in the same manner utilized for books-in-series.
- to facilitate user/access services all indexing/abstracting services are encouraged to utilize the co-indexing entry note indicated at the bottom of the first page of each article/chapter/contribution.
- this is intended to assist a library user of any reference tool (whether print, electronic, online, or CD-ROM) to locate the monographic version if the library has purchased this version but not a subscription to the source journal.
- individual articles/chapters in any Haworth publication are also available through the Haworth Document Delivery Service (HDDS).

DEDICATION

This volume is dedicated in honor of Dr. Gurdev Singh Khush, the world's foremost rice breeder and architect of the Green Revolution in rice farming, for his lifelong contributions to increase the quality, quantity, and availability of food for billions of people. Dr. Khush was born on 22 August 1935 at Rurkee, Punjab, India, and was raised on a wheat farm. After graduating in 1955 with a BS in Agriculture from the renowned Punjab Agricultural University of India, he went to the University of California, Davis, receiving his PhD in Genetics in 1960 under the direction of G. Ledyard Stebbins. Dr. Khush then worked on the faculty of this university as an Assistant Geneticist (1960-1967), and carried out research on the cytogenetics of tomatoes in collaboration with C. M. Rick. As a result of these investigations, all the linkage groups of tomato were associated with their respective chromosomes, all the centromeres were mapped, 129 genes were delimited to respective chromosome arms, and orientation of all the linkage groups were determined.

In 1967, Dr. Khush joined the International Rice Research Institute (IRRI) as a plant breeder, became head of Plant Breeding Department in 1972, and was promoted to Principal Scientist of the Institute in 1986, a distinction conferred on only seven of more than 500 scientists who are either currently working or have earlier worked at IRRI. He has since played a key role in developing more than 300 rice varieties with a dedicated effort to keep rice production ahead of burgeoning population growth, particularly in the developing world.

It is a measure of Dr. Khush's stature as the world's foremost rice breeder that, in any rice field, anywhere in the world, there is a 60 percent chance that the rice was either bred at IRRI under his leadership or developed from IRRI varieties. One of them, IR36, released in 1976 became the most widely planted variety of rice, or of any other food crop, the world has ever known. It was planted on 11 million hectares in Asia in the 1980s, yielding an additional five million tons of rice a year, boosting rice farmers' incomes by US$1 billion, and, because of its resistance to pests, saving an estimated US$500 million a year in insecticide costs. IR64 later replaced IR36, as the world's most popular rice variety and IR72, released in 1990, became the world's highest-yielding rice variety. Dr. Khush's final work, the creation of a new plant type of rice, is almost complete. The plants are already yielding strongly in temperate areas of China, and are expected to be ready for farmers in tropical Asia by 2005.

Dr. Khush has trained numerous rice breeders, supervised 20 postdoctoral fellows, and served as a major professor of 15 PhD degree candidates, 25 MSc students, and 60 non-degree trainees. He has visited 50 rice growing countries to observe rice research and production systems and served as a consultant to 15 national rice improvement programs.

Dr. Khush has written three books, edited five books, and authored 25 review articles, 70 book chapters, and 155 research papers in refereed journals. His book on *Cytogenetics of Aneuploids* published by Academic Press, New York, is used widely as a text.

Dr. Khush has become one of the world's most decorated scientists, winning the Japan Prize in 1987, International Service in Agronomy Award in 1989, the World Food Prize in 1996, and both the Wolf Prize from Israel and the Padma Shri Award from his native India in 2000, and the China International Scientific and Technological Cooperation Award for 2001. He is a Fellow of the Royal Society of London, The Third World Academy of Sciences, Indian National Science Academy, American Society of Agronomy, and Foreign Associate of US National Academy of Sciences. He is a member of the Scientific Advisory Committee to International Foundation for Science, Sweden, The Third World Academy of Sciences, Italy, Department of Biotechnology, Government of India, and the Rockefeller Foundation's International Program on Biotechnology.

The Nobel laureate, Dr. Norman Borlaug, has summed up Dr. Khush's career by saying, "The impact of Dr. Khush's work upon the lives of the world's poorest people is incalculable."

Quality Improvement in Field Crops

CONTENTS

ABOUT THE EDITORS

A. S. Basra, PhD, is an eminent plant physiologist at the Punjab Agricultural University in Ludhiana, India. He is currently a visiting scientist at the University of California, Davis, USA. His outstanding work on seed quality and cotton fiber quality has been internationally recognized. Dr. Basra has more than 80 research publications to his credit, and is the lead editor of 11 books on topical subjects in plant/crop science. He is the Founding Editor-in-Chief of the *Journal of Crop Production* and the *Journal of New Seeds* (Food Products Press). He is a member of the American Society of Agronomy, the Crop Science Society of America, the American Society of Plant Biologists, the American Association for the Advancement of Sciences, the American Institute of Biological Sciences, the New York Academy of Sciences, the Australian Society of Plant Physiologists, the American Society for Horticultural Science, and the International Society of Horticultural Science. A decorated scientist who has received several coveted awards and honors, Dr. Basra has made scientific visits to several countries fostering cooperation in agricultural research at the international level.

L. S. Randhawa, PhD, is a plant breeder with Sutter Seeds of Yuba City, California. Previously, he has worked as Professor of Plant Breeding at the Punjab Agricultural University in Ludhiana, India; as Head of the Central Institute for Cotton Research, Regional Station in Sirsa, India; and as Commonwealth Academic Staff Fellow at the University of Nottingham in the United Kingdom. He has developed 7 varieties and 2 hybrids of cotton including the world's first leaf curl virus resistant American cotton hybrid, LHH 144. He has been invited to present papers at 19 national and international conferences including World Cotton Research Conference I (Brisbane, Australia) and II (Athens, Greece). Dr. Randhawa has also served as Co-Chair of the Asian Cotton Research and Development Association of the International Cotton Advisory Committee and as a member of its working group on Cotton Biotechnology.

Preface

Crop yields have increased dramatically in the 20th century, powered mainly by changes in the genetic potential of the crop, and the way in which it has been agronomically managed. Nevertheless, the challenge to feed a world population that is expected to increase by about 2.5 billion people during the next 25 years is formidable. The majority of this projected population growth is expected to occur in developing countries, which are already witnessing increased pressure on unit land and reduction in the capacity of farmers to produce sufficient quantity of nutritive food for their families. Crop improvement efforts have often been limited to higher yields rather than improved nutritional quality.

Today, we face a new challenge in crop production because the rate of improvement (yield and/or composition) is linear while the rate of demand increase has become exponential. Recent analyses suggest that the rate of increase in yields of several crops may have dropped over the last decade. There has been a progressive decline in the annual rate of increase in cereal yield, so that at present, the annual rate of yield increase is below the rate of population increase. Despite significant achievements in food production, more than 800 million people consume less than 2,000 calories a day and are chronically undernourished. Nutritional security must become an essential component of food security worldwide.

On the other hand, with increased crop yields over the past fifty years, the end-uses have proliferated requiring special quality characteristics. With rising incomes, there is a growing demand from an increasingly informed, affluent and selective consumer for the availability of foods with improved flavor, safety, health or technological charac-

[Haworth co-indexing entry note]: "Preface." Basra, A. S., and L. S. Randhawa. Co-published simultaneously in *Journal of Crop Production* (Food Products Press, an imprint of The Haworth Press, Inc.) Vol. 5, No. 1/2 (#9/10), 2002, pp. xxiii-xxiv; and: *Quality Improvement in Field Crops* (ed: A. S. Basra, and L. S. Randhawa) Food Products Press, an imprint of The Haworth Press, Inc., 2002, pp. xv-xvi. Single or multiple copies of this article are available for a fee from The Haworth Document Delivery Service [1-800-HAWORTH, 9:00 a.m. - 5:00 p.m. (EST). E-mail address: getinfo@haworthpressinc.com].

teristics. New cultivars must match the defining quality characteristics of the targeted market class.

Significant progress in crop quality improvement has been achieved using conventional breeding methods, or the new biotechnologies, or a combination of both. The development of the new technologies for trait modification has opened the possibility of engineering a wide range of new food products with enhanced nutritional and functional value, which will have a tremendous impact on human and animal health in both developed and developing countries.

In this context, the primary aim of this volume is to provide an intelligent synthesis of the available information based on exciting research that has been going on to discover, characterize, modify and manipulate genes controlling key quality traits, and point out directions for future research. A group of leading authorities in the field has contributed their specialist knowledge highlighting the use of agronomic management, conventional plant breeding, and modern biotechnologies in quality improvement of important food, feed, and fiber crops. We wish to sincerely thank these authors who accepted our invitation to contribute to this special issue and succeeded in making an outstanding contribution.

We hope that this volume will provide the much needed state-of-the-art information about the important aspects of crop quality improvement, and will serve to stimulate crop scientists and students to enter in this exciting field in a common effort to understand and manipulate the quality traits to produce better quality food, feed and fiber products for enhanced quality of life in the 21st century.

A. S. Basra
L. S. Randhawa

Quality (End-Use) Improvement in Wheat: Compositional, Genetic, and Environmental Factors

R. J. Peña
R. Trethowan
W. H. Pfeiffer
M. van Ginkel

SUMMARY. Wheat provides nutrients and the raw materials for industrialized food production. Recent global economic trends and increases in urban population growth have led to an increased demand for wheat-based convenience foods (fast, ready-to-eat, frozen foods, etc.) and for new wheat-based products. These factors have resulted in a greater emphasis than ever on the end-use quality of wheat. This paper reviews grain compositional aspects influencing the processing and quality attributes of the main foods produced with wheat, as well as the breeding strategies and methodologies used to achieve germplasm with desirable end-use quality. Common wheat (*Triticum aestivum*) is used in bread (leavened, flat, and steamed), noodles, biscuits, and cakes. Durum wheat (*T. turgidum* L. var. *durum*) is used globally in alimentary pasta and regional foods (flat breads, couscous, and burghoul) in North Africa and West Asia. Grain characteristics (grain hardness, protein content/qual-

R. J. Peña is Cereal Chemist, R. Trethowan is Bread Wheat Breeder, W. H. Pfeiffer is Durum Wheat Breeder, and M. van Ginkel is Bread Wheat Breeder, International Maize and Wheat Improvement Center Institute (CIMMYT, INT), Apartado Postal 6-641, 06600 Mexico, D. F., Mexico.

Address correspondence to: R. J. Peña, CIMMYT, INT, Apartado Postal 6-641, 06600 Mexico, D. F., Mexico (E-mail: j.pena@cgiar.org).

[Haworth co-indexing entry note]: "Quality (End-Use) Improvement in Wheat: Compositional, Genetic, and Environmental Factors." Peña, R. J. et al. Co-published simultaneously in *Journal of Crop Production* (Food Products Press, an imprint of The Haworth Press, Inc.) Vol. 5, No. 1/2 (#9/10), 2002, pp. 1-37; and: *Quality Improvement in Field Crops* (ed: A. S. Basra, and L. S. Randhawa) Food Products Press, an imprint of The Haworth Press, Inc., 2002, pp. 1-37. Single or multiple copies of this article are available for a fee from The Haworth Document Delivery Service [1-800-HAWORTH, 9:00 a.m. - 5:00 p.m. (EST). E-mail address: getinfo@ haworthpressinc.com].

ity, enzymatic activity, etc.) play a moderate to important role in the processing and end-use quality of wheat-based products. Among these, gluten strength and extensibility, which are determined by glutenin (HMW and LMW) and gliadin composition, are two of the main factors that determine quality. The complex and generally additive nature of inheritance of most quality traits has led to the development of several indirect tests used in early and advanced generations to increase the frequency of high yielding lines with desirable quality attributes. Additionally, characterization of HMW and LMW glutenins and gliadins allows breeders to combine protein content and quality more effectively. The use of molecular-marker-assisted selection and genetic transformation is expected to accelerate the tailoring of new wheat varieties to meet specific end-use quality requirements. Accumulating desirable quality genes will help reduce genotype × environment effects on quality–presently among the major challenges confronting breeders. *[Article copies available for a fee from The Haworth Document Delivery Service: 1-800-HAWORTH. E-mail address: <getinfo@haworthpressinc.com> Website: <http://www. HaworthPress.com> © 2002 by The Haworth Press, Inc. All rights reserved.]*

KEYWORDS. Biotechnology in wheat quality, breeding methodology, common wheat, durum wheat, end-use quality, genotype × environment interaction, gluten proteins, grain quality, wheat-based foods

INTRODUCTION

Wheat was an important catalyst in the establishment of permanent settlements that gave rise to civilization. Wheat adapts to almost all climatic conditions, with the exception of the very warm tropics. Its wide adaptation and unique property of forming gluten, a viscoelastic storage protein complex that allows the production of diverse foods, makes wheat the most important food crop in the world. Although there are several wheat species within the genus *Triticum,* the most widely cultivated ones are the hexaploid *Triticum aestivum* L., and the tetraploid *Triticum turgidum* L. var. *durum,* known commercially as common wheat and durum wheat, respectively. The two species differ in genomic structure, grain composition, and end-use quality attributes. These differences affect the type of use given the grain from each species: durum wheat is used globally to make pasta and regionally in the production of flat breads and other important foods; common wheat flour is used for making bread and biscuits (cookies). There are also large differences in grain composition among cultivars of the same wheat species, and both quantitative and qualitative compositional differences can make a cultivar suitable or unsuitable for a given baking process or food type.

Wheat breeders must identify specific quality traits to satisfy market and processing demands. These demands are driven more than ever by the consumer. This article reviews the main uses of wheat and the improvement of grain compositional characteristics that affect quality.

CURRENT IMPORTANCE AND TRENDS IN END-USES OF WHEAT

Approximately 90% of the wheat produced in the world–about 585 million metric tons (World Grain, 1999)–is common wheat. Durum wheat production amounts to approximately 27 million tons (Morancho, 2000). Common wheat is generally milled into flour (refined and whole meal) and used for the production of diverse leavened and flat breads, biscuits (cookies), noodles, and other baked products. In contrast, durum wheat is generally milled into semolina (coarse grits) to manufacture alimentary pasta and couscous (cooked grits) in Arab countries. Some durum wheat flour is used in the production of medium-dense breads in Mediterranean and Middle Eastern countries (Quaglia, 1988; Qarooni, 1994).

For many years, several important wheat-based foods were specific to certain regions (e.g., Arabic flat breads, couscous, and oriental noodles) and unknown outside their region of origin. However, migrations from rural to urban areas have permitted the exchange of cultural and dietary habits (Montague, 1998). Today's consumers, particularly those living in urban areas, look for more healthy, nutritious foods and/or convenience foods (frozen foods, instant noodles, etc.). Newly marketed wheat-based foods, such as noodles and flat breads in Europe, the Americas and Australia, or leavened breads and wheat-based fast foods in Asia, are easily accepted by urban populations. For example, the "flour tortilla," a wheat flat bread from Mexico, has become one of the fastest growing products of the baking industry in the U.S.A. (Kohn, 2000).

CLASSIFICATION AND GRAIN COMPOSITIONAL CHARACTERISTICS

Wheat Types and Classes

For trading purposes, wheat is classified into distinct categories of endosperm hardness (soft, semi-hard, and hard) and grain color (red, white, and amber). Wheat can be further divided into subclasses based on growing habit (spring or winter). Each wheat subclass may be grouped into grades that are used to adjust the basic price of a wheat stock, by applying premiums or penalties based on physical characteristics of the wheat lot. Wheat grades are determined by the purity of a wheat class or subclass, by the effects of external factors on grain soundness (rain, heat, frost, insect, and mold damage), and by the cleanliness (dockage, foreign material) of the wheat lot. All these grading factors affect both flour and semolina milling quality. Grain protein content and alpha-amylase activity–enzymatic activity associated with grain germination–are frequently considered grading factors in wheat trading. These two factors are important in determining the end-use properties of wheat and can be tested rapidly.

Grain Hardness

Grain hardness is determined by the packing of grain components in the endosperm cells. It is often referred to as the resistance of the grain to an applied fracturing force or to the energy required to reduce the grain sample into fine particles (whole meal, semolina, refined flour). In general, durum wheat has a harder endosperm than hard-grained common wheat. Grain hardness is a quality trait associated with the milling properties of wheat (Miller et al., 1982; Finney et al., 1987), with the water absorption capacity of flour/semolina, and with the baking quality of the resulting dough.

Milling duration, milling energy requirements, and the level of starch damage produced in the milled flour are all influenced by grain hardness. Hard wheat requires longer milling time, more milling energy and produces a larger amount of damaged starch compared to soft wheat.

A 15 KD protein attached to the surface of the starch granule is associated with grain hardness in common wheat; starch from soft wheat tends to have more of this protein than starch from hard wheat (Greenwell and Schofield, 1986). The actual role of the 15 KD protein (controlled by a gene on chromosome 5DS) in determining grain hardness is not well understood (Rahman et al., 1991).

Starch

Native starch, the main component of the wheat grain (70-75%, dry weight), has little influence on the differences in functional properties existing among flours used in bread, cookie, and cake making; however, it plays an important role in the textural properties of flour noodles (Lee et al., 1987). The amylose and amylopectin components of starch damaged mechanically during flour milling interact with other constituents of the baking formula, thereby influencing the water absorption and fermentation time requirements of bread-making dough and the staling and crumb textural properties of bread. A small amount of damaged starch is desirable in bread-making flours but highly undesirable in cookie and cake-making flours, as it may considerably reduce the expansion capacity of cookie dough (Miller and Hoseney, 1997) and cake batter during baking. For this reason, the cookie and cake industries use flour from soft-grained wheat, which has very low levels of mechanically damaged starch and, consequently, low flour water absorption.

On the other hand, the swelling and pasting properties of native starch influence the eating quality of wheat flour noodles, particularly white noodles that are smooth, soft, and slightly elastic (Lee et al., 1987; Bhattacharya and Corke, 1996). High starch swelling and desirable noodle softness have been associated with the absence of the granule bound starch synthase (GBSS) protein controlled by genes on chromosome 4A, commonly referred to as Null4A (Ross et al., 1996). Other starch and protein related factors also influence starch swelling and noodle softness and may mask the quality effects of Null4A (Ross et al., 1996). Thus screening for Null4A may not be effective in improving starch properties for noodle making.

Proteins

Protein content in wheat varies between 8 and 17%. The factors that affect protein content are the genetic make-up of the variety, permanent and variable environmental factors, and crop management practices. Most (78-85%) endosperm protein is gluten, a very large complex primarily comprised of polymeric (multiple polypeptide chains linked by disulfide bonds) and monomeric (single-chain polypeptides) proteins known as glutenins and gliadins, respectively (MacRitchie, 1994). Glutenins, through inter-peptide disulfide bonding, confer elasticity, whereas gliadins, with their globular structure, confer viscous flow to the gluten complex (see Shewry and Tatham, 1997, for a review). Gluten is, therefore, responsible for most viscoelastic properties of wheat flour and semolina dough, and is the main factor determining the use of a wheat variety by the baking and pasta-making industries. Although variations in grain protein (gluten) content may significantly influence the dough strength of a wheat variety (Simmonds, 1989), protein quantity alone cannot explain differences in quality characteristics among wheat cultivars. Therefore, protein quality related to the polymeric/ monomeric protein ratio and the size of the aggregated protein polymer are important (see Weegels et al., 1996, for a review).

Wheat flour contains similar amounts of glutenins and gliadins, and any change in the glutenin/gliadin ratio may change its viscoelastic properties. The glutenin fraction is, however, the major protein factor responsible for variations in dough strength among wheat varieties (Fu and Sapirstein, 1996).

Genes in the complex *Glu-1* and *Glu-3* loci located in the group 1 chromosomes control glutenins (Table 1). In the *Glu-A1*, *Glu-B1*, and *Glu-D1* loci there are genes coding for 0 or 1, 1 or 2, and 2 high molecular weight (HMW) subunits of glutenin, respectively. The relationships between frequently found HMW subunits and dough strength and bread-making properties are well established (Payne, 1987; Shewry et al., 1992). Quality scores for several HMW subunits of common wheat are shown in Table 2. Low molecular weight (LMW) subunits of glutenin, controlled by genes of the *Glu-3* complex loci, are also important in determining gluten viscoelasticity (see Weegels et al., 1996, for a review). However, much remains unknown regarding the relationship between LMW glutenin subunit composition and gluten strength in common wheat. This is in part due to the larger number of LMW subunits in bread wheat compared to durum wheat (D genome proteins are not present in durum wheat). The large number of components (up to 7) that make up a single LMW subunit makes it practically impossible to resolve them by using the conditions for determining HMW composition, particularly if the gliadins that co-migrate with the LMW are not removed as a first step (Gupta et al., 1990). A new sequential protein ex-

TABLE 1. Wheat gluten proteins and their genetic control.

Proteins	Chromosome Arm			Locus		
Glutenins						
High molecular weight glutenins	1AL	1BL	1DL	*Glu-A1*	*Glu-B1*	*Glu-D1*
Low molecular weight glutenins	1AS	1BS	1DS	*Glu-A3*	*Glu-B3*	*Glu-D3*
Gliadins						
γ- and ω-gliadins	1AS	1BS	1DS	*Gli-A1*	*Gli-B1*	*Gli-D1*
α- and β-gliadins	6AS	6BS	6DS	*Gli-A2*	*Gli-B2*	*Gli-D2*

TABLE 2. Quality scores assigned to HMW glutenin subunits based on gluten quality-related parameters.[a]

Subunit	Score[b]	
	SDS-Sedimentation[c]	Alveograph W
Glu-A1		
2*	3	5
1	3	3
Null	1	2
Glu-B1		
17 + 18	3	6
7 + 9	2	5
7 + 8	3	-
6 + 8	1	1
7	1	2
20	-	1
Glu-D1		
5 + 10	4	6
5 + 12	-	2
2 + 12	2	2
3 + 12	2	-
4 + 12	1	1

[a]: Source: Pogna et al. 1992.
[b]: Higher value indicates better quality effect.
[c]: Source: Payne 1987.

traction procedure combined with ID SDS-PAGE has been developed to obtain well-resolved patterns of HMW and LMW subunits in a single gel (Singh et al., 1991). The procedure has facilitated the study of the relationship between genetic control and impact on grain quality of LMW subunits. This notwithstanding, it remains more difficult to determine LMW subunit composition than HMW composition.

In durum wheat, gluten quality is more closely associated with specific LMW subunits and gliadins than with specific HMW glutenins (Ruiz and Carrillo, 1995; Liu et al., 1996; Peña, 2000). However, some HMW subunits associated with strong gluten type and good bread making quality in durum wheat have been identified (Pogna et al., 1988; Peña, 2000). The role of specific LMW subunits on gluten strength in durum wheat has been determined; LMW-2 and its variants confer stronger gluten characteristics than LMW-1 (Pogna et al., 1988; Peña et al., 1994).

Genes in the *Gli-1* and *Gli-2* loci located in the group 1 and group 6 chromosomes, respectively, control gliadins (Table 1). Some allelic variants at the complex gliadin loci *Gli-1* have been found to influence gluten properties in common wheat (Sozinov and Poperelya, 1980; Branlard and Dardevet, 1985). However, there is still controversy regarding the effects of individual *Gli-1* alleles (Nieto-Taladriz et al., 1994). In durum wheat, gliadin γ-45 is associated with high gluten strength, while gliadin γ-42 tends to be associated with weaker gluten and poorer viscoelastic properties (Kosmolak et al., 1980). These associations are thought to be related to the tight linkage between *Glu-B3* and *Gli-B1*; the causal effects are attributed to allelic variations (LMW-1 and LMW-2) at *Glu-B3* (Payne et al., 1984).

Flour/Semolina Pigments

In many countries, flours used for bread and noodle making must be white with no discoloration or yellow pigment. In contrast, the yellow pigment (xantophylls or luteins) of semolina is extremely important in assessing durum wheat for pasta quality.

Enzymatic Activity

High alpha-amylase activity has a large negative effect on baking dough properties, as it hydrolyzes flour starch excessively. Grain with very high levels of amylase activity may be rejected as a food product and downgraded to feed grain in the market. Polyphenol oxidase (PPO), also referred to as tyrosinase, is an enzyme complex located in the bran layers of wheat. PPO may cause product discoloration in both oriental noodles and durum pasta through the oxidization of phenols into polyphenols, which are dark brown. Low levels of PPO activity ensure that fresh flour noodles and fresh pasta will maintain their luster. Common wheat generally has higher PPO levels than durum wheat (Bernier and Howes, 1994).

The yellowness of pasta may also be significantly reduced due to the oxidative degradation of yellow pigment (lutein) by the enzyme lipoxygenase (Irvine and Winkler, 1950). Lipoxygenase (LOX) oxidizes semolina yellow pigments during pasta making. Low LOX activity is more important than high grain yellow pigment content in retaining the yellow color in pasta (Borrelli et al., 1999).

Minor Grain Constituents

Lipids, pentosans, soluble proteins, and other minor grain constituents also play a role in determining wheat flour quality. Their effect on flour and dough functionality can be corrected, generally by adjusting ingredients (e.g., use of additives, improvers) before baking. Thus, they are not considered major factors in grain quality improvement.

WHEAT-BASED FOODS AND GRAIN QUALITY REQUIREMENTS

Flour and Semolina

Milling yield is an extremely important characteristic for millers, particularly in countries where leavened bread is a key product. This is because higher rates of flour extraction result in higher financial returns from the same volume of grain. Breeding lines with potential for release are often rejected in the final round of evaluation because milling yield is a percentage point less than the industry target. Acceptable flour extraction rates may vary from ~70 to 80%, depending on the country/region and the intended use of the flour. Refined flour specifications may also vary widely regarding ash content (.040-0.60%), granulometry, protein content, and levels of starch damage, depending on the intended use and, in some instances, local legislation. In countries where non-leavened breads (e.g., chapatis) are common, such as in the Indian Sub-Continent, whole wheat is milled with very little bran discarded. Grain test weight, thousand-kernel weight, and hardness are the main factors influencing milling yield (Finney et al., 1987). Soft wheat suffers less starch damage during milling than hard wheat and generally yields less flour. Durum wheat is harder than common wheat, resulting in higher semolina yields. Semolina is defined as the purified middling that will pass through a No. 20 U.S. sieve, of which not more than 3% will pass through a finer No. 100 U.S. sieve (Finney et al., 1987). Of 75% milled product, 80-85% should be semolina (semolina yield > 60%, on average about 65%). Ash values of durum semolina are economically important and higher (0.60-0.70%) than those of common wheat refined flours (0.40-0.60%); in countries such as Italy ash content in semolina is specified by law. One particularly important grading factor in semolina milling is the grain discoloration known as black point. This is caused by pathogens such as *Bipolaris* spp. and *Alternaria* spp., and can add undesirable black specks (particles from black point, bran pieces, or other visible foreign materials) to semolina. Up to 15% black point is acceptable. The presence of starchy or yellow berry kernels also reduces semolina (not flour) milling yields and negatively affects end-product quality, because starchy parts of the kernel have lower protein content than the vitreous portion. Minimum requirements for vitreousness vary from 75% to much higher proportions for top grades.

Bread

Wheat bread provides more nutrients to the world population than any other single food source (Pomeranz, 1987). Bread consumption is increasing globally. In developed countries, people consume bread, particularly bread prepared with whole-grain and multi-grain flours, as an inexpensive source of complex carbohydrates, dietary fiber, and proteins (Faridi and Faubion, 1995; Seibel, 1995). In developing countries, bread consumption increases with urban population growth, increased income, and the adoption of processed, convenience foods, particularly in China and Southeast Asia (Owens, 1997) and in North African and Middle Eastern countries (Prior, 1997). Bread consumption in Sub-Saharan Africa is low and varies widely from country to country. It is greater in wheat-producing countries (South Africa, Ethiopia, Sudan, Kenya, Zimbabwe, Tanzania, and Zambia) in Eastern and Southern Africa, and lower in those that rely on wheat imports.

All wheat-based bread types (leavened, flat, and steamed) are made from viscoelastic and cohesive dough prepared from refined or wholemeal flour. There are differences in specific grain quality requirements, processing conditions, and end-product properties within each bread type.

Leavened Breads

Leavened breads are popular in almost all parts of the world. They are made with wheat (and/or wheat-rye blends in Central and Eastern Europe) viscoelastic dough, which expands by the action of gas produced by yeast and/or other fermenting agents. Leavened breads are characterized by a small crust to crumb ratio. The combination of bread formula, dough viscoelastic properties, length of the fermentation stage, and oven-stage conditions determines size, volume, crust thickness, and crumb structure of a given type of bread. Pan breads (pan-type bread, hamburger, hot dog buns, etc.) generally have a thin to medium thin light-brown crust and a light, uniformly distributed crumb structure. Their shape is determined by the panning mold. In contrast, hearth breads, baked on the oven floor (French-type baguette, sweet rolls, croissants, etc.) generally have a semi-hard to hard crust and an unevenly distributed crumb structure. The shape given by hand to the dough determines the shape of hearth breads.

Hard to medium-hard grain is preferred for leavened breads. This is because the levels of damaged starch from these wheat classes result in the high dough water absorption desired by bakers. The type of bread and the bread-making process used, on the other hand, determines flour (or dough) strength requirements. In general, mechanized, high-speed mixing requires stronger dough than manual or semi-mechanized mixing. Hard to medium-hard wheat with strong gluten is more suitable for mechanized production of leavened breads such as pan type bread and hamburger and hot dog buns (Wrigley, 1991). Those with medium-strong gluten are suitable for semi-mechanized or manual production of hearth breads.

To develop wheat cultivars with suitable pan bread quality, breeders must select hard to medium hard-grained wheats with high milling yield, medium to high flour protein, and low yellow pigment. Flour water absorption should be high. Medium-strength and extensible doughs are also favored by most users (Finney et al., 1987). Breeders can use grain hardness estimates, flour protein (> 13%, Wrigley, 1994), medium to high values of SDS-sedimentation, and low yellow pigment scores to favorably skew their breeding populations in the early generations. At an intermediate stage in the breeding process (F4 to F7), when the number of lines is reduced, small-scale milling tests can be applied to further truncate the material. When advanced materials with suitable disease resistance and good yield potential are identified, a more exhaustive examination of physical dough and baking properties is justified.

Steamed Breads

Steamed bread is popular in China and throughout East Asia (Pomeranz, 1978; Lin et al., 1990). Steamed bread is spherical and possesses a spongy crumb and no crust. It is made with yeast-fermented viscoelastic dough that is steamed rather than baked. Steaming prevents the formation of the brown crust characteristic of baked breads. Steamed bread may be plain or filled with sweet bean paste, or with meat and/or vegetables (Nagao, 1995b). There are three main types of steamed bread; northern-style and southern-style Chinese steamed breads, and Cantonese and/or Southeast Asia steamed bread. Northern-style Chinese steamed bread is denser and chewier, while Cantonese style is softer and less cohesive (Crosbie et al., 1998). Southern-style Chinese and Cantonese and/or Southeast Asia steamed breads are consumed in southern China, Southeast Asia, Korea, and Japan.

To produce good steamed breads, the grain should be white in color, semi-hard with bright white flour. Milling yield should be high and flour protein intermediate (10-13%, Wrigley, 1994). Hard to semi-hard wheat possessing medium strong and extensible gluten is more suitable for northern Chinese steamed bread. Soft wheat with medium to low gluten strength is better suited to southern Chinese steamed bread, and steamed breads consumed in Southeast Asia (Lin et al., 1990; Nagao, 1995a; Huang et al., 1996; Crosbie et al., 1998). There are visual and texture requirements for steamed bread, but these cannot be selected for with any certainty in the early generations. Breeders should essentially apply the same early generation selection regime as for leavened bread. However, slightly softer grain can be retained and flours must have a very low level of yellow pigment. Final evaluation will require customer "feedback" on texture and appearance.

Flat Breads

Flat breads can be single layered (e.g., the Indian chapati, Mexican tortilla and Arabic tanoori) or double layered (e.g., the Egyptian baladi and Arabic or pita bread). Flat breads are characterized by a high crust to crumb ratio and are generally round or oval-shaped, and less than 1 cm to around 4-6 cm thick. All flat breads are made from viscoelastic doughs prepared with 100% wheat flour or with composite flours that include wheat and other cereals. Some flat breads are made of wheat flour (whole or refined), water, and salt. Others are fermented using sourdough starters or yeast, and still others contain chemical leavening agents. Flat breads may be thin and crisp (e.g., the Israeli matzo and the Iranian lavosh) or soft and flexible such as the Mexican flour tortilla and the Indian naan (Faridi, 1988; Qarooni, 1996; Singh and Kulshresta, 1996).

Flat breads are consumed in northern Europe (bannock, lefse, plank, etc.), North Africa (baladi, Arabic or pita, Moroccan, etc.), the Middle

East (barbari, bazlama, lavosh, tanoori, etc.), the Indian Sub-Continent (chapati, naan, etc.) and North and Central America (tortilla, etc.). Faridi (1988) published an excellent review of the various types of flat breads all over the world. Some flat breads remain ethnic foods, made at home using traditional clay or brick ovens, or baked on a hot clay or metal plate. Others have become very popular, given that they fit the needs of the fast food and/or convenience food industry. This is the case of the Mexican flour tortilla, Indian naan, and Egyptian baladi and Arabic or pita breads (Faridi, 1988; Steinberg, 1996). Mechanized, large-scale production of these breads satisfies the demand of urban consumers (Faridi, 1988; Kohn, 2000).

Breeders aiming to develop good flat bread products target white (red is at times acceptable), hard-grained wheat with high milling extraction and intermediate flour protein (11-13%, Wrigley, 1994). Hard to medium-hard wheats with medium-strong and extensible gluten are suitable for flat-type breads such as the two-layered Arabic baladi, and the single-layered Indian chapati, Mexican flour tortilla, etc. (Qarooni, 1996; Singh and Kulshresta, 1996). Flat bread doughs should be medium in strength and extensible. Breeders can essentially utilize the same early generation selection criteria as those used for leavened bread. The suitability of a variety for flat bread production will be determined once physical dough properties are determined.

Flour Noodles

Wheat flour noodles, a staple food in northern China, are widely consumed all over East Asia (Huang, 1996; Kim, 1996). To make noodles, a stiff, crumbly dough is developed by hand-kneading or slow mixing of the ingredients and then letting the dough rest for 30 min. A dough sheet is then made, cut into noodle strands, and boiled (or dried). Mechanized noodle production predominates in Japan, while most noodles in China are handmade (Nagao, 1995a). There are three main types of noodles: white salted (WSN), yellow alkaline (YAN), and instant (bag and cup noodles). White salted noodles (fresh, dried, boiled and steamed and dried) are made of flour, salt and water. In addition to the ingredients of WSN, YAN includes alkali (roughly 1% of alkaline salts, flour weight basis) to develop their characteristic yellow color. In Japan WSN are known as *udon*, and YAN as *ra-men* (Nagao, 1995b). WSNs are bright, creamy-white, smooth and soft, but slightly elastic to permit a clean bite. In contrast, YAN are light yellow to yellow, smooth but firmer and more elastic than WSN, and offer slightly more resistance to the bite.

The greater firmness and elasticity of YAN is due to the use of flours with higher protein content and stronger gluten (Huang, 1996; Nagao, 1995a). Instant noodles are steamed and fried, or steamed and dried. In instant noodle production, fresh noodles are waved and steamed in a steaming tunnel for 1-3 min prior to cutting. Non-fried instant noodles require longer steaming time than fried ones (Nagao, 1995b). Drying by frying requires the steamed noodles to be passed through a frying tunnel for approximately 1-2 min at 140-150°C for bag instant noodles and 1.5-2.0 min at 160°C for cup instant noodles (Kim, 1996; Nagao, 1995b).

Desirable noodle brightness and whiteness are achieved when low extraction (below 70%) flour having low ash content (less than 0.5%) is used. White salted noodles require intermediate flour protein, whereas YAN will range from high (> 13%) in Japan to low (> 9.5%) in southern China and Indonesia (Nagao, 1995a; Ross et al., 1996). Grain hardness also varies; in WSN soft to semi-hard grain is acceptable, whereas the grain must be hard to produce acceptable YAN. Starch properties are also important, particularly in WSN. Starch paste viscosity must be high (Lee et al., 1987); however, there is more flexibility in the production of yellow alkaline types where medium levels are acceptable. The reaction of flour to the addition of alkali is also important in making YAN, where a clear yellow color is sought. As with steamed breads, noodle texture and taste are very important (Finney et al., 1987), and doughs should generally be medium to strong.

In selecting for flour-noodle quality, breeders can use the same tests as those used for leavened breads; however, their interpretation will depend on the product being sought. This is particularly important when using protein and grain hardness as indicators of eventual noodle quality. The swelling power test may also be applied in preliminary screening, as it correlates well with starch pasting viscosity (McCormick et al., 1991). The role of flour pigment in determining suitable reaction to the addition of alkali is not well defined, so selection for increased yellow pigment may or may not be beneficial in the case of alkaline noodles.

Soft Wheat Products (Cookies and Cakes)

Cookies (biscuits), cakes, and other soft wheat products, such as those made with batter, are produced around the world in a very large variety of shapes, textures, sizes, and flavors. They are generally sweet goods where wheat flour represents 1/3 to 1/2 of the formula. Cookies are usually made with inelastic and stiff dough, but some cookies, cakes, pancakes, and waffles require thick, viscous batters. The development of viscoelasticity in cookie dough is prevented by the inclusion of large amounts of sugar and fat in the baking formula (sugar may equal as much as 2/3 and fat as much as 1/2 the amount of flour used in the formula). Both cookie spread and batter expansion are largely determined by the viscosity of the system and the availability of free water (water acts as a leavening agent as it evaporates) during the baking stage. Soft wheat flour (not hard wheat flour) is suitable for the production of cookies, cakes, and other products from batters, because it has limited ability to form viscoelastic dough coupled with low water absorption capacity (Miller and Hoseney, 1997). Air and chemicals are, in addition to water vapor, the leavening agents in these foods.

To make suitable cookies and cakes, wheats with low flour protein (8-10%, Wrigley, 1994) and weak, extensible dough, possessing low water absorption, are sought. Low water absorption can best be achieved with soft wheat flours characterized by low levels of damaged starch and low protein content. Breeders can use NIR analysis to select for soft grain and low flour protein. Low protein in combination with low SDS-sedimentation values can be used to skew populations favorably in the early generations. Eventually a cookie test is required to determine suitability for specific cookie and cake manufacture (Finney et al., 1987).

Durum Wheat Products

Durum wheat is eaten as alimentary pasta in most countries. The most commonly known pasta products are spaghetti and macaroni. Durum wheat is also a major staple food in the West Asia/North Africa (WANA) region where most of the world's durum wheat area is concentrated. In WANA durum wheat is used to make a range of local and processed products, such as pasta and couscous, made with semolina; burghoul, made with boiled and dried grain; flat bread, and several other minor local foods. Other uses of durum wheat include flat bread production in India (chapati) or Mexico (tortilla). In the Andean Region, durum wheat is consumed as a local dish called mote (boiled whole grain).

Alimentary Pasta

Although alimentary pasta originated in Italy, it is now consumed around the world. Pasta is made primarily (and, in several countries, exclusively) from durum wheat semolina. Semolina and water (~30% of semolina weight) are mixed to form a stiff dough that is extruded at high pressure and in a vacuum to produce long and short pasta products of diverse sizes and shapes. Fresh pasta products are dried to 12-12.5% moisture content using short or long drying regimes. Dry pasta is boiled for 6-10 min, depending on pasta type, and then consumed. Good quality cooked pasta should be smooth, firm, and slightly elastic, offering some resistance to the bite ("al dente" texture). Filled pasta ("tortellini," "ravioli," etc.) is made with fresh pasta sheets cut to the desired shape, filled with meat, cheese and/or vegetables, sealed and then dried or canned fresh, with or without sauce. Filled pasta is particularly popular in Italy.

Flat Breads, Couscous, and Burghoul (Bulgur)

Durum wheat flour, alone or blended with other flours, is widely used in Mediterranean countries for making dense leavened breads (Quaglia, 1988). In the WANA region, 50-60% of durum wheat is consumed as single- or double-layered flat bread, and as dense, leavened bread (Williams, 1985). Durum wheat flour flat breads, although similar in processing and general attributes to those prepared with common wheat, are characterized by an appealing yellowish color and slower staling compared to common wheat. The latter is because durum wheat flour dough has higher water absorption capacity than common wheat.

Couscous and burghoul are two of the most important traditional foods in the WANA region. Couscous is a national dish in North Africa, where it is consumed two or three times a day (Qarooni, 1994). Couscous consists of steam-cooked small granules of agglomerated semolina and is consumed with vegetables and/or meat stews. It is prepared by moistening semolina while mixing and rubbing to agglomerate wet semolina particles. Small granules of agglomerated semolina are then steamed and dried. Couscous may be steamed and used, or steamed and dried (to 10-12% moisture) for later use. Couscous is generally dried under the sun during the hot summer season, but large-scale automated production of couscous is now common in large urban areas (Qarooni, 1994). Burghoul (also known as bulgur and boulgur) is a popular food in West Asia (Williams, 1985; Qarooni, 1994). Durum grain is soaked and then boiled in excess water. The cooked wheat is dried (10-11% moisture) under the sun. The dried grain is rubbed to remove outer bran layers and cracked to produce the coarse pieces called burghoul. Burghoul is the basic ingredient of several regional dishes containing vegetables and/or meats. Large-scale automated production of burghoul takes place in large urban areas in the region as well as in the USA (Qarooni, 1994). Durum burghoul possesses a desirable characteristic yellow color and a hard texture that cannot be obtained with common wheat.

Processing requirements of flat breads prepared with durum flours are practically the same as those prepared with common wheat, apart from the stronger yellow color, higher water absorption, and better keeping properties associated with durum wheat. For this reason, flat breads will not be considered further in this section. Pasta and couscous are the main uses of semolina. For reasons as yet not well understood, the grain quality characteristics required to produce high quality couscous are the same, or similar, to those required for high quality pasta production. Therefore, in the following text we will refer only to pasta quality-related factors. Grain factors associated with quality generally include kernel vitreousness, protein content, gluten strength, and yellow pigment concentration. High quality pasta is characterized by a uniform, bright golden yellow color, free of specks, non-stickiness and low cooking loss and maintains the correct firmness or chewiness ("al dente" trait) after some overcooking. The minimum grain protein content required by the pasta processing industry is roughly 12.5% (11.5% semolina protein), but higher concentrations are preferred and often rewarded by higher prices. Protein content is critical in durum wheat, since it can account for 30-40% of the variability in pasta cooking quality.

BREEDING FOR END-USE QUALITY

To satisfy the processing and quality requirements of traditional and new wheat-based products, the wheat industry requires several wheat types, each characterized by specific grain quality attributes that should be consistent from lot to lot. More and more frequently, the value and acceptability of a wheat crop on the market is determined by how much its grain attributes satisfy specific processing and end-use quality requirements. In the current free trade environment, the wheat industry can find on the export-market wheat quality types not encountered locally. Hence, to remain competitive, local farmers look for wheat varieties that satisfy their grain yield expectations and the quality requirements of their target market.

Improving yield potential without negatively affecting grain quality is difficult, mainly because increases in grain yield are generally accompanied by a decrease in grain protein content, which is strongly associated with bread-making quality. Therefore, wheat breeders must give grain quality the same level of importance as yield potential and disease resistance. To develop wheat cultivars for specific food markets, breeders should:

- Understand the genetic control of specific grain components,
- Understand the relationship between grain composition and processing qualities, and
- Achieve rapid identification and manipulation of quality-related traits by using quick, reliable, low-cost methodologies for testing quality.

Table 3 summarizes the major characters sought by breeders in developing common and durum wheat products.

Breeding Methodology

Parameters Considered in Breeding for Quality

The complex and generally additive nature of inheritance of most quality traits (Wrigley, 1994) has led to the development of a range of indirect tests. By applying these tests in the early generations, the population mean is favorably shifted and increased frequency of homozygous lines with the desired quality characteristics can be expected at the end of the breeding process. Tests commonly used for the determination of grain quality are described below. The sample size and approximate time required for each test are provided in Table 4.

Flour and Semolina Milling Quality

A range of small-scale milling equipment to estimate flour yield potential is available, from micro-mills (Finney et al., 1987) to Brabender Quadrumat Jr. and Sr. mills. To obtain a more precise estimation of flour and semolina yield potential, it is necessary to mill larger samples (1-2 kg) using laboratory pilot mills such as the Buhler mill (Simmonds, 1989). These mills contain two to three sets of break and reduction rolls, similar to those used in industrial milling. In addition, they have better flour sifting mechanisms and, in the case of semolina milling, semolina purifiers. Near-infrared reflectance spectroscopy (NIR), if calibrated properly, can provide estimates of potential flour extraction (Burridge

TABLE 3. A rough guide to the key quality characteristics sought by breeders when developing cultivars suitable for some major products.

Product	Grain Color	Flour Protein %	Grain Hardness	Flour Pigment	Dough Strength	Dough Extensibility
Leavened Bread	Red/white	High	Hard/semi-hard	Low	Strong/medium	High
Flat Bread	White	Intermediate	Hard/semi-hard	Low	Medium/strong	High
Steamed Bread	White/red	Intermediate	Semi-hard	Low	Medium	High
White, Salted Noodle	White	Intermediate	Soft/semi-soft	Low	Medium	Intermediate
Yellow, Alkaline Noodle	White/red	Low/High	Hard/semi-hard	not defined	Medium/strong	Intermediate
Biscuits (cookies) and Cakes	Red/white	Low	Soft	Low	Weak	Intermediate
Alimentary Pasta (Durum wheat)	Amber	Intermediate/high	Vitreous	High	Strong	Low

TABLE 4. Approximate sample size and time required for small-scale quality tests.

Test Name	Sample (g)	Time/Sample (min)	Test Name	Sample (g)	Time/Sample (min)
Milling (grain)			**Enzymatic Activity (whole meal /flour)**		
Micro-mills (whole mill, UDY, Tecator)	5	1	Falling Number (alpha amylase)	10	1-5
Brabender Quadrumat Junior mill	100-500	10-15	Dyed-Substrate Test (alpha amylase)	2	60*
Brabender Quadrumat Senior mil	500-1000	15-25	Polyphenol Oxidize (PPO)	1g	30*
Buhler or Similar mill	> 500	25-40	**Starch Pasting Properties (whole meal/flour)**		
NIR Analysis (whole meal/flour)			Flour Swelling Volume Test	10	20-30*
Grain hardness, Moisture, Protein, etc.	10-15	< 1	Rapid Visco Analyzer (Newport, Scientific)	10	5
NIR Analysis (grain)	>10	< 1	**Dough Rheology (flour)**		
Gluten Strength (whole meal/flour)			Mixograph (National Mfg.)	2-35	5-10
SDS-Sedimentation (whole meal flour)	1-5	20*	Farinograph (Brabender)	10-60	15-30
Gluten Index (Glutomatic, FN)	10	10	Alveograph (Chopin)	60-250	40-60
Flour Color (whole meal/flour)	10	< 1	Extensograph (Brabender)	60-200	100-130
Pigment Content			**Baking Tests (flour)**	2-250	100-300*
(whole meal/flour/semolina)	1-10	80-120*	**Pasta Making and Pasta Cooking Quality**	1000-5000	2 days

*Indicates many test can be run concurrently

et al., 1994). Wishna (1998) examined back illumination imaging, and light and near-infrared transmission as improved predictors of semolina milling parameters.

Small-Scale Tests to Determine Physical, Chemical, and Functional Grain/Flour Quality Factors

Protein content. Several methods can be used to estimate protein content, including the Kjeldahl and Dumas methods, calorimetric/spectrophotometric assessments, and near infrared reflectance (NIR). Of these, the NIR analysis using either milled flour or whole grain is the most versatile and provides the fastest estimate (Williams et al., 1982; Delwiche et al., 1998).

Grain hardness. Rapid small-scale tests (based on grinding time, grinding volume, or particle size distribution) used to determine grain hardness make it relatively easy to screen for hardness as early as the F3 generation. NIR analysis of the particle size distribution of whole grain flour or analysis of intact grain samples are very quick and useful in early generation screening. Rough estimates of hardness can also be made visually or by simply biting the grain.

Flour/semolina yellow pigment. Easy-to-use reflectance and pigment extraction methods are available to breeders. Semolina color is determined by xanthophylls, especially lutein, measured as pigment concentration or by calorimetric instruments. Semolina color heritability is high (Johnston et al., 1983; Clarke et al., 2000) and largely additive. Color can be determined visually by comparison with standard samples or directly with a reflectance colorimeter. The pigment may also be extracted and pigment concentration measured by a spectrophotometer. NIR has been used more recently to assess pigment (McCraig et al., 1992) and is a rapid, inexpensive screening tool. In early generations, pigment is frequently determined from whole meal samples (Clarke et al., 2000) to reduce costs and turnover time, while in later stages of the breeding process, semolina and pasta color are measured. In the latter case, the CIE 1976 L*, a*, b* color system is widely used in breeding programs and standard in the pasta manufacturing industry.

Enzymatic activity. Rainfall at or prior to harvest can cause seeds to germinate on the spike, significantly reducing grain quality. Tolerance to preharvest sprouting is linked to grain dormancy (Mares, 1989). Dormancy inhibits the production of germinative enzymes, particularly alpha-amylase. Breeders can indirectly measure alpha-amylase levels using the Falling Number test (AACC, 1983) to measure the viscosity of a flour suspension, or dye-labeled starch substrates that measure alpha-amylase activity by change in the dye color over time to indicate tolerance to preharvest sprouting (Meredith and Pomeranz, 1985). A calorimetric microtiter plate assay is available for detection of polyphenol-oxidase (PPO) activity (Bernier and Howes, 1994).

Sedimentation test (Zeleny and SDS). The Zeleny (Zeleny, 1947) and SDS-sedimentation (Axford et al., 1979) tests can be used to obtain a semi-quantitative estimation of the amount of glutenin (or indirectly, of general gluten strength). These tests are based on the expansion of glutenins (also known as gel proteins) in isopropanol/lactic or SDS/lactic acid solution and currently the most rapid and reliable single small-scale tests (see Weegels et al., 1996, for a review). The tests are widely used to screen early generation wheat lines for gluten strength (strong to weak). Heritability of SDS-sedimentation is intermediate to high (Clarke et al., 2000). Correlation between protein concentration and predictors of gluten strength differs among methods (Ruiz and Carrillo, 1995; Peña, 2000), while SDS-sedimentation, used widely in early generation selection, and industrial standards, such as Alveograph and Mixograph parameters, are high (Clarke et al., 2000; Peña, 2000). Selection in early generations using SDS sedimentation is frequently practiced from F2 onwards (Clarke et al., 1997).

Glutenins (HMW and LMW) and gliadins. Sodium dodecyl sulfate-polyacrylamide gel electrophoresis (SDS-PAGE) of whole protein extracts can be used in early generations to select lines with desirable HMW subunit composition and in advanced stages to define desirable HMW combinations in the progeny of new crosses. Selecting for HMW-glutenins has proved effective in improving the gluten strength of both bread and durum wheat. Low molecular weight (LMW) glutenins, comprising approximately 80% of total glutenin, have for some time been neglected by breeders because assay methods were unreliable. The sequential protein extraction procedure (to separate gliadins from glutenins) combined with 1D SDS-PAGE provides well-resolved patterns of HMW and LMW subunits in a single gel (Singh et al., 1991). Although this procedure has facilitated the study of the relationship between genetic control and quality of LMW subunits, it is not simple enough to rapidly screen for desirable LMW glutenins.

Screening for gliadin bands γ-42 and γ-45 (by APAGE) and LMW gluten subunit LMW-2 has been implemented in many durum breeding programs (Federmann et al., 1994; Clarke et al., 1997). Early generation selection for pasta making quality using monoclonal antibodies (Howes et al., 1995) to identify desirable gamma gliadins (γ-45) or LMW glutenin subunits (LMW-2) has been proposed. The linkage of γ-42 with bronze (brown) glume color and γ-45 with white glume color (Leisle et al., 1985) can be exploited for selection in early generations, although recombinants do exist. Furthermore, indirect selection for gluten strength via direct selection for SDS-sedimentation has resulted in high frequencies of LMW-2; also, LMW-1 was nearly eliminated and favorable HMW-glutenin allelic variants were accumulated (Peña, 1995; Pfeiffer, 2000).

Starch pasting properties. Starch pasting viscosity is very important in determining the quality of flour-noodles (Lee et al., 1987) and of Arabic flat breads (Quail et al., 1990). The Amylograph/Viscograph and, more recently, the Rapid Visco Analyzer (RVA) are used to obtain a complete profile of starch pasting properties. While the first requires a large sample size and a considerably longer testing time, the RVA requires a 3-4 g sample and only a few minutes to reveal the pasting profile of the tested material. Therefore, the RVA is now considered a rapid test suitable for the early selection of wheat lines with desirable starch pasting viscosity for noodle-making (Panozzo and McCormick, 1993). As an alternative to the RVA test, starch-pasting properties can be measured inexpensively using the flour swelling power test (McCormick et al., 1991).

Measures of dough strength and extensibility. Direct measures of dough rheology (mixing properties, strength, and extensibility) are much more time-consuming and require more flour compared to the indirect tests described above. However, the Mixograph (Finney and Shogren, 1972) potentially offers breeders the most versatility because it allows evaluation of more samples per day, using smaller quantities of flour, than alternatives like the Brabender Farinograph and Extensograph, and the Chopin Alveograph. The correlation between the Mixographic parameters, mixing time and tolerance to over mixing, and more specific measures of strength and extensibility obtained with the Chopin Alveograph or the Brabender Extensograph is high (Finney et al., 1987). For the same reasons indicated above for the Mixograph, the Alveograph is considered a better alternative than the Extensograph in determining dough strength and extensibility.

Small-scale baking. To properly evaluate loaf volume and crumb structure, baking tests must be performed. Finney et al. (1987) describe the process in some detail. Baking tests are performed only on breeding materials deemed to have, from other measures of grain quality, suitable bread-making quality. The tests range from micro-bakes using 2-5 g of flour (Grass et al., 1997) to larger traditional bakes using 100 g of flour.

Traditional Breeding Approaches

Breeding Methods

Selection and testing for quality begins in early generations. Crop management × quality interactions are of critical importance, and agronomic practices suitable for both visual field selection and identification of quality attributes must be employed. If a breeding program employs a pedigree or modified pedigree breeding method, early generation quality tests are potentially very valuable. However, in a bulk system where selected progenies are carried to homozygosity in a bulk before head rows are taken, early generation testing will be of little value, as the lines with the desired quality will be obscured in the mean of the bulk population. This does not preclude the use of such tests to remove bulks with low mean quality parameters. Obviously, a much higher frequency of derived lines must be screened in a bulk system once the bulk is finally broken up into its constituent homozygous lines.

Selection of Parents

Plant breeders select at least one parent with the desired quality when designing their crossing strategies, particularly as end-use requirements frequently determine the fate of potential new cultivars.

All potential progenitors should be characterized for the requisite quality parameters before crossing. Exploration of genetic variation for quality traits present in wild relatives and alien species may require pre-breeding or progenitor building before such species are crossed widely in the breeding program. The value of new glutenin subunits obtained from crossing durum wheat with *T. tauschii*, the donor of the D-genome in hexaploid wheat, which results in man-made synthetic hexaploid wheat, has yet to be determined. However, evidence suggests that some of these new alleles may enhance quality when optimally recombined with the appropriate glutenin and gliadin combinations found in hexaploid wheat (Peña et al., 1995).

The use of wild emmer wheat (*T. turgidum* L. var. *dicoccoides*) has been investigated and proposed to enhance grain protein concentration and protein quality in durum wheat (Levy and Feldman, 1989; Joppa et al., 1991). Research on quality improvements in durum wheat via common wheat *Glu-D1* HMW glutenin subunits (D-genome), *Glu-D1* alleles 5 + 10 and 2 + 12 suggests there is scope for further quality enhancement (Liu et al., 1996).

Application of Quality Tests in the Breeding Process

Genetics, environmental conditions, and crop management practices influence protein concentration in the grain. In contrast to the low heritability of protein content, grain hardness, yellow pigment, and bread and pasta processing quality are highly heritable and can be readily improved through conventional breeding. Protein heritability estimates vary from low to moderate (Clarke et al., 2000) and selection for protein is complicated by a negative association with grain yield. For instance, if a population is too severely truncated on the basis of grain protein, then the probability of selecting high yielding lines with the desired quality is limited. Hence, selection for protein is not efficient in early generations (Legg et al., 1991) and should be restricted to intermediate segregating generations from F4 onwards. Heritability of protein content can be improved by using statistical procedures to remove environmental trends. Adjustment of protein for environmental trends using the moving mean procedure tended to improve realized heritability for protein concentrations (Clarke et al., 1998). However, this technique may be difficult to apply in early generations, since selection intensities are high and not all entries are harvested. Higher heritability estimates can be realized via replications, locations, and years of testing; however, these options are impractical in early generations. Consequently, molecular markers may have great potential (Humphreys et al., 1998).

Plant breeders must also cull populations with some care on the basis of parameters such as the SDS-sedimentation index (sedimentation/protein content) and HMW-glutenin subunits, if grain yield potential is to be maintained (O'Brien et al., 1989; Trethowan et al., 2000). The stage in the breeding process at which quality determination takes place will influence which tests are applied. If a series of F2-derived F3 or F4 populations is to be screened, then the large number of lines generally found in most breeding programs at this stage and the limited seed amounts will restrict the choice of test. In these instances, the breeder should consider NIR analysis of grain hardness and protein, SDS-sedimentation (rough measurement of gluten strength), HMW-glutenins, grain yellow color/pigment (durum wheat), Falling Number (amylase activity), PPO (noodle wheat), and RVA analysis of starch (noodle wheat). These tests are repeated in the F5 and F6 generations. Advanced lines entering multi-location yield trials are screened for specific quality attributes using more time-consuming procedures. This can also be achieved by screening a considerably smaller set of advanced materials for flour/semolina yield, yellow color, and dough viscoelasticity. Gluten Index and the Mixographic parameters, mixing time, and mixing tolerance/stability are particularly useful for identifying gluten strength. Time-consuming, yet accurate, measures of dough rheology (Farinograph, Extensograph or Alveograph) and quality (bread, cookie, pasta, etc., using laboratory-scale methods) must be applied on relatively small numbers of elite advanced lines before considering them for varietal release. In durum wheat, elite advanced lines and progenitors are tested for milling parameters (e.g., semolina yield, ash) and for actual pasta processing and cooking quality attributes, including spaghetti water absorption, cooked weight, cooking loss, over-cooking features, firmness or "al dente" trait, stickiness, speck count, pasta color, and brightness. Applying pressure for quality traits at both the segregating and advanced stages allows breeders to develop wheat varieties that are not only adapted to cropping conditions, but also have the quality attributes required by the baking and pasta industries.

Hybrid wheat may also offer breeders the opportunity to tailor end-use quality through the selection of parents with high or complementary quality characteristics while exploiting potential heterosis for both quality and yield. Most authors report an intermediate expression of grain quality traits in hybrid wheat somewhere between the two parents (Borghi and Perenzin, 1994; Cukadar et al., 2000). However, since this intermediate quality is obtained at higher yield levels, hybrid wheat may still offer farmers a considerable advantage in some environments.

BIOTECHNOLOGY IN QUALITY IMPROVEMENT

Application of Genetic Transformation

Biotechnology offers the possibility of investigating the genetic and biochemical basis of individual protein subunits and other molecules contributing to the end-use quality of wheat. Using genetic transformation, new cultivars with improved quality can be developed through the insertion of genes coding for key grain quality attributes. Some researchers have reported improved functional properties from the transformation of wheat with HMW glutenin subunits (Anderson et al.,

1996; Blechl and Anderson, 1996; Barro et al., 1997). These authors found improvements in dough strength among plants expressing the transgenes; however, the level of expression of the inserted genes and their contribution to quality in wheat advanced lines suitable for varietal release are yet to be determined (Anderson et al., 1996; Blechl and Anderson, 1996; Blechl, 1998).

The functional properties of starch can also be modified using transformation. Li et al. (1999) have cloned a gene encoding wheat starch synthase 1, responsible for extending starch polymers (hence the balance between amylose and amylopectin). This gene, if expressed in transgenic plants, may affect starch quality. The proportion of amylose to amylopectin and starch granule size affect water adsorption and pasting viscosity (Dengate, 1984). Anti-sense gene constructs may also be used to suppress expression of some deleterious characters such as the rye secalins associated with the 1B/1R translocation in wheat (Blechl, 1998). Transformed plants have been produced using this construct; however, the effects on grain quality have not yet been determined (P. Langridge, personal comm). Other potential applications include neutralization of undesired yellow flour pigment associated with a yield enhancing gene-complex (*Lr-19*) located on chromosome 7D (Knott, 1984).

Application of Molecular Markers

Molecular marker technology is expected to increase the efficiency and speed of breeding for quality. Instead of selecting progeny by determining gene action (determination of their effect on quality), breeders will use marker assisted selection (MAS) to select germplasm carrying desirable genes. Although MAS has been useful in selecting plants carrying genes associated with resistance to pests and diseases, it has not yet proved to be an efficient alternative to traditional chemical and biochemical small-scale testing. The identification of molecular markers in wheat is hampered by a general lack of polymorphism (Helenjaris et al., 1995). It will be difficult to identify markers for milling and dough handling properties of complex heritability (Parker et al., 1996). However, there may be some application for specific traits controlled by one or two genes. For example, in bread wheat a single gene located on chromosome 5D controls grain hardness. A restriction fragment length polymorphism (RFLP) has been developed that allows discrimination among some hard and soft wheats (Jolly et al., 1993). A marker for high grain protein identified from *T. dicoccoides* has been used to select durum and common wheats with higher grain protein (Howes et al.,

1994). Other studies have suggested that markers can be applied to simply inherited traits like flour color (Parker et al., 1996).

EFFECTS OF GENOTYPE × ENVIRONMENT INTERACTIONS ON QUALITY IMPROVEMENT

The interaction of genotype with environment has been and still is one of the major challenges confronting plant breeders. Peterson et al. (1998) reported that the interaction variance (G × E) for flour protein in their study of 30 genotypes across 17 locations was lower than that for genotypes. Similarly, Lukow and McVetty (1991) reported significant, but considerably smaller, G × E for flour protein compared to the variance attributed to genotypes. Although a cultivar may have the potential to produce a specified end-use quality, realizing this quality is dependent upon flour protein content and quality. Yellow berry, closely related with vitreosity in durum wheat (Dexter et al., 1989), is strongly affected by crop management practices such as rate and timing of N application, irrigation, and planting date (Sombrero and Monneveux, 1989; Shashi-Madan and Rajendra-Kumar, 1998). Several characters contributing to good quality have high heritabilities and relatively small G × E effects, including SDS-sedimentation (Lukow and McVetty, 1991; Peterson et al., 1998), HMW subunits (MacRitchie et al., 1990), flour pigment (Parker et al., 1996), flour yield (Lukow and McVetty, 1991; Fenn et al., 1994), and grain hardness (Fenn et al., 1994). However, previous studies have shown G × E effects to be significant for many quality tests, although generally smaller than genetic and location effects (Lukow and McVetty, 1991; Robert and Denis, 1996). Robert and Denis (1996) found the Alveograph gave relatively small G × E effects for W (strength) and P (tenacity) but significantly larger effects for P/L (extensibility). Peterson et al. (1998) found the variation attributed to environment was greater than that for genotype for the Mixograph and many baking parameters, even though the G × E interaction variance was generally lower than that of genotype. In contrast, Lukow and McVetty (1991) reported genotypic variance to be much greater than that of environment for the same characters. Reports of G × E associated with loaf volume have been inconsistent, with authors finding large (Peterson et al., 1998) to relatively small (Lukow and McVetty, 1991; Fenn et al., 1994) interactions. In the Lukow and McVetty (1991) and Fenn et al. (1994) studies, the G × E effect on flour protein was relatively small; since protein and loaf volume are highly correlated, this result is not surprising.

The implications of G × E on selection for quality, while significant for plant breeders, are less than those associated with yield. Many of the key quality traits measured by breeders as indirect indications of potential end-use quality are not greatly affected by G × E. While location effects can be large, genotypes tend to rank the same way across locations. Selection for SDS, grain hardness, grain or flour protein, flour pigment, and seed appearance in early generations can greatly assist breeders in identifying wheats with good quality. The SDS-sedimentation/flour protein ratio allows correction for variable protein levels associated with particular locations and correlates well with physical dough properties and baking quality while maintaining variability for yield potential (Trethowan et al., 2000). The relatively low G × E associated with Mixograph and Alveograph measures of physical dough properties also indicates that these more time-consuming measurements can be conducted on samples collected from a small range of representative test sites. Therefore, while most breeding programs conduct yield evaluations over many locations, a subset of locations will provide an adequate representation of end-use requirements.

CONCLUSIONS

The globalization of wheat markets, improved communications, and demographic changes are having an impact on the various uses of wheat. Plant breeders need to better predict these changes in regional demands if they are to position their breeding programs to meet those demands. In countries currently exporting wheat and particularly in countries that traditionally import wheat, breeders must also be conversant with global trends in wheat consumption. Understanding the basic compositional characteristics of the desired end product will enable researchers to apply the most appropriate indirect quality tests in early breeding generations. Application of these tests will depend heavily on their expense, correlation with key end-use characteristics, and speed and ease of use. The potential interaction with environment of key quality characteristics, measured using both indirect and direct techniques will also influence the breeders' choice of test locations.

Researchers working in isolation cannot develop wheats with the quality characteristics necessary to satisfy changing market requirements. More than ever, multidisciplinary approaches involving cereal chemists, plant breeders, agronomists, and market economists will be fundamental to meeting changing consumer demands.

REFERENCES

AACC. (1983). *Approved Methods of the AACC*, 8th ed., St. Paul, MN: American Association of Cereal Chemists, method 26-30; method 14-10.

Anderson, O.D., F. Bekes, P. Gras, J.C. Kuhl, and A. Tam. (1996). Use of bacterial expression system to study wheat high-molecular-weight (HMW) glutenins and the construction of synthetic HMW-glutenin genes. In: *Gluten '96*, ed. C.W. Wrigley. Victoria, Australia: Royal Australian Chemical Institute, pp. 195-198.

Axford, D.W.E., E.E. McDermott, and D.G. Redman. (1979). Note on the sodium dodecyl sulfate test of bread-making quality: Comparison with Pelshenke and Zeleny tests. *Cereal Chemistry* 56:582-584.

Barro, F., L.F. Rooke, Gras, P. Bekes, A.S.Tatham, R. Fido, P.A. Lazzeri, P.R. Shewry, and P. Barcelo. (1997). Transformation of wheat with high molecular weight subunit genes results in improved functional properties. *Nature Biotechnology* 15:1295-1299.

Bernier, A.-M. and N.K. Howes. (1994). Quantification of variation in tyrosinase activity among durum and common wheat. *Journal of Cereal Science* 19:157-159.

Bhattacharya, M. and H. Corke. (1996). Selection of desirable pasting properties in wheat for use in white salted or yellow alkaline noodles. *Cereal Chemistry* 73: 721-728.

Blechl, A.E. (1998). Gene transformation: A new tool for the improvement of wheat. *Wheat Yearbook*, March 1998, Economic Research Service/USDA, pp. 30-32.

Blechl, A.E. and O.D. Anderson. (1996). Expression of a novel high-molecular-weight glutenin in transgenic wheat. *Nature Biotechnology* 14:875-879.

Borghi, B. and M. Perenzin. (1994). Diallel analysis to predict heterosis and combining ability for grain yield, yield components and bread-making quality in bread wheat (*T. aestivum*). *Theoretical and Applied Genetics* 89:975-981.

Branlard, G. and M. Dardevet. (1985). Diversity of grain protein and bread wheat quality. I. Correlation between gliadin bands and flour quality characteristics. *Journal of Cereal Science* 3:329-343.

Brunori, A., G. Galterio, and G. Mariani. (1991). Relationship between gluten components in durum wheat and pasta quality. In: *Gluten Proteins*, eds. W. Bushuk and R. Tkachuk, St. Paul, MN: American Association of Cereal Chemists, pp. 253-260.

Burridge, P.M., G.A. Palmer, and G.J. Hollamby. (1994). Developing a strategy for rapid wheat quality screening. In: *Proceedings of the 7th Assembly of the Wheat Breeding Society of Australia*, eds. J. Paull, I.S. Dundas, K.J. Shepherd, and G.J. Hollamby. Adelaide, Australia, pp. 57-60.

Carrillo, J.M., J.F. Vazquez, and J. Orellana. (1990). Relationship between gluten strength and glutenin proteins in durum wheat cultivars. *Plant-Breeding* 104: 325-333.

Clarke, F.R., J.M. Clarke, and R.M. DePauw. (1998). Improving selection for protein by adjusting for environmental trends. In: *Proceedings of the Ninth International Wheat Genetics Symposium* V. 4, ed. Slinkard, A.E. University of Saskatchewan, Saskatoon: University Extension Press, pp. 142-144.

Clarke, J.M., F.R. Clarke, N.P. Ames, T.N. McCaig, and R.E. Knox. (2000). Evaluation of predictors of quality for use in early generation selection. In: *Durum Wheat*

Improvement in the Mediterranean Region: New Challenges, eds. C. Royo, M.M. Nachit, N. Di Fonzo, J.L. Araus. Serie A: Séminaires Méditerranéennes No. 40. Zaragoza, España: Instituto Agronomico Mediterraneo de Zaragoza,, pp. 439-446.

Clarke, J.M., B.A. Marchylo, M.I.P. Kovacs, J.S. Noll, T.N. McCaig, and N.K. Howes. (1997). Breeding durum wheat for pasta quality in Canada. In: *Wheat: Prospects for Global Improvement*, eds. H.-J. Braun, F. Altay, W.E. Kronstad, S.P.S. Beniwal, and A. McNab. The Netherlands: Kluwer Academic Publishers, pp. 223-228.

Crosbie, G.B., S. Huang, and I.R. Barclay. (1998). Wheat quality requirements of Asian foods. *Euphytica* 100:155-156.

Cukadar, B., R.J. Peña, and M. vanGinkel. (2000). Yield potential and bread-making quality of bread wheat hybrids produced by a chemical-hybridizing agent, Genesis. In: *Proceedings of the 6th International Wheat Conference*, June 2000, Budapest, Hungary (in press).

Delwiche, S.R., R.A. Graybosch, and C.J. Peterson. (1998). Predicting protein composition, biochemical properties, and dough-handling properties of hard red winter wheat flour by near-infrared reflectance. *Cereal Chemistry* 75:412-416.

Dengate, H.N. (1984). Swelling, pasting and gelling of wheat starch. In: *Advances in Cereal Science and Technology*, Vol. 6, ed. Y. Pomeranz. St. Paul, MN: American Association of Cereal Chemists, pp. 49-82.

Edwards, I.B. (1997). A global approach to wheat quality. In: *International Wheat Quality Conference*, eds. J.L. Steele and O.K. Chung. Manhattan, KS: Grain Industry Alliance, pp. 27-37.

Dexter, J.E., B.A. Marchylo, A.W. MacGregor, and R. Tkachuk. (1989). The structure and protein composition of vitreous, piebald and starchy durum wheat kernels. *Journal of Cereal Science* 10:19-32.

Fares, C., G. Novembre, G. Galterio, N.E. Pogna, and N. di-Fonzo. (1997). Relationship between storage protein composition and gluten quality in breeding lines of durum wheat (*Triticum turgidum* spp. *durum*). *Agricoltura Mediterranea* 127: 363-368.

Faridi, H. (1988). Flat breads. In: *Wheat Chemistry and Technology*, ed. Y. Pomeranz. St. Paul, MN: American Association of Cereal Chemists, pp. 457-506.

Faridi, H. and J.M. Faubion, (1995). Wheat usage in North America. In: *Wheat End Uses Around the World*, eds. H. Faridi, and J.M. Faubion. St. Paul, MN: American Association of Cereal Chemists, pp. 1-41.

Federmann, G.R., E.U. Goecke, A.M. Steiner, and P. Ruckenbauer. (1994). Biochemical markers for selection towards better cooking quality in F2-seeds of durum wheat (*Triticum durum* Desf.). *Plant Varieties and Seeds* 7:71-77.

Fenn, D., O.M. Lukow, W. Bushuk, and R.M. DePauw. (1994). Milling and baking quality of 1BL/1RS translocation wheats. 1. Effects of genotype and environment. *Cereal Chemistry* 71:189-195.

Finney, K.F. and M.D. Shogren. (1972). A ten-gram Mixograph for determining and predicting functional properties of wheat flours. *Baker's Digest* 46:32-35, 38-42, 77.

Finney, K.F., W.T. Yamazaki, V.L. Youngs, and G.L. Rubenthaler. (1987). Quality of hard, soft and durum wheats. In: *Wheat and Wheat Improvement*, ed. E.G. Heyne. 2nd ed. Madison, Wisconsin, pp. 677-748.

Fu, B.X. and H.D. Sapirstein, (1997). Fractionation of monomeric proteins, soluble and insoluble glutenin, and relationships to mixing and baking properties. In: *Gluten '96*, ed. C.W. Wrigley. Victoria, Australia: Royal Australian Chemical Institute, pp. 340-344.

Galterio, G., L. Grita, and A. Brunori. (1993). Pasta-making quality in *Triticum durum*. New indices from the ratio among protein components separated by SDS-PAGE. *Plant Breeding* 110:290-296.

Gras, P.W., F.W. Ellison, and F. Bekes. (1997). Quality evaluation on a micro-scale. In: *International Wheat Quality Conference*, eds. J.L. Steele and O.K. Chun. Manhattan, KS: Grain Industry Alliance, pp. 161-172.

Greenwell, P. and J.D. Schofield, (1986). A starch granule protein associated with endosperm softness in wheat. *Cereal Chemistry* 63:379-380.

Gupta, R.B., F. Bekes, C.W. Wrigley, and H.J. Moss. (1990). Prediction of wheat quality in breeding on the basis of LMW and HMW glutenin subunit composition. In: *Sixth Assembly of the Wheat Breeding Society of Australia.* Tamworth, NSW, pp. 217-225.

Helenjaris, T., M. Slocum, S. Wright, A. Schaefer, and J. Neinhuis. (1986). Construction of genetic linkage maps in maize and tomato using restriction fragment length polymorphism. *Theoretical and Applied Genetics* 72:761-769.

Howes, N.K., M.I. Kovacs, D. Leisle, M.R. Dawood, and W. Bushuk. (1989). Screening of durum wheats for pasta-making quality with monoclonal antibodies for gliadin 45. *Genome* 32:1096-1099.

Howes, N.K., J.M. Clarke, and D. Leisle. (1994). Selection for high grain protein derived from *T. dicoccoides* in durums using molecular markers. *Annual Meeting of the American Agronomy Society*, 13-17th November, Seattle, WA.

Huang, S. (1996). A look at noodles in China. *Cereal Foods World* 41:199-204.

Huang, S., S.H. Yun, K. Quail, and R. Moss. (1996). Establishment of flour quality guidelines for Northern style Chinese steamed bread. *Journal of Cereal Science* 24:179-185.

Irvine G.N. and C.A. Winkler. (1950). Factors affecting the color of macaroni. II: Kinetic studies of pigment destruction during making. *Cereal Chemistry* 27:205-218.

Humphreys, D.G., J.D. Procunier, W. Mauthe, N.K. Howes, P.D. Brown, and R.I.H. McKenzie. (1998). Marker-assisted selection for high protein concentration in wheat. In: *Wheat Protein Production and Marketing: Proceeding of the Wheat Protein Symposium*, eds. D.B. Fowler, W.E. Geddes, A.M. Johnston, and K.R. Preston. Saskatoon, Saskatchewan: University Extension Press, University of Saskatchewan, pp. 255-258.

Johnston, R.A., J.S. Quick, and J.J. Hammond. (1983). Inheritance of semolina color in six durum wheat crosses. *Crop Science* 23:607-610.

Jolly, C.J., S. Rahman, A.A. Kortt, and T.J.V. Higgins. (1993). Characterization of wheat Mr 15 000 'grain softness protein' and analysis of the relationship between its accumulation in the whole seed and grain softness. *Theoretical and Applied Genetics* 86:589-97.

Joppa, L.R., G.A. Hareland, and R.G. Cantrell. (1991). Quality characteristics of the Langdon durum-dicoccoides chromosome substitution lines. *Crop Science* 31: 1513-1517.

Kaan, F., B. Chihab, C. Borries, P. Monneveux, and G. Branlard. (1995). Prebreeding and breeding durum wheat germplasm (*Triticum turgidum* var. *durum*) for quality products. In: *Durum Wheat Quality in the Mediterranean Region*. Options-Mediterraneennes, Serie A, no. 22. Zaragoza, España: Instituto Agronomico Mediterraneo de Zaragoza, pp. 159-166.

Kim, S.-K. (1996). Instant noodle technology. *Cereal Foods World* 41:213-218.

Kohn, S. (2000). An update of the U.S. Baking Industry. *Cereal Foods World* 45:94-97.

Knott, D.R. (1984). The genetic nature of mutations of a gene for yellow pigment linked to *Lr 19* in 'Agatha' wheat. *Canadian Journal of Genetics and Cytology* 26:392-393.

Kosmolak, F.G., J.E. Dexter, R.R. Matsuo, D. Leisle, and B.A. Marchylo. (1980). A relationship between durum wheat quality and gliadin electrophoregrams. *Canadian Journal of Plant Science* 60:427-432.

Lee, C.H., P.J. Gore, H.D. Lee, B.S. Yoo, and S.H. Hong. (1987). Utilization of Australian wheat for Korean style dried noodle making. *Journal of Cereal Science* 6:283-97.

Legg, W.G., D. Leisle, P.B.E. McVetty, and L.E. Evans. (1991). Effectiveness of two methods of early generation selection for protein content in durum wheat. *Canadian Journal of Plant Science* 71:629-640.

Leisle, D., M.I. Kovacs, and N. Howes. (1985). Inheritance and linkage relationships of gliadin proteins and glume color in durum wheat. *Canadian Journal of Genetics and Cytology* 27:716-721.

Levy, A.A. and M. Feldman. (1989). Location of genes for high grain protein percentage and other quantitative traits in wild wheat *Triticum turgidum* var. *dicoccoides*. *Euphytica* 41:113-122.

Li, Z., S. Rahman, B. Kosar-Hashemi, G. Mouille, R. Appels, and M.K. Morell. (1999). Cloning and characterization of a gene encoding wheat starch synthase 1. *Theoretical and Applied Genetics* 98:1208-1216.

Lin, Zuo-Ji, D.M. Miskelly, and H.J. Moss. (1990). Suitability of various Australian wheats for Chinese-style steamed bread. *Journal of the Science of Food and Agriculture* 53:203-213.

Liu, C.Y., K.W. Shepherd, and A.J. Rathjen. (1996). Improvement of durum wheat pasta making and bread making qualities. *Cereal Chemistry* 73:155-166.

Lukow, O.M. and P.B.E. McVetty. (1991). Effect of cultivar and environment on quality characteristics of spring wheat. *Cereal Chemistry* 68:597-601.

MacRitchie, F. (1994). Role of polymeric proteins in flour functionality. In: *Wheat Kernel Proteins: Molecular and Functional Aspects*. Bitervo, Italy: Universita degli studi della Tuscia, pp. 145-150.

MacRitchie, F., D.L. duCros, and C.W. Wrigley. (1990). Flour polypeptides related to wheat quality. In: *Advances in Cereal Science and Technology*, ed. Y. Pomeranz. St. Paul, MN: American Association of Cereal Chemists, 10:79-146.

Madan, S. and R. Kumar. (1998). Effect of sowing time and NPK fertilization on yellow berry and yield of durum wheat (*Triticum durum*) varieties. *Indian Journal of Agricultural Sciences* 68:371-372.

Mares, D.J. (1989). Preharvest sprouting damage and sprouting tolerance: Assay methods and instrumentation. In: *Preharvest Field Sprouting in Cereals*, ed. N.F. Derera. Boca Raton, Florida: CRC Press Inc., pp. 129-170.

Matsuo, R.R. and J.E. Dexter. (1980). Relationship between some durum wheat physical characteristics and semolina milling properties. *Canadian Journal of Plant Science* 60:49-53.

McCaig, T.N., J.G. McLeod, J.M. Clarke, and R.M. DePauw. (1992). Measurement of durum pigment with an NIR instrument operating in the visible range. *Cereal Chemistry* 69:671-672.

McCormick, K.M., J.F. Panozzo, and S.H. Hong. (1991). A swelling power test for selecting potential noodle quality wheats. *Australian Journal of Agricultural Research* 42:317-323.

McMaster, G.J. and J.T. Gould. (1995). Wheat usage in Australia and New Zealand. In: *Wheat End Uses Around the World*, eds. H. Faridi and J.M. Faubion. St. Paul, MN: American Association of Cereal Chemists, pp. 267-285.

Miller, B.S., S. Afework, Y. Pomeranz, B. Bruinsma, and G.D. Booth. (1982). Measuring the hardness of wheat. *Cereal Foods World* 27:61-64.

Miller, R.A. and R.C. Hoseney. (1997). Factors in hard wheat flour responsible for reduced cookie spread. *Cereal Chemistry* 74:330-336.

Montague, D. (1998). Global grain perspectives. *World Grain*, September 1998, pp.13, 14, 16, 18.

Morancho, J. (2000). Production et commercialisation du blé dur dans le monde. In: *Durum Wheat Improvement in the Mediterranean Region: New Challenges*, eds. C. Royo, M.M. Nachit, N. Di Fonzo, J.L. Araus. Serie A: Séminaires Méditerranéennes No. 40. Zaragoza, España: Instituto Agronomico Mediterraneo de Zaragoza, pp. 29-33.

Nagao, S. (1995a). Wheat usage in East Asia In: *Wheat End Uses Around the World*, eds. H. Faridi and J.M. Faubion. St. Paul, MN: American Association of Cereal Chemists, pp. 167-189.

Nagao, S. (1995b). Wheat Products in East Asia. *Cereal Foods World* 40:482-487.

Nieto-Taladriz, M.T., M.R. Oerretant, and M. Rousset. (1994). Effect of gliadins and HMW and LMW subunits of glutenin on dough properties in the F6 recombinant inbred lines from a bread wheat cross. *Theoretical and Applied Genetics* 88:81-88.

O'Brien, L., J.F., Panozzo, and J.A. Ronalds. (1989). F3 response to f2 selection for quality and its effect on F3 yield distribution. *Australian Journal of Agricultural Research* 40:33-42.

Owens, G. (1997). China: Handled with care. *Cereals International*, Sept.-Oct. 1997, pp. 14-16.

Panozzo, J.F. and K.M. McCormick. (1993). The rapid viscoanalyser as a method of testing for noodle quality in a wheat breeding programme. *Journal of Cereal Science* 17:25-32.

Parker, G.D., K.J. Chalmers, A.J. Rathjen, and P. Langridge. (1996). Identification of molecular markers linked to protein, flour, color and milling yield in wheat. In: *Proceedings of the 8th Assembly of the Wheat Breeding Society of Australia*, eds. R. Richards, C.W. Wrigley, H.M. Rawson, G.J. Rebetzke, J.L. Davidson and R. Brettell. Canberra, Australia, pp. 105-108.

Payne, P.I. (1987). Genetics of wheat storage proteins and the effect of allelic variation on bread making quality. *Annual Reviews in Plant Physiology* 8:141-153.

Payne, P.I., E.A. Jackson, and L.M. Holt. (1984). The association between gliadin γ-45 and gluten strength in durum wheat varieties. A direct causal effect or the result of genetic linkage? *Journal of Cereal Science* 2:73-81.

Peña R.J. (1995). Quality improvement of wheat and triticale. In: *Wheat Breeding at CIMMYT: Commemorating 50 Years of Research in Mexico for Global Wheat Improvement*, eds. S. Rajaram. and G.P. Hettel, 1995. Wheat Special Report No. 29. Mexico, D. F.: CIMMYT, pp. 100-111.

Peña, R.J. (2000). Durum wheat for pasta and bread making. Comparison of methods used in breeding to determine gluten quality-related parameters. In: *Durum Wheat Improvement in the Mediterranean Region: New Challenges*, eds. C. Royo, M.M. Nachit, N. Di Fonzo, J.L. Araus. Serie A: Séminaires Méditerranéennes No. 40. Zaragoza, España: Instituto Agronomico Mediterraneo de Zaragoza, pp. 423-430.

Peña, R.J., J. Zarco-Hemandez, A. Amaya-Celis, and A. Mujeeb-Kazi. (1994). Relationships between 1B-encoded glutenin subunit compositions and bread making quality characteristics of some durum wheat (*Triticum turgidum*) cultivars. *Journal of Cereal Science* 19:243-249.

Peña, R.J., J. Zarco-Hernandez, and A. Mujeeb-Kazi. (1995). Glutenin subunit compositions and bread-making quality characteristics of synthetic hexaploid wheats derived from *Triticum turgidum* × *Triticum tauschii* (coss.) Schmal Crosses. *Journal of Cereal Science* 21:15-23.

Peterson, C.J., R.A. Graybosch, D.R. Shelton, and P.S. Baenziger. (1998). Baking quality of hard winter wheat: Response of cultivars to environment in the Great Plains. In: *Wheat: Prospects for Global Improvement*, eds. H.-J. Braun, F. Altay, W.E. Kronstad, S.P.S. Beniwal, and A. McNab. The Netherlands: Kluwer Academic Publishers, pp. 223-228.

Pfeiffer, W.H. (2000). CIMMYT's past, present and future breeding goals for durum wheat. *FAO Book*, (in press).

Pogna, N.E., D. Lafiandra, P. Feillet, and J.C. Autran. (1988). Evidence for a direct causal effect of low molecular weight subunits of glutenins on gluten viscoelasticity in durum wheats. *Journal of Cereal Science* 7:211-214.

Pogna, N.E., R. Radaelli, T. Dackevitch, A. Curioni, and A. Dal Belin Perufo. (1992). Benefits from genetics and molecular biology to improve the end use properties of cereals. In: *Cereal Chemistry and Technology: A Long Past and a Bright Future*, ed. P. Feillet. Montpellier, France: INRA, pp. 83-93.

Pomeranz, Y. (1978). Wheat in China. In: *Advances in Cereal Science and Technology*, vol. 2, ed. Y. Pomeranz. St. Paul, MN: American Association of Cereal Chemists, pp. 387-413.

Pomeranz, Y. (1987). Bread around the world. In: *Modern Cereal Science and Technology*, ed. Y. Pomeranz. New York, NY: VCH Publishers, Inc., pp. 258-333.

Prior, D. (1997). The cradle of civilization: A snapshot of the flour milling industry. *Feed & Grain*, October 1997, pp. 15-17.

Quaglia, G.B. (1988). Other durum wheat products. In: *Durum Wheat: Chemistry and Technology*, eds. G. Fabriani and C. Lintas. St. Paul, MN: American Association of Cereal Chemists, pp. 263-282.

Qarooni, J. (1994). Historic and present production, milling and baking industries in the countries of the Middle East and North Africa. Manhattan, KS: Department of Grain Science and Industry, Kansas State University.

Qarooni, J. (1996). Wheat Characteristics for fiat breads: Hard or soft, white or red? *Cereal Foods World* 41:391-395.

Quail, K., G. McMaster, and M. Wootton. (1990). The role of flour components in the production of Arabic bread. In: *Proceedings of the 40th Cereal Chemistry Conference*. Victoria, Australia: RACI, pp. 113-116.

Rahman, S., C.J. Jolly, and T.J. Higgins. (1991). The chemistry of wheat-grain hardness. *Chemistry in Australia* 58:397.

Robert, N. and J.B. Denis. (1996). Stability of baking quality in bread wheat using several statistical parameters. *Theoretical and Applied Genetics* 93:172-178.

Ross, A.S., K.J. Quail, and G.B. Crosbie. (1996). An insight into structural features leading to desirable alkaline noodle texture. In: *Cereals '96*, ed. C.W. Wrigley. Victoria, Australia: Royal Australian Chemical Institute, pp. 115-119.

Ruiz, M. and J.M. Carrillo. (1995). Relationships between different prolamin proteins and some quality properties in durum wheat. *Plant Breeding* 114:40-44.

Seibel, W. (1995). Wheat usage in Western Europe. In: *Wheat End Uses Around the World*, eds. H. Faridi and J.M. Faubion. St. Paul, MN: American Association of Cereal Chemists, pp. 93-125.

Shewry, P.R., N.G. Halford, and A.S. Tatham. (1992). High molecular weight subunits of wheat glutenin. *Journal of Cereal Science* 15:105-120.

Shewry, P.R. and A.S. Tatham. (1997). Disulfide bonds in wheat gluten proteins. *Journal of Cereal Science* 25:207-227.

Simmonds, D.H. (1989). Inherent quality factors in wheat. In: *Wheat and Wheat Quality in Australia*. © CSIRO: William Brooks Qld., 2:37-38.

Singh, N.K., K.W. Shepherd, and G.B. Cornish. (1991). A simplified SDS-PAGE procedure for separating LMW subunits of glutenin. *Journal of Cereal Science* 14: 203-208.

Singh, R.B. and V.P. Kulsherestha. (1996). Wheat. In: *Fifty Years of Crop Science Research in India*. New Delhi, India: Indian Council of Agricultural Research, pp. 219-249.

Sosinov, A.A. and F.A. Poperelya. (1980). Genetic classification of prolamins and its use for plant breeding. *Annales of Technological Agriculture* 29:229-245.

Steinberg, I. (1996). The rise of the flat bread. *Milling Flour and Feed Cereals International* (Summer Supplement, 1996) 190:8-11.

Sombrero, A. and P. Monneveux. (1989). Yellow berry in durum wheat (*Triticum durum* Desf.): The effect of nitrogen and water supply and cultivar. *Agricoltura Mediterranea* 119:349-360.

Trethowan, R.M., R.J. Peña, and M. vanGinkel. (2000). Breeding for grain quality: A manipulation of gene frequency. In: *Proceedings of the 6th International Wheat Conference*, June 2000, Budapest, Hungary (in press).

Weegels, P.L., R.J. Hamer, and J.D. Schofield. (1996) Critical Review: Functional properties of wheat glutenin. *Journal of Cereal Science* 23:1-18.

Williams, P.C. (1985). Survey of wheat flours used in the Near East. *Rachis* 4:17-20.

Williams, P.C., K.H. Norris, and W.S. Zarowski. (1982). Influence of temperature on estimation of protein and moisture in wheat by near-infrared reflectance. *Cereal Chemistry* 59:473-477.

Wishna, S. (1998). HVAC, back illumination imaging, light and near-infrared transmission as predictors of durum wheat milling parameters. *ASAE Annual International Meeting*, Orlando, Florida, USA, 12-16 July, 1998, p. 6.

Wrigley, C.W. (1991). Improved tests for cereal-grain quality based on better understanding of composition-quality relationships. In: *Cereals International*, eds. D.J. Martin, and C.W. Wrigley. Victoria, Australia: Royal Australian Chemical Institute, pp. 117-120.

World Grain. (1999). World grain review. Where does the grain go? *World Grain*, November, 1999, pp. 32-34, 36-37.

Yupsanis, T. and M. Moustakas. (1988). Relationship between quality, color of glume and gliadin electrophoregrams in durum wheat. *Plant Breeding* 10:30-35.

Zeleny, L. (1947). A simple sedimentation test for estimating the bread making and gluten qualities of wheat flour. *Cereal Chemistry* 24:465-475.

Breeding Wheat for Improved Milling and Baking Quality

E. J. Souza
R. A. Graybosch
M. J. Guttieri

SUMMARY. The molecular determinants of flour functionality and the targeted end-uses for flour define the methods used for improving wheat (*Triticum aestivum* L.) quality. We review major biochemical systems (gluten, grain hardness, starch, pentosans, lipids, and pigments) affecting wheat milling and baking quality. In early segregating generations (F_1 to F_4), breeding programs select for the basic criteria that define a market class (e.g., grain hardness, color, kernel shape, and gluten strength). When desired and possible, improvement of highly heritable traits (e.g., PPO, GBSS mutations, and glutenin sub-units) in early generations is practiced through selection for desired seed or molecular phenotype. In advanced generations (F_5 to cultivar release), bake tests and rheological measures identify breeding lines fitting the subtle characteristics of a market class. Breeding programs favor rapid measures of end-use quality to rank breeding lines for relative end-use quality rather than time consuming protocols that would precisely measure the rheology of flour.

E. J. Souza is Associate Professor, Plant Breeding and Genetics, University of Idaho, P.O. Box AA, Aberdeen, ID 83210 (E-mail: esouza@uidaho.edu).

R. A. Graybosch is Research Geneticist, Agricultural Research Service, USDA, University of Nebraska, Lincoln, NE 68583 (E-mail: rag@unlserve.unl.edu).

M. J. Guttieri is Senior Support Scientist, University of Idaho, P.O. Box AA, Aberdeen, ID 83210 (E-mail: mgttri@uidaho.edu).

[Haworth co-indexing entry note]: "Breeding Wheat for Improved Milling and Baking Quality." Souza, E. J., R. A. Graybosch, and M. J. Guttieri. Co-published simultaneously in *Journal of Crop Production* (Food Products Press, an imprint of The Haworth Press, Inc.) Vol. 5, No. 1/2 (#9/10), 2002, pp. 39-74; and: *Quality Improvement in Field Crops* (ed: A. S. Basra, and L. S. Randhawa) Food Products Press, an imprint of The Haworth Press, Inc., 2002, pp. 39-74. Single or multiple copies of this article are available for a fee from The Haworth Document Delivery Service [1-800-HAWORTH, 9:00 a.m. - 5:00 p.m. (EST). E-mail address: getinfo@haworthpressinc.com].

Integration of genetic, biochemical, and rheological factors with breeding goals and realistic selection protocols results in the improvement of end-use quality for new cultivars of wheat. *[Article copies available for a fee from The Haworth Document Delivery Service: 1-800-HAWORTH. E-mail address: <getinfo@haworthpressinc.com> Website: <http://www.HaworthPress. com> © 2002 by The Haworth Press, Inc. All rights reserved.]*

KEYWORDS. Wheat quality, flour, gluten, waxy wheat, polyphenol oxidase, breeding

ABBREVIATIONS. AACC–American Association of Cereal Chemists, ABA–abcisic acid, AWRC–alkaline water retention capacity, FN–falling number, GBSS–granule bound starch synthase, *Ha*–Hardness locus, NIR–near infrared reflectance, PPO–polyphenyl oxidase, *pin*A and *pin*B–puroindoline A and B, QTL–quantitative trait loci, RVA–rapid visco-analyser, SDS sedimentation–sodium dodecyl sulfate sedimentation test for gluten strength, SRC–solvent retention capacity, SKCS–single-kernel characterization system

INTRODUCTION

Working definitions of wheat (*Triticum aestivum* L.) quality vary through the wheat marketing chain. To the producer, wheat quality implies grain cleanliness and purity, grain yield and test weight, and, in many cases, protein content (depending upon whether premiums are being offered). The miller also is concerned with cleanliness, purity, test weight, and protein content. Yet, the miller focuses on flour yield, as well as the ability of the flour to satisfy customer manufacturing requirements. Manufacturers, the bakers, noodle and pasta makers, are most interested in performance of a particular flour lot in a specific food application.

Wheat moves through commercial channels from farm to manufacturer based upon systems of market class and grain grade. The kernel appearance characteristic of a market class in domestic and international trade represents an array of quality functions associated with the market class. New cultivars must match the defining quality characteristics of the targeted market class. The wheat breeder's task, therefore, is to develop high-yielding, disease resistant cultivars with quality characteristics that satisfy customer needs and generally are consistent with the expectations for the market class by each link of the marketing chain. For example, hard red winter wheat should bake bread, and soft white wheat should make pastries and cakes, etc.

Underlying genetic systems and macromolecules define dough rheology across the spectrum of wheat food products and define the characteristics of wheat classes. The genetic systems of these macromolecules are largely independent in their inheritance, producing a wide range of genetic combinations in wheat flour functionality. Environmental conditions also enhance or suppress expression of these flour components in combinations to increase the range of variation that millers and bakers observe in flours. Quality traits are subject to significant environmental and genotype × environmental variation (Peterson et al., 1992; 1998; Baenzinger et al., 1985). Breeders selecting improved quality genotypes must understand the genetic components of quality, their interaction with environment, and then use genetic based selection strategies to achieve end-use targets.

BIOCHEMICAL DETERMINANTS OF QUALITY

Simply described, wheat manufacturing quality is defined by kernel hardness and endosperm protein. In international trade, wheat typically is classified as either hard or soft (Giroux and Morris, 1998). Kernel hardness affects the milling characteristics and levels of starch damage from milling. Grain protein content and quality of a wheat class are major determining factors for assigning value grain in the marketplace. Gluten, the functional matrix of endosperm proteins in wheat, is unique to wheat among the cereal grains and provides the basis for the viscoelastic properties of dough, mechanical mixing character, extensibility, and elasticity. Beyond hardness and gluten, structural composition of starch (amylose:amylopectin ratio) regulates pasting properties of batters when heated and the firming of cell structure following baking. Penotsans, starch, and proteins condition the water absorption and viscosity in doughs. The primary biochemical components of wheat grain, protein, starch, and lipids interact to determine manufacturing quality. The relative amounts of these compounds, their composition, and their interactions, are the primary determinants of flour quality.

Flour proteins: Both flour protein concentration and composition affect processing quality. Flour protein is comprised of gluten and non-gluten proteins. Non-gluten proteins (often termed albumins and globulins) primarily are enzymes and other proteins soluble in aqueous solutions. Gluten proteins biologically function as seed storage proteins to be digested and utilized by germinating embryos. Gluten proteins initially are deposited in membrane-bound protein bodies. In soft wheats, protein bodies retain their identity (Stenvert and Kingswood, 1977), but in hard wheats the protein bodies coalesce, forming a continuous matrix that envelopes the starch grains (Bechtel et al., 1982).

Gluten proteins are comprised of two groups of molecules, gliadins and glutenins, glutamine and proline are the dominant amino acids, and disulfide bonds occur (Shewry et al., 1984). Gliadins contain only intramolecular disulfide bonds, exist in their native state as monomers, and are soluble in aqueous-alcohol solutions such as 70% ethanol. Four classes of gliadins are recognized, based on apparent molecular weights (*Mr*) as defined by gel electrophoresis and the chromosomal locations of the genes from which they originate. γ-Gliadins have the largest *Mr*, and arise from genes on the short arms of group 1 chromosomes. They are followed in decreasing *Mr* by γ-gliadins, also arising from 1AS, 1BS and 1DS, and the α- and β-gliadins, which are encoded by genes on 6AS, 6BS and 6DS. *Mr* of the gliadins ranges from approximately 25,000 to 75,000. Gliadins provide viscous flow to dough systems, and likely interact with native and added lipids during baking.

Glutenins form intermolecular disulfide bonds and are extracted with aqueous-alcohol only in the presence of reducing agents such as β-mercaptoethanol or dithiothreitol. In their native states, glutenins interact to form very large molecular weight polymers. Electrophoretic mobilities divide glutenins into two classes. High-molecular-weight (HMW) glutenin subunits arise from genes on the long arms of group 1 chromosomes, and have *Mr* ranging from 80,000-120,000. Low-molecular-weight (LMW) glutenin subunits have *Mr* similar to gliadins, and are encoded by genes on the short arms of group 1 chromosomes. A large, intermolecular disulfide linked matrix of glutenin proteins provide elasticity and strength in dough systems.

All gluten protein encoding genes occur in complex loci, composed of tightly linked genes (Payne et al., 1984). The HMW glutenin subunit clusters (*Glu-A1, Glu-B1* and *Glu-D1* loci) each contain two structural genes, one encoding a larger *Mr* x-subunit, the second encoding a smaller *Mr* y-subunit. In most hexaploid wheats, the *Glu-A1y* gene is inactive. Null, or non-functional genes also occur at other loci. Thus, hexaploid wheats carry 0-5 functional genes, with 4 and 5 being most commonly encountered. The *Mr* of the encoded gene products is genetically variable, with a multiple alternate alleles at each functional locus.

LMW glutenins arise from complex loci on the short arms of group 1 chromosomes. Various gliadins arise from the same complex loci. DNA sequencing of LMW glutenin- and gliadin-encoding genes demonstrates close sequence similarity to gliadins, suggesting the LMW glutenins and gliadins share a common evolutionary origin (Colot et al., 1989). However, LMW glutenins have the capacity to form inter-molecular disulfide bonds while LMW gliadins do not. Genetic recombination within these complex loci is rare; hence, all protein subunits arising from closely linked genes tend to be co-inherited (Payne et al., 1984).

Numerous reports link the presence/absence of specific gluten protein alleles, especially those encoding HMW glutenin subunits, with wheat breadmaking quality (Payne, 1987). Certain glutenin subunits, especially specific HMW glutenin subunits, increase dough strength. Other reports find the total glutenin polymer, expressed either as a proportion of total flour protein or as proportion of flour weight, is related to increased dough strength (Weegels et al., 1996). Loss of glutenin genes, from mutation, chromosomal deletion, or the presence of wheat-rye (*Secale cereale* L.) chromosomal translocation lines (in the latter wheat genes encoding LMW glutenins and gliadins are replaced by genes producing monomeric rye secalin proteins), reduces glutenin content and dough strength (Graybosch et al., 1990). In contrast, over-expression of the *Glu-D1* 1Dx5 allele in transgenic wheat significantly increased gluten strength beyond normally accepted quality parameters (Rooke et al., 1999). Selection for gluten strength can evaluate protein subunit composition summation like the "Payne" system of scoring advantageous HMW glutenin subunits (Payne, 1987). The voluminous study of glutenin proteins concludes simply that certain subunits interact efficiently in polymer formation, resulting in glutenin macro-polymers of both larger size and greater abundance, conditioning greater dough strength (Weegels et al., 1996). Glutenin content alone, however, is not the only protein factor determining wheat quality. Both total flour protein concentration, and the ratio of gliadins to glutenins, contributes to flour functionality (Graybosch et al., 1996). Gliadin concentration is linked to water absorption (Graybosch et al., 1993a), and gliadins likely interact with native and added lipids to form gas bubbles. In short, hard wheat flour needs adequate protein and gliadin to produce loaves of desired size, but also need glutenin to provide tolerance to overmixing, and to maintain loaf size after baking.

Kernel hardness: Hardness is the physical measure of kernel resistance to crushing force. Harder wheat requires greater force to mill for flour due to stronger binding of starch granules to the endosperm protein matrix. Hard texture in turn leads to greater damaged starch during the milling process as the endosperm fractures in planes through starch granules rather than breaking into smaller irregular aggregates containing intact starch granules (Pomeranz and Williams, 1990). Biochemically, softness is associated with a complex of starch granule membrane bound proteins collectively known as friabilin (Giroux and Morris, 1998). A lipid binding protein, friabilin, is composed of purolindoline a and b that are encoded by the Hardness locus (*Ha*) on the short arm of chromosome 5D (Gautier et al., 1994; Campbell et al., 1999). Puroindolines are unique in cereals for their emulsifying characteristics. In hard wheats, the emulsifying characteristics of the puroindolines is lost, and starch granules are bound tightly within the endosperm starch-protein matrix (Lillemo and Morris, 2000). Commonly, either of two mutations condition the transition of the *Ha* locus from the dominant (soft) to the recessive (hard) state: (1) a serine for glycine substitution at position 46 in puroindoline b (*Pinb-D1b*), or (2) the loss of expression of the puroindoline a (*Pina-D1b*) (Giroux and Morris, 1997, 1998). Rarer mutations of leucine to proline at position 60 (*Pinb-D1c*) and tryptophan to arginine at position 44 (*Pinb-D1d*), also induce the hard phenotype (Lillemo and Morris, 2000). The exact mutation to hardness is significant for breeders of hard wheat because the *Pinb-D1b* mutation typically produces softer hard kernels with higher flour extraction than *Pina-D1b* mutations. The major effects of the *Ha* 5DS locus are modified by other quantitative factors. Campbell et al. (1999) attributed 60% of the genetic variation for kernel hardness to the *Ha* locus with at least one other significant QTL contributing to hardness.

Soft texture derives in part from open space within the protein matrix surrounding the starch granules of non-vitreous soft wheats. In hard wheats the open spaces are filled, limiting the easy disassociation of one starch granule from another and from the protein matrix (Glen and Sanders, 1990). Pentosan quantities increase as hardness increases across market classes (Bettge and Morris, 2000). Membrane bound pentosans of the endosperm are associated with both increasing hardness and deterioration of soft wheat quality through increased water absorption. The membrane bound pentosans, primarily water-soluble arbinoxylans, are partially causal to increased hardness. As the membrane bound pentosans increase they may help bind the starch granules more tightly to the protein matrix, filling the intercellular airspaces (Bettge and Morris, 2000).

Direct selection toward a specific hardness phenotype is conducted using near-infra red (NIR) instrumentation and single kernel characterization system (SKCS). The SKCS, because it tests individual kernels, identifies breeding selections that are heterogeneous for the *Ha* locus. This is a distinct improvement over NIR for selection in hard × soft crosses, allowing discarding or reselection of breeding lines that are not true to market class. Two standard soft wheat measures of quality are the alkaline water retention tests (AWRC) and the sugar snap cookie test. Both are sensitive to the effects of damaged starch. The AWRC directly measures water absorption by the flour. By contrast, in a sugar-syrup environment of the sugar snap cookie dough, damaged starch draws water away from the sugar syrup, producing a more viscous dough. The flow of the dough is constrained with the increasing viscosity producing a smaller cookie diameter (Slade and Levine, 1994). Substituting damaged starch for flour in a sugar snap cookie formulation or for native prime starch in an all-starch cookie model system decreased cookie diameter and increased AWRC (Donelson and Gaines, 1998). In both sugar snap cookies and AWRC tests other components of the dough can mimic the effects of damaged starch. Gluten can both constrain expansion flow of cookie dough and increase water absorption of flours (Souza et al., 1994). Water-soluble pentosans are strongly hydrophilic, absorbing 10 times their weight in water, mimicking the effects of damaged starch in simple evaluation methods (Kulp, 1968; Jelaca and Hlynka, 1971). Modifying the cookie formulation using the wire-cut cookie can focus the selection toward the specific effects of damaged starch (Gaines et al., 1996, AACC method 10-53 and 10-54). Slade and Levine (1994) have proposed specific solvent tests for the individual components of water absorption. The solvent retention capacity (SRC) test partitions the absorptive capacity of flours using differential solvents (Gaines, 2000). The four common solvents are: water for a global measure of absorption, sodium carbonate to quantify damaged starch, sucrose solutions to measure pentosans, and lactic acid to assess the effects of high molecular weight glutenins.

Both damaged starch and pentosans are elevated by increasing kernel hardness (Gaines, 2000). Preliminary tests with the SRC evaluation indicate that it has potential to address a battery of flour functionalities for both hard and soft wheats by quantifying individual components of rheology (Gaines, 2000). In particular, this evaluation appears to be able to separate the effects of damaged starch and pentosans. Pentosans are structural carbohydrates, primarily arabinoxylans, with xylose, arabinose, and protein-bound galactose. Within the seed, approximately 70% of pentosans are found within the seed coat and pericarp (Hoseney, 1986). Poor milling genotypes may have excessively high flour pentosan content because of poor bran separation. Within the endosperm, the pentosan composition derives from remnant cell walls, which typically are composed of pentosans, hemi-celluloses, β-glucans, but no celluloses (Hoseney, 1986). Endosperm pentosan is composed primarily of xylose polymers with branches of arabinose (D'Appolonia and Rayas-Duarte, 1994). High water-soluble and total pentosan contents correlate with environment and genotypes that produce small diameter cookies (Kaldy et al., 1991; Bettge and Morris, 2000). Greater pentosan content may be desirable in some cases for hard wheat flours to increase the moisture content of baked bread. However, water absorption derived from pentosan is of less value in the bread baking process than water absorption from gluten. Pentosans may degrade bread appearance, yet may also interact with amylose to slow amylose related staling (D'Appolonia and Rayas-Duarte, 1994).

Starch composition: Until the last decade, starch quality research focused on the effects of damaged starch. Little variation in starch biochemical properties was suspected. Wheat starch is composed of two types of glucose polymers, unbranched amylose and branched amylopectin molecules. Typical amylose content ranges from 25-30% of starch. Amylose is synthesized by the granule-bound starch synthase (GBSS), also known as the waxy protein. Spontaneously occurring waxy (amylose-free) wheats have never been observed, no doubt due to the hexaploid nature of bread wheats. Any spontaneous mutations of the genes (*wx* loci) encoding GBSS were not detected, as two functional copies remain. In the early 1990s, Nakamura and colleagues separated by gel electrophoresis the gene products of the three *wx* loci (Nakamura et al., 1993). Using this technique, non-functional or "null" alleles were found to be quite common at the *Wx-A1*, and *Wx-B1* loci, and were detected in at least one source at the *Wx-D1* locus (Yamamori et al., 1994). Combining the three null alleles produces waxy (amylose-free) wheat starch (Nakamura et al., 1995).

Waxy wheats are only just beginning to be characterized. However, wheats carrying one or two null alleles occur among advanced breeding lines and cultivars. Such wheats, often termed "partial waxy," produce starch with reduced amylose content, generally in the range of 15% for double-null lines to 20-25% in single null lines (Nakamura et al., 1993). Reduced amylose content is advantageous in certain Asian noodle products, especially in udon noodles, typically produced from soft wheats (Miura and Tanii, 1994). The increased starch swelling associated with elevated amylopectin softens noodles, decreasing the noodle's hardness and elevating oil absorption in fried noodles. High amylose content is undesirable for high protein noodles where firm texture is desirable. Lines with single null alleles do not suffer in breadmaking applications. However, the presence of two null alleles may reduce loaf internal crumb grain and lead to partially collapsed loaves (unpublished observations, University of Nebraska Wheat Quality Lab). The firming of crumb texture due to amylose is reduced in high amylopectin wheats.

There are some reports of high amylose content wheats, especially among durum (*Triticum turgidum* var. *durum* L.; Watanabe and Miura, 1998). In addition, genetic polymorphism for genes encoding starch branching enzymes has been detected, suggesting that some genetic complementation could result in the production of more wheats with elevated amylose contents (Nagamine et al., 1998). High amylose wheats are not yet common in breeding populations. However, in the near future, wheat breeders may have the genetic materials available to develop wheats with starch amylose contents ranging from 0 to more than 40%.

Preharvest sprouting: In addition to starch composition, starch quality is affected by preharvest sprouting. Preharvest sprouting elicits synthesis of α-amylase in the aluerone tissue, degrading the endosperm starch granule structures even before radicle emergence is obvious. Textures of bread, sponge cakes, cookies, and sheeted noodles are sensitive to even low levels of preharvest sprouting, causing excessively extensible and sticky doughs (Rasper et al., 1993). The effects of preharvest sprouting are obvious in the rapid reduction of falling number (FN) measurements and the correlated Rapid Viscoanalyser (RVA) assays (Mathewson and Pomeranz, 1978; Panozzo and McCormick, 1993). The grain trade uses FN to measure preharvest sprouting damage and exclude sprouted grain from food grade wheat. Breeders develop wheats resistant to preharvest sprouting to reduce farmer's risk of economic loss in the case of untimely rains at harvest. Red wheats are less prone to preharvest sprouting than white wheats, with the red pigmentation alleles (*R*) positively linked with tolerance to preharvest sprouting (Flintham, 1993). Red wheats traditionally are planted in areas, such as the eastern U.S. and northern Europe, with high probabilities of rain at harvest.

Preharvest sprouting resistance is anecdotally and quantitatively assigned to a number of older white seeded cultivars, such as 'Brevor' and 'Clark's Cream' (DeMacon and Morris, 1993). Expression of the trait typically has a large interaction between the environment and genotype (Anderson and Sorrells, 1993) limiting gains from selection. Seed dormancy fades quickly after harvest; timely assessment of preharvest sprouting, which adjusts for maturity differences among wheat genotypes, is laborious and can detract from other breeding considerations that occur at harvest. Some evaluation can be deferred by storing mature spikes below 20°C to prevent loss of dormancy (Noll and Czarnecki, 1980). Standardized dormancy tests include measures of sprouting in spikes placed in artificial mist chambers and germination of threshed seeds (DeMacon and Morris, 1993; Paterson and Sorrells, 1990). Direct selection dormancy tests can improve the tolerance of wheat genotypes to preharvest sprouting (Sorrells and Paterson, 1986; Morris and Paulsen, 1989).

Marker-assisted selection is an alternative to direct selection for preharvest sprouting. Anderson and Sorrells (1993) studied two single cross populations involving white winter wheats and found 10 QTLs for preharvest sprouting tolerance. Different markers were identified in the two populations suggesting the possibility of pyramiding genes for tolerance. Additional work has investigated the genes specifically involved in mediating the germination response. Abscisic acid (ABA) suppresses gibberellic acid (GA)-mediated synthesis of α-amylase. At least one source of preharvest sprouting in white wheat, 'Clark's Cream', has elevated sensitivity to ABA. In wheat, a number of ABA-induced genes appear to be related to dormancy (Morris et al., 1991; Rayfuse et al., 1993). Synteny between wheat and maize (*Zea mays* L.) suggests that several wheat QTLs may correspond to the maize *Viviparous* locus (Sorrells, personal communication), an ABA-regulated gene controlling precocious germination (Vasil et al., 1995; Schultz et al., 1998). The ABA-mediated regulation of dormancy may be sufficiently general as to allow direct selection for ABA receptors or for transgenic modification of the genes inhibiting germination.

Lipids: Wheat does not store appreciable amounts of lipid within the endosperm. Flour contains approximately 1% (by weight) lipid (Morrison, 1988). Flour lipids are classified as starch or nonstarch lipids. Starch lipids, largely phospholipids, are integral parts of the starch granule. Nonstarch endosperm lipids largely represent remnant plastid membranes, or derive from germ and aleurone layers during milling.

Nonstarch endosperm lipids are approximately a 50:50 mix of nonpolar lipids (including free fatty acids) and polar lipids (glyco- plus phospholipids). Increased polar lipid content may be related to improved breadmaking qualities, especially loaf volume and internal loaf appearances (Panozzo et al., 1993). Indeed, lipids are common components of dough "improvers" and are used to improve shelf-life duration, crumb texture, and loaf volume. The genetics of wheat lipid composition are not well understood. On chromosome 5D, a gene linked to the *Ha* hardness gene cluster conditions increased polar lipid content (Morrison et al., 1989). The enhanced polar lipid content may be a consequence of hardness, a cause of hardness, or merely an associated trait. Genetic diversity in lipid composition does not seem great enough to warrant selection for the trait as part of a quality improvement program.

Flour pigmentation and product color: Specific genetic systems, primarily polyphenol oxidase (PPO) in the aleurone of wheat, condition the brightness and color stability of Asian noodles (Hatcher and Kruger, 1993). Flour pigmentation (mainly yellowness) is independent of the browning effects of PPO (Miskelly, 1984). Yellow flour pigmentation is due primarily to xanthophyll deposited in the endosperm (Miskelly, 1984).

PPO enzymes degrade noodle brightness through time and are the primary factors determining noodle brightness 24 h after sheeting. Factors such as high protein and starch damage also can dull brightness (Miskelly, 1984), and may be important components of within-cultivar environmental variation in noodle color. However, genetic differences between cultivars are due primarily to PPO (Baik et al., 1995). The browning from PPO enzymes will darken whole grain doughs (Miskelly, 1984) and chapati made with high extraction flour (Singh and Sheoran, 1972). Milling technique can reduce PPO from flour (Hatcher and Kruger, 1993). As extraction rates increase, the PPO level in the flour fraction also increases dramatically (Hatcher and Kruger, 1993). Shifting from 75% flour extraction to a 50% extraction patent flour reduces flour PPO levels 88 to 100% in hard wheats and causes a corresponding improvement in flour brightness (Kruger et al., 1994). Wheats with genetically low PPO activity can be milled to higher extraction levels and, therefore, have greater value in milling and noodle manufacture.

Genetic variation exists among cultivars for levels of PPO in the aleurone (Bernier and Howes, 1994) and the degree to which it can be removed by milling (Kruger et al., 1994). The trait appears to be highly heritable with limited genotype × environment interactions (Windes et al., 1995). Selection in Idaho for improved noodle color resulted in the release of the hard white noodle wheat 'Idaho 377s', which has approximately 50% lower whole grain PPO activity compared to the hard white wheat cultivar 'Klasic' (Souza et al., 1997). However, a base level of PPO activity always exists within the *T. aestivum*. In contrast, cultivars of durum wheat commonly have null activity for PPO (Mahoney and Ramsey, 1992). Therefore, the D genome is speculated to be the source of baseline PPO activity in bread wheat and may be conditioned by a locus on chromosome 2DS (Howes et al., 1995). Segregation studies with synthetic hexaploids in wheat suggest at least two genes are present in the D genome conditioning temporal dulling typically attributed to PPO (Souza, personal observation).

PPO has been studied in other plant species. PPO is a copper binding metaloprotein that catalyzes the hydroxylation of monophenols to *o*-diphenols and *o*-dihydroxy phenols to *o*-quinones. The *o*-quinones are highly electrophilic and cross-links cellular substituents. The reactivity of PPO enzyme products produces the browning discoloration common in food products from apples to potatoes. Wheat likely has multiple gene copies of PPO. PPO enzyme isolated from wheat exists as a 115,000 kD tetramer with two types of subunits, a 30,000 kD and a 23,500 kD polypeptides (Interesse et al., 1983). The activity of the PPO enzymes recovered from wheat leaves had higher activity toward di- and polyphenols than monophenols (Interesse et al., 1980). Kruger (1976) identified 12 isozymes in developing seeds of wheat. Two of the isozymes were unique to the maternal seed tissues and remained at maturation. However, Kruger (1976) was unable to find PPO isozymes particular to the endosperm in the mature endosperm. The developmentally regulated activity in wheat is consistent with studies in other species where activity of PPO is induced and lost during specific stages of growth and development (Thygesen et al., 1995).

Flour pigmentation is independent of the PPO browning effects (Miskelly, 1984). Yellowness of wheat products is due primarily to xanthophyll deposited in the endosperm (Miskelly, 1984). Flour chromaticity tends to increase with flour extraction rate (Kruger et al., 1994). Moderately pigmented wheat flours, typical of US bread wheats, would have appropriate chromic character for white salted noodles, Korean-style fried noodles, and fried noodles of Northern China (Crosbie et al., 1990). Higher levels of pigmentation may be desirable for alkali noodles and Cantonese-style fried noodles (Azudin, 1998). Specific cultivars such as 'Batavia-AUS', 'Krichauff', and 'Rosella' have high levels of flour xanthophylls (Azudin, 1998). Among North American hard wheats 'Idaho 377s' and 'Manning' have favorable combinations of low PPO and moderate to high levels of yellow pigmentation. Generally though, pronounced yellow pigmentation is uncommon in US bread wheats because it discolors the loaf interior. Genetic control of yellow pigmentation in euploid wheat is poorly documented. Mapping work with the recombinant inbred population 'Clark's Cream' × NY6432-18 suggests that yellow pigmentation is controlled by several loci that are strongly affected by environment, yet have limited genotype × environment interaction (Souza and Udall, unpublished data). In Intermountain U.S. spring wheat trials, yellow pigmentation appears to have a large environmental component with some genotypes expressing yellow pigmentation more often than others (Souza et al., 1993). Late season moisture stress is one environmental factor that elevates the level of flour pigmentation (Guttieri et al., in review) and may be related to the drought stress response of wheat, which elevates synthesis of xanthophylls (Loggini et al., 1999).

SELECTION FOR QUALITY

Commercial mills and bakeries stringently specify flour based both on flour composition (moisture, protein, ash) and rheological functionality (mixograph, farinograph, and alveograph) tests (Faridi et al., 1994). Commercial measures of soft wheat flour quality attempt to precisely determine rheological characteristics of the flour specified for bakery formulations. Genetic selection assays, in contrast, attempt to correctly rank different wheat selections for their relative quality merits. Because the exact measure of quality in genetic selection is relatively unimportant (the phenotype may be completely different in the next environment), genetic selection typically uses hardness tests, micro-milling assays, particle size index, SDS sedimentation, small-scale bake tests, and quick measures of solvent absorption rather than the time-consuming farinograph and alveograph tests. Genetic selection typically emphasizes testing with high throughput and low per-unit assay cost. The integrated selection of quality measures significantly improved the quality of hard wheat cultivars in the U.S. (Cox et al., 1989; Souza et al., 1993). Although less rigorously studied, selection of soft wheat cultivars has produced more uniform and better quality pastry and cracker wheats than existed in the North America prior to World War II.

Wheat improvement programs typically use populations segregated by grain hardness. Commonly, hard and soft wheat improvement is handled in different programs breeding for specific geographic regions. Each breeder tailors their crop improvement program based on goals and available resources. Nonetheless, hard and soft wheat improvement programs often follow parallel selection channels, with Table 1 representing a breeding program ideotype. Parents are selected for favorable complementation of genes, with at least one parent having adapted manufacturing quality. Following crossing, selfing and segregation produce an array of inbred lines. Typically only limited and unreplicated seed quantities are available for testing during inbreeding. Destructive tests, like milling tests, limit the ability to advance germplasm to the next generation. Also, early generations often are produced in environments, such as greenhouses or off-season winter nurseries, that are unrepresentative of target production environments. Therefore, selection for additive genetic traits with minimal environmental interaction, such as PPO reaction, or marker-assisted selection are likely to be most profitable.

Early selection among inbred lines typically emphasizes meeting minimal class quality requirements that are easy to assess, such as gluten strength, protein quantity, hardness, and flour water absorption (Table 1, F_4). As the breeding line numbers decrease and replication through time and space increases, selection emphasizes destructive tests that more finely discriminate between acceptable and unacceptable genotypes (Table 1, F_5 to F_7). Typically, breeders channel materials into target uses. If breeding materials do not fit predetermined categories of quality, for better or for worse, the lines are discarded. Regional or multistate cooperative testing evaluates breeding lines across multiple growing regions. Regional quality laboratories perform levels of quality testing that are typically beyond the scope of a single breeding program (Table 1, F_8 to F_9). Finally, prior to, or at the time of, cultivar release, industry panels of end-users conduct quality testing to verify the smaller scale testing conducted in breeding programs. Each step of selection must be optimized based on the number of lines evaluated and the relative merit, cost, and time for the assay.

Breeding for hard wheat quality: Leavened bakery products such as breads, rolls, bagels, baguettes, etc., and certain types of Asian noodles traditionally use flour of hard wheat. Pasta noodles typically are produced from durum (tetraploid) wheats. The precise quality characteristics desired in a breeding program vary somewhat, depending upon the goals of the program. Programs designed to develop wheats for large-scale mechanized bakeries using sponge-and-dough technology or the Chorleywood process aim for wheats with strong tolerance to over-

TABLE 1. Comparison of selection for hard wheat quality and soft wheat quality in the context of a breeding program developing improved cultivars for production.

Generation	Hard Wheat Quality Scheme	Soft Wheat Quality Scheme
Parental generation: Characterize parents for important quality traits. Use information to develop complementary matings	**Important traits of interest:** 1RS translocation, GBSS alleles, PPO levels, and HMW glutenin composition	**Important traits of interest:** GBSS alleles, PPO levels, and HMW glutenin composition, Ha character, pentosan levels
F₁-F₃ generations: Advance materials using breeding methods (pedigree, bulk, etc.). Limited selection for disease resistance, height, and winter hardiness. Head selections made from F₃ bulks.	**Limited quality selection:** Some progeny testing for gluten strength and assessment of highly heritable traits such as seed color and PPO.	**Limited quality selection:** Some progeny testing for hardness and assessment of highly heritable traits such as seed color and PPO.
F₄ generation: F₃₋₄ headrows are grown at a single location. Select rows to advance based on disease resistance, winter survival, lodging resistance, etc. Retain 100 grams seed from each selected row.	**Selection for minimal quality requirements:** Test grain samples for hardness (using NIR hardness or single kernel hardness tester), PPO activity, falling numbers and protein concentration. Mill 50 grams to flour in a laboratory scale (micro) mill. Test flour samples for flour protein content, flour pasting for stored product discoloration, starch swelling power (or GBSS allelic status) and some measure of dough or gluten strength (SDS sedimentation or mixograph). Discard samples that do not meet or exceed check cultivars in quality characteristics.	**Selection for minimal quality requirements:** Test grain samples for hardness (using NIR hardness), PPO activity. Discard small seed genotypes. Club wheats with non-club kernel confirmation should be discarded. Discard wheats with excessively high protein content. Select club wheats for very low SDS-sedimentation values. Cracker type wheats: discard wheats with very low SDS-sedimentation or values similar to hard winter wheats. Micro-milling for gross estimate of milling performance, starch swelling power test.

TABLE 1 (continued)

Generation	Hard Wheat Quality Scheme	Soft Wheat Quality Scheme
F₅ generation: Preliminary yield testing with limited replication within locations planted at one to several locations. Evaluate yield potential, agronomic type, general area of adaptation, and quantitative disease resistance traits (e.g., adult plant rust resistance, foot rots and virus resistance). Use irrigation applied at maturity in one set of trials to assess preharvest sprouting resistance.	*Intermediate scale quality assessment:* Test harvested grain from irrigation study for susceptibility to preharvest sprouting using the falling number test. Bulk grain from multi-location trials (or, if resources allow, test independent environments). Test grain for protein content and grain hardness. Mill grain to flour on an experimental laboratory mill (Buhler or other similar type of mill). Test flour samples for protein concentration, starch swelling (omit if GBSS allelic status is known) gluten strength (mixograph or SDS sedimentation test). Bake wheats with acceptable gluten strength using a small-scale, 100-gram pup-loaf (straight-dough) procedure. Alkaline or Chinese (neutral pH) noodles for color and color stability measures. Segregate lines based on results; low-mid protein content with mellow gluten and lack of product discoloration become "noodle" wheats. Higher protein, medium to strong gluten types with good loaf characteristics (volume, internal and external appearance) become "bread" wheats.	*Intermediate scale quality assessment:* Bulk grain from multi-location trials (or, if resources allow, test independent environments). Test grain for protein content and grain hardness. Mill grain to flour on an experimental laboratory mill (Buhler or other similar type of mill). Select for high break flour yield. Test flour samples for protein concentration, flour ash, starch swelling (omit if GBSS allelic status is known). Use micro-tests to select for low water absorption capacity using either AWRC or SRC tests. Bake sugar snap cookies, measure diameter and top-grain. Alkaline or white salted noodles for color and color stability measures.

*F₅ generation:*

F₆ and F₇ generations:
Grow in multiple locations with additional levels of replications. Assess yield adaptation and relative stability of performance.

F₈ and F₉ generations:
Enter genotypes into regional testing and continue testing within local testing programs. Assess regional adaptation and allow other researchers to validate disease resistance, yield performance, and quality traits.

Replicated quality testing:
Repeat quality tests as per F_5 generation. If resources allow, test locations independently to establish environmental stability.

Select more intensive for quality based on multi-year and multi-location averages, adding selection pressure for traits of lower heritability such as mixing tolerance, crumb grain and noodle texture analysis (particularly noodle hardness).

Advance to the F_7 and F_8 generations only the lines that pass more stringent quality testing in the previous generation.

Large scale quality testing:
Repeat quality tests as per F_5-F_7. In the U.S., one of four USDA-ARS Regional Quality Laboratories conduct such tests on regional entries.

Arrange for large-scale milling (50-60 kg) with noodle and bake tests by commercial partners. In the U.S., this is coordinated under the aegis of the Wheat Quality Council, and noodle testing may be arranged through the Wheat Marketing Center, Portland, Oregon.

Make release decisions based on agronomic performance and quality tests. Use lines for the next round of matings in the breeding program.

Replicated quality testing:
Repeat quality tests as per F_5 generation. If resources allow, test locations independently to establish environmental stability.

Select more intensive for quality based on multi-year and multi-location averages, adding selection pressure for traits of lower heritability such as flour ash, break flour yield, and top-grain score, and noodle texture analysis.

If the soft line is targeted towards, udon or other noodle application, RVA analysis should be used to select for high initial and final flour viscosity.

Large scale quality testing:
Repeat quality tests as per F_5-F_7. In the U.S., one of four USDA-ARS Regional Quality Laboratories conduct such tests on regional entries. Add additional finished product testing: wire-cut cookies, Japanese sponge cakes, Chinese steam breads, batter viscosity, and pancakes.

Arrange for large-scale noodle (50-60 kg) and bake tests by commercial partners. In the U.S., this is coordinated under the aegis of the Wheat Quality Council, and noodle testing may be arranged through the Wheat Marketing Center, Portland, Oregon.

Make release decisions based on agronomic performance and quality tests. Use lines for the next round of matings in the breeding program.

mixing. Longer mixing times can be tolerated in sponge-and-dough plants, while the Chorleywood process requires short mixing times, but still requires tolerance to overmixing. Programs producing wheats for developing countries target mellow gluten types beneficial for bread produced at home or in small-scale bakeshops. In such applications, high protein content and long mixing times can be detrimental.

Thick Asian noodles (Fried or Hokkien) require lower protein contents and mellower gluten than typical for leavened bakery products (Miskelly, 1996). Thin noodles such as bammee or ghioza commonly use high protein, strong gluten wheats such as Dark Northern Spring or Canadian Western Red Spring. Because flour protein dulls the appearance of Asian noodles, manufacturers are limited in their ability to increase flour protein content to improve flour functionality. Therefore, wheats developed for noodles need good gluten strength per unit of protein.

For use in U.S. commercial bread making procedures, hard wheat varieties should have adequate protein concentration (flour protein after milling of greater than 11.5%), good mixing tolerance, adequate loaf volume, and should produce loaves with good internal appearance. Flour water absorption should be high enough to meet product specifications. Tolerance to overmixing is defined in commercial settings as the range of mix times above and below peak dough development, at which a given lot of flour will function properly in a baking procedure. This is an important variable, as, in mechanized plants, under- or overmixed doughs can result in poor final product or production stoppage ("down-time"). Proper grain hardness and grain conformation (test weight or thousand-kernel weight) are needed for the milling process. Low levels of PPO is linked to discoloration of refrigerated and frozen doughs, making reduced levels of PPO more desirable than high.

Wheat breeding programs historically selected early generation materials on the basis of protein concentration, combined with some measure of gluten strength, either the SDS sedimentation test (AACC method 56-63 modified by Dick and Quick, 1983) or some type of physical dough testing procedure (primarily mixograph, extensigraph, and alveograph; AACC methods 54-21, 54-10, and 54-30A, respectively). While numerous schemes have been proposed for the selection of wheats based either on HMW glutenin subunit composition (Payne, 1987) or via direct measurement of glutenin content (Singh et al., 1990), such procedures have not gained favor in U.S. hard wheat breeding programs. This situation partially derives from the relatively high frequency of "optimal" subunits already present in U.S. hard wheat gene pools (Graybosch, 1992). The same type of information also derives from simpler and less expensive tests such as mixograph analysis and SDS sedimentation tests. Various glutenin-based assays have been correlated with small-scale tests. Yet, their prediction of commercial-scale performance essentially is unknown.

Worldwide, wheat-rye chromosomal translocations in breeding gene pools probably are the single most important negative factor influencing quality. Lines carrying either 1AL.1RS or 1BL.1RS have reduced glutenin content, more water-soluble protein, and, especially among 1BL.1RS wheats, reduced mixing tolerance. Cytogenetic, biochemical, immunological and molecular tests detect 1RS in wheat with some tests applicable to large population sizes (Andrews et al., 1996). The effects of 1RS vary with both its location in the genome (1AL.1RS wheats are less severely impacted than 1BL.1RS wheats) and with the genetic background in which it resides (Graybosch et al., 1993b). Poor quality 1RS wheats readily are detected by use of SDS sedimentation tests (Lee et al., 1995); those that pass the minimal values established by each breeding program may possess quality characteristics permitting their commercial use.

In later generations, small-scale (100 gram flour, AACC method 10-10B) "pup-loaf" bake procedures are used (Table 1, F_5 and higher). All tests used by U.S. wheat breeding programs are limited in that there are doubts as to how accurately they can predict commercial-scale quality control test bakes (Graybosch et al., 1999). The best predictor of overall commercial quality potential is large pup-loaf volume, representing the combined favorable effects of high protein concentration and high glutenin content. Most programs lack the resources to evaluate breeding lines for commercial bread performance, leaving breeders with the available small-scale tools and the hope that the selected lines will prove suitable in large-scale systems.

The processing quality of nearly all hard wheats improves with increased flour protein concentration. Protein concentration is somewhat amenable to selection. Safe, accurate, and simple tests, especially those based on near-infrared reflectance spectroscopy (NIR; AACC method 39-10 and 39-11), are available as selection tools. Protein concentration is, however, highly subject to environmental conditions, with as much as 50% of the overall variation in protein concentration being due to environment (Peterson et al., 1992). Genetic sources of increased protein concentration are known and, more importantly, seem to be conditioned by a number of independent loci. Thus, gene pyramiding may increase protein concentration. Two notable sources of higher flour protein are

'Atlas 66' and 'Plainsman V'. 'Atlas 66' contributed to the development of the hard winter wheat cultivars 'Lancota' and 'Vista'. 'Plainsman V' served as a parent for 'Norkan', 'McGuire', 'Karl', 'Nuplains', and 'Wesley'. The two sources differ in their effects on dough strength. 'Atlas 66' types, such as 'Lancota', have shorter mix times, but good mixing tolerance and good loaf volume potential. 'Plainsman V' types have long mixing times, strong tolerance to overmixing and high loaf volume potential, producing superior baking quality wheats (Peterson et al., 1993).

Breeders in the Great Plains of North America have concentrated on the development of improved hard red winter and spring wheats. Current market forces are driving an increased emphasis on hard white wheats. Hard white is the wheat of choice in many international markets, and in the U.S. for use in whole-grain bakery products. International uses of hard white wheats include Asian-type "wet" noodles, flatbreads, and tortillas. Breeding-level tests and commercial noodle quality are disconnected to a greater degree than in bread quality. Noodle texture and mouth-feel are assessed by large-scale tests and taste panels. However, some simply measured traits are important starting points for selection. Noodle hardness as measured by TA-TX2 Texture Analyzer is closely related to the texture characteristic often called "bite," or the force to incise the noodle. Harder bite is preferred for most noodles, except for the soft wheat-based Japanese udon noodle. Bite hardness is highly correlated to loaf volume (Lang et al., 1998). Higher protein content increases loaf volume and noodle hardness. Yet, increased protein content also dulls noodle color (Baik et al., 1995), limiting noodle manufacturers' flexibility to correct poor gluten strength with increased flour protein. Selection for loaf volume per unit of protein may provide a window for improvement of both noodle and bread quality. Later segregation of breeding lines may be necessary to channel mellow gluten wheats for noodles, and stronger gluten types for breads. Environmental effects on dough strength can be significant, and stronger gluten wheats tend to have more stable quality under severe heat stress, an effect common in the Great Plains. Yet, the low protein hard wheat production, common in higher rainfall zones and irrigated production, may be ideally suited for Asian export quality if noodle product color can be improved.

Product discoloration also is a concern, but small-scale tests can address this issue. Flour PPO enzyme activity is associated with discoloration of many food products derived from white wheats, especially those that are refrigerated between formulation and cooking. Thus, newly developed white wheats should be selected for low, or, if possible, nil levels of PPO (a trait also of potential value in red wheats) using simple PPO assays (Kruger et al., 1994; Bernier and Howes, 1994). Other, unidentified factors may also contribute to product discoloration. Dough sheeting tests, in which flour is mixed with alkaline solutions, can be used to evaluate lines for color stability (Kruger et al., 1992; Morris et al., 2000).

In general, hard red wheats are more tolerant to preharvest sprouting than hard white wheats (Mares and Ellison, 1989). In breeding programs, the FN test (AACC method 56-81B), which is an indirect measurement of α-amylase activity, can be used in early generations to identify white wheats with preharvest sprouting tolerance. An important limitation, however, is that the FN test may be used to differentiate lines only if sprouting has occurred. When environmental conditions are not "favorable" for expression of sprouting susceptibility, irrigation systems might be needed to induce the trait. Alternatively, germination tests on grain harvested at physiological maturity, and subsequently dried, have been explored as a means of estimating sprouting susceptibility (Wu and Carver, 1999).

Starch swelling potential, the ability of heated starch gel to absorb and retain water, contributes to Asian noodle quality when softer texture is desired (Azudin, 1998). Starch swelling is inversely correlated with amylose content. Partial waxy lines, or those carrying 1 or 2 GBSS null alleles, typically have both reduced starch amylose and increased starch swelling potential. The linkage is tenuous between the presence of GBSS null alleles and desired texture of noodles typically made from hard wheats, such as alkaline, Chinese, or wonton noodles. One report claims a negative relationship (Ross et al., 1996). However, additional reports show positive effects of elevated starch swelling potential on alkaline noodle quality (Konik et al., 1994). This clearly is an area that requires additional investigation. Starch properties could be measured either by direct determination of swelling power (Crosbie et al., 1992; Konik et al., 1993), or indirectly by determining GBSS genotypes through gel electrophoresis. The latter is unambiguous, but does require some technical expertise.

Grain hardness probably is the single most important quality factor, as it ultimately governs the type of end-use quality applications in which a wheat cultivar may be used. Due to partitioning of gene pools, many programs rarely screen for hardness in early generations. In addition, wheats of opposite hardness to that desired are easily detected during milling for small-scale tests. The availability of SKCS makes hardness screening during early generations relatively easy. In addition, the SKCS instrument can determine seed weight and size, and more recent models have been outfitted with NIR technology for protein concentration determination. The instrument also calculates the deviation within a sample for each trait, useful information if one is attempting to develop more uniform cultivars.

Selection for wheat quality traits within hard wheat breeding programs increases in complexity, as the goals of hard wheat breeding programs diversify. An integration of quality screening in a breeding program follows (Table 1). This scheme is based on the discussion above, and on the long established programs at the University of Nebraska/USDA-ARS program, Lincoln, Nebraska, USA and the University of Idaho, Aberdeen, Idaho, USA. It also reflects the authors' personal bias, in that some subjectivity has been used in assigning assays to specific generations. This is based, however, on the authors' experience with the various tests. The scheme also assumes that financial resources are not infinite. No differentiation based on grain color is included, as it is assumed that, at present, either red or white wheats could be sold and used in either bread or noodle applications. It also assumes, as is the case in most U.S. breeding programs at this writing, that small-scale noodle-testing equipment is not available.

This program should gradually improve quality, or maintain quality characteristics, and diversifying end-use quality among elite lines for a range of hard wheat uses. Many early generation tests are simple, with limited requirements of capital equipment and technical expertise. Multi-year and location testing helps establish the environmental stability of quality within selected lines.

Breeding for soft wheat quality: Soft wheat end-uses range from air-classified, chlorinated cake flours to whole-wheat flours for artisan breads. Good soft wheats tend to have high break flour yields with high overall flour extraction rates. Because of the breadth of soft wheat uses, the diversity of flour functionality may be greater than among hard wheats. Common among soft wheat breeding goals is selection of physically soft wheats that require low milling energy, producing low levels of damaged starch and pentosans in the finished flour. Most pastry, cracker, and snack foods benefit from the low water absorption characteristics of typical soft wheats. Batters and pancake mixes are exceptional in requiring soft wheats with relatively weak gluten and high water absorption due to damaged starch or pentosans. Soft wheats grown in moisture stressed environments have the phenotype desired for batters. Therefore, breeders typically target the dominant quality use (low water absorption flours) and allow normal environmental variation to provide need variability for the marketplace.

Core selection methods for improving soft wheat quality are: (1) grain hardness (NIR and SKCS), (2) protein concentration, (3) milling yield and measures of milling quality such as particle size index (AACC method 55-30) and flour ash (AACC method 08-01), (4) water absorption (AWRC, AACC method 56-10, and SRC, AACC method 56-11; Gaines, 2000), and (5) sugar snap cookie bake (AACC method 10-50D). Secondary measures of quality that are used in regional programs or for specific end-uses are: (1) mixograph and alveograph for measuring gluten strength and water absorption, (2) flour swelling test or RVA for measuring flour pasting viscosity (Gaines et al., 2000), (3) sheeted noodle color (Kruger et al., 1992), and (4) MacMichael viscosity (AACC method 56-80). Japanese sponge cake is the definitive test for club wheat quality, yet is of secondary importance for other soft classes (Nagao et al., 1976). For each additional trait selected the power to make breeding progress for any one trait is decreased. Fortunately, many of the core quality characteristics of soft wheats are positively correlated (Gaines, 1985). Of the secondary quality characteristics some are relatively independent to moderately negatively correlated to the core quality measures. Noodle color falls into the first category with initial brightness associated with lower protein content and better milling yield (Kruger, 1994) while change in noodle color due to PPO is generally independent of other quality measures (Baik et al., 1995). Gluten strength (Souza et al., 1994) and elevated amylopectin content associated with greater flour swelling volume are genetically independent of hardness characteristics, yet may be negatively associated to other core quality measures (Souza and Guttieri, personal observations at Idaho Wheat Quality Laboratory). However, the association is slight enough for gain from selection to occur for both core and secondary measures of quality.

In addition to core and secondary selection criteria, a number of evaluations exist to characterize finished product quality. Some, such as the wire-cut cookie (AACC method 10-53 and 10-54) are exceptionally time consuming (5 to 10 samples per day) while other tests, such as Brookfield viscosity (AACC method 56-81) have high genotype × environment components. Chinese steamed bread represents a particular challenge for North American breeders; the evaluation is still fairly subjective (Huang et al., 1995) and the preferred color and texture of Chinese steamed bread varies among geographic and ethnic populations. Genotype × environment variation may also highly significant for steamed bread quality (Hou et al., 1991). Time consuming soft wheat evaluations that are poorly heritable or represent emerging markets may be important components of wheat quality research and market development efforts. However, these tests are external to core selection protocols until a combination of market demand and mechanization justify their routine application to breeding line selection.

Breeders commonly use tandem selection, sequentially 'building' a target genotype by selecting for grain softness in early generations, then incorporating more costly intermediate and advanced testing such as milling, AWRC, and cookie bakes (an idealized selection incorporating author biases is presented in Table 1). Grain hardness is controlled by a single locus with high heritability (Campbell et al., 1999) and minor modifying genes. Therefore, one to two selection cycles realizes the softness characteristics measured by NIR or SKCS. Additional gain from selection with NIR/SKCS is limited because genetic variation among selected lines is slight relative to environment and G × E interaction effects. More important than hardness selection after initial NIR/SKCS characterization is the physical softness measured by milling yield, particle size index, and damaged starch estimated through AWRC and SRC.

Researchers often need to accommodate field methodology problems in assessing milling quality and milling softness (or more commonly softness equivalent; Finney and Andrews, 1986). Late season biotic stress shrivels grain, reducing flour yield and increasing damaged starch due to milling. Late season rains or drought commonly reduce milling yield in eastern U.S. soft wheat programs. To measure true milling performance, some programs aspirate grain samples to remove environmentally damaged kernels (Gaines et al., 1998). In western soft white programs late-season rains are less important than late season drought. The degree of grain shriveling in droughted trials may measure abiotic stress tolerance rather than innate milling performance. Extreme drought eliminates useful soft wheat selection because test weights and protein contents are beyond tolerances used by mills and the relative differences among breeding lines are often masked by the effects of stress. If other locations are available for evaluation of quality, the drought location may be discarded for quality evaluation or reduced in weight at the time of selection.

Milling selection within early testing is critical to improving soft wheat quality, because it quickly boosts the average quality of the population. Elevated break flour yields are common among cultivar with better soft wheat quality and is a defining characteristic of club wheat cultivars. In our breeding program (Souza and Gutteri), we routinely begin Braebender Quadrumat Senior milling selection in the F_5 generation, with selection pressure for high break flour yield and recently added selection using the SRC test to reduce milling induced damaged starch levels. The USDA-ARS Western Wheat Quality Laboratory, Pullman, WA, USA pushes the selection of soft wheats to earlier generations with a modified Braebender Quadrumat Junior (Jeffers and Rubenthaler, 1977) for preliminary yield trials. This system is further modified to mill 50-g samples from F_3 headrows using a relative humidity/moisture equilibration cabinet as opposed to tempering with water (C.F. Morris personal communication). The USDA micromill (Shoup et al., 1957) mills as little as 5-10 g effectively removing any generational limits to evaluating milling quality. Gaines et al. (2000) have developed predictive models for the total long-flow milling yield of soft breeding lines based on SKCS evaluation coupled to 10 g grain samples milled in a Quadrumat junior mill (AACC method 26-50). The Gaines protocol also does not require tempering, reducing the assay time.

Grain or flour protein content is commonly measured in soft wheat breeding lines for adjusting individual bake results of tests like sugar snap cookies that are sensitive to high protein contents. Yet, selection for yield and test weight in conjunction with selection for cookie quality normally identifies genotypes with acceptable protein contents. Additional selection for protein may be unnecessary as abnormally high protein content among soft wheat genotypes is often a sign of agronomic problems, such as low yield or sensitivity to biotic stress.

Heritability of core quality measures is relatively high. Bergman et al. (1998) found heritabilities in a hard × soft recombinant inbred line population to exceed 80% for kernel hardness, flour yield, and AWRC. Heritabilities in crosses between two adapted soft wheat parents typically are lower than those described by Bergman et al. Yet, the proportion of variation in soft wheat quality measures found among genotypes often greatly exceeds genotype × environment interactions and errors associated with the measures (Bassett et al., 1989; Baenziger et al., 1985). In the authors' experience (Souza and Guttieri), several site-years of testing definitively separates poor from average check cultivars and six to eight site-years are needed to confirm the superiority of a breeding line over average check cultivars. The recently adopted solvent retention capacity (SRC) AACC protocol (Gaines, 2000) is attractive because it has relatively small genotype × environmental errors and much greater range in variation between advanced breeding lines. Among soft white spring wheat breeding lines, the ratio of genotype variation to genotype × environment variation for the SRC solvents is two to four times greater than for sugar snap cookie diameter (Souza and Guttieri, unpublished data). Because the SRC segregates the effects of damaged starch, pentosans, and HMW-glutenin, it can identify superior quality lines more quickly and inexpensively than AWRC or bake tests. It also identifies complementary genotypes, lines with different patterns of desirable quality, for future cycles of crossing and selection.

Secondary measures of wheat quality are less well integrated into U.S. selection methodologies than the core methodologies. The poor development derives in part from their newness to North American breeders, yet also from the diversity among genotypes within North American soft wheat classes, which have not been selected extensively for gluten strength, PPO activity, or starch paste viscosity.

Of the secondary measures gluten strength is probably the best characterized genetically. Gluten strength varies among soft wheats, with soft red wheats stronger on average than soft white and club wheats having characteristically weak gluten strength. Modern club cultivars typically carry the 1Dx2-1Dy12 allele at the *Glu-D1* locus. The recent exception, 'Hyak' carried the 1Dx5-1Dy10 *Glu-D1* allele and was rejected by the industry as atypical of the club class. Most techniques for assessing wheat gluten strength are applicable to soft wheat. In western U.S. classes of wheat the mixograph and SDS-sedimentation are more commonly used than the alveograph, which is more characteristic of quality evaluations in the eastern U.S. Strong gluten wheats are typically not segregated within the classes of soft wheat, yet some level of gluten strength within the class is desired for correct machining of many snack foods and essential for leavened crackers. To meet gluten strength requirements for machining, hard red winter wheat is commonly blended with soft classes in amounts varying depending on the target product and the gluten strength of the soft grain being blended. Within the breeding strategy, SDS-sedimentation or lactic acid SRC evaluation are relatively inexpensive assays for gluten strength and can be applied in early generation selection (Carter et al., 1999). Club wheats should be selected for very low SDS-sedimentation values, common soft wheats targeted at SDS-sedimentation levels somewhat below hard red winter wheats, and wheats with high SDS-sedimentation values (similar to hard red winter or stronger) should be discarded except for some specialty purposes. Although the best application for stronger gluten soft wheat is in leavened crackers, breeders lack a good small-scale cracker test that can be used at this time. Additional selection may lie in using the alveograph to select for the extensible doughs needed for crackers (high l/p values) or through SRC evaluation targeting low water absorption (below 500 g kg^{-1}) and moderate gluten strength as measured by lactic acid (850 to 950 g kg^{-1}).

Partial waxy genotypes (high amylopectin starch) are common among soft white spring wheats, rarer among soft white and soft red winter wheats, and non-existent in club wheat cultivars. GBSS mutations are a significant source of variation in flour and starch pasting, yet quantitative variation also exists among soft wheats (Udall et al., 1999). The GBSS null alleles reduce the time and temperature required to reach an initial peak viscosity, elevating the initial peak viscosity (Gaines et al., 2000b). Direct selection for the GBSS mutation through gel electrophoresis (Nakamura et al., 1993) may be sufficient to identify soft cultivars useful for batters and soups where high temperature viscosity is desirable (Gaines et al., 2000b). Final viscosity of the cooled gel may be more important than initial peak viscosity for Asian products such as udon noodles, Korean style noodles, and steamed breads. Although generally associated with the initial peak viscosity, the final viscosity is conditioned by additional factors beside the GBSS mutations (Gaines et al., 2000a). A firm link between reduced amylose content, starch swelling power and enhanced noodle quality has been established for Japanese udon noodles (Miura and Tanii, 1994), typically produced from flours of soft textured wheats. Australian breeding programs targeting udon noodle products use the flour swelling volume test to select for high final gel volume, which is related to final viscosity (Crosbie et al., 1992; Konik et al., 1993). Subsequent microtests of flour swelling requiring approximately 30 mg of flour are readily applied to early generation evaluation and less specialized equipment than the standard, large sample, flour swelling test (Fu et al., 1998). Breeders selecting for flour swelling to service Asian markets should practice joint selection for product color, as many high moisture products (e.g., noodles and steamed breads) also have stringent requirements for pigmentation and rate of discoloration conditioned by PPO.

As a class soft white wheats tend to have lower PPO levels than red wheats. Yet, significant variation exists within both soft white and soft red wheat classes. Selection for PPO may occur in almost any generation as the tests using tyrosine salts or L-DOPA are non-destructive, rapid, inexpensive, and can use one seed to several grams of seeds (Bernier and Howe, 1994). We (Souza and Guttieri) routinely evaluated thousands of breeding lines each year for PPO level, typically in segregating populations and headrows. Good response to selection for qualitative differences between genotypes occurs within crosses segregating for low PPO activity. Sources of low PPO activity among soft wheats include 'Eltan', 'Stephens', 'Treasure', 'Whitebird', and 'Tincurrin'. Yet, Asian product color is more complex than PPO. In as early a generation as a patent flour can be produced, selection should occur using a Pekar flour slick (AACC method 14-10) or a fresh noodle (Lang et al., 1998 and Gutteri et al., in review). Selected genotypes should have bright sheets without harsh colors (grays, greens, and pinks) and limited dulling after 24 hours. Moderate yellow pigmentation is desired for some products (Southern Chinese steamed bread and cream colors for udon). However, excessive xanthophylls may cause marketing problems because of poor acceptance in some North American and international markets. Both environment and genotype are important sources of variation for noodle color. Genotype × environmental variation is relatively limited for most noodle color attributes (Lang et al., 1998 and Gutteri et al., in review). Therefore, two to three site-years testing is sufficient to discard genotypes with unacceptable color, if a locally adapted standard cultivar is included within each breeding trial to correct for environmental variation.

FUTURE DIRECTIONS

Our outline of quality selection posed for the different wheat classes is largely based on direct selection for physical characteristics of the grain, flour, and food products. Increasing emphasis on molecular biology applications in crop improvement suggests that indirect selection for quality improvement by selection of genetic phenotype will be increasingly used. Current methodology for quality improvement emphasizes many low cost assays that integrate the functionality of multiple genetic systems. This raises the question, then, of whether molecular markers should be used rather than quality tests. Molecular markers could theoretically be developed to select lines with the optimal characteristic for each trait described above, and this all could be done with a single procedure. For example, PCR markers have been developed to discriminate genotypes carrying the good bread-making quality *Glu-D1* 1Dx5-1Dy10 allele from the poor bread-making quality 1Dx2-1Dy12 allele (D'Ovidio and Anderson, 1994; Varghese et al., 1996). This approach has the advantage of being nondestructive, rapid, and technically easier than traditional PAGE analysis of seed proteins. PCR markers also have demonstrated utility in evaluating interspecific introgressions. For example, PCR-based markers associated with rye (*Secale cereale* L.) chromatin have been developed (Koebner, 1995; Francis, 1995; Lee et al., 1996) for detection of rye chromatin in wheat-rye introgression lines. PCR based markers have also been used in our laboratory (Souza and Guttieri) to select for GBSS alleles. Giroux and Morris have routinely characterized wheats for the *PinA* and *PinB* mutations associated with grain hardness and milling yield differences among hard wheats (Giroux and Morris, 1997, 1998).

While in theory marker assisted selection sounds attractive, aspects of wheat quality limit its application. Many quality traits are complicated, likely are governed by multiple genes, and are subject to environmental variation. In addition, multiple sources of improvement exist for traits such as protein content, dough strength, loaf volume, etc. The use of markers for these traits established in one or a few genetic backgrounds will limit genetic diversity by selection against lines carrying alternate sources of improvement for the trait in question. Finally, the establishment of markers for large-scale quality performance (commercial bread and noodle manufacturing) would be extremely difficult, if not impossible. The large samples sizes needed, coupled with the need to conduct initial experiments in multiple populations, each composed of approximately 100 individuals, would sour even the wealthiest granting agency. Ultimately, though, molecular markers could replace many early generation tests, or assays useful in characterizing parents, but end-use performance tests in later generations likely will continue to be necessary.

Finally, this review has emphasized improvement of milling and flour performance as measured by its ability to produce a manufactured product. Increasing emphasis in the food trade on functional foods, foods with enhanced nutritional or health benefits, suggests increased breeding emphasis be placed in modification of flour constituents unrelated to rheology. Qualitative increase in provitamin A content of rice through genetic transformation provides preview of wheat modifications (Ye et al., 2000). Enhancements to the level of antioxidants in flour may also be possible through this same pathway (Guerinot, 2000). The integrating theme for breeding wheat quality, though, is to identify the target market parameters (whether rheology or functional), understand the underlying mechanisms conditioning the desired qualities, and use the most inexpensive technology possible to measure the desired trait on as many samples as possible, and then use the knowledge to select improved genotypes with improved quality while accomplishing the other goals of the wheat breeding program.

REFERENCES

American Association of Cereal Chemists. 2000. *Approved Methods of the AACC*, 10th ed. The Association: St. Paul, MN.

Anderson, J., and M. Sorrells. 1993. RFLP analysis of genomic regions associated with resistance to pre-harvest sprouting in wheat. *Crop Sci.* 33:453-459.

Andrews, J.L., M.J. Blundell, and J.H. Skerritt. 1996. Differentiation of wheat-rye translocation lines using antibody probes for *Gli-B1* and *Sec-1*. *J. Cereal Sci.* 23: 61-72.

Azudin, M.N. 1998. Screening of Australian wheat for the production of instant noodles. In: *Pacific People and Their Food*, eds. A.B. Blackeney and L. O'Brien, 101-122, Amer. Assoc. of Cereal Chem., St. Paul, MN, USA.

Baenziger, P.S., R.L. Clements, M.S. McIntosh, W.T. Yamazaki, T.M. Starling, D.J Sammons, and J.W. Johnson. 1985. Effect of cultivar, environment, and their interaction and stability analyses on milling and baking quality of soft red winter wheat. *Crop Sci.* 25:5-8.

Bassett, L.M., R.E. Allan, and G.L. Rubenthaler. 1989. Genotype × environment interactions on soft white winter wheat quality. *Agron. J.* 81:955-960.

Baik, B.-K., Z. Czuchajowska, and Y. Pomeranz. 1995. Discoloration of dough for oriental noodles. *Cereal Chem.* 72:198-205.

Bechtel, D.B., R.L. Gaines, and Y. Pomeranz. 1982. Protein secretion in wheat endosperm–formation of the matrix protein. *Cereal Chem.* 59: 336-343.

Bernier, A.-M. and N.K. Howes. 1994. Quantification of variation in tyrosinase activity among durum and common wheat cultivars. *J. Cereal Sci.* 19: 157-159.

Bettge, A.D. and C.F. Morris. 2000. Relationship among grain hardness, pentosan fractions, and end-use quality of wheat. *Cereal Chem.* 77:241-247.

Campbell, K.G., C.J. Bergman, D.G. Gualberto, J.A. Anderson, M.J. Giroux, G. Hareland, R.G. Fulcher, M.E. Sorrells, and P.L. Finney. 1999. Quantitative trait loci associated with kernel traits in a soft × hard wheat cross. *Crop Sci.* 39:1184-1195.

Carter, B.P., C.F. Morris, and J.A. Anderson. 1999. Optimizing the SDS sedimentation test for end-use quality selection in a soft white and club wheat breeding program. *Cereal Chem.* 76:907-911.

Colot, V., D. Bartels, R. Thompson, and R.B. Flavell. 1989. Molecular characterization of an active wheat LMW glutenin gene and its relation to other wheat and barley prolamin genes. *Mol. Gen. Genet.* 216: 81-90.

Cox, T.S., M.D. Shogren, R.G. Sears, T.J. Martin, and L.C. Bolte. 1989. Genetic improvement in milling and baking quality of hard red winter wheat cultivars, 1919 to 1988. *Crop Sci.* 29:626-631.

Crosbie, G.B., W.J. Lambe, H. Tsutsui, and R.F. Gilmour. 1992. Further evaluation of the flour swelling volume test for identifying wheats potentially suitable for Japanese noodles. *J. Cereal Sci.* 15:271-280.

D'Appolonia, B.L. and P. Rayas-Duarte. 1994. Wheat carbohydrates: structure and functionality. In: *Wheat: Production, Properties, and Quality*, eds. W. Bushuk and V.F. Rasper, 107-127, Chapman and Hall, London, UK.

DeMacon, V.L. and C.J. Morris. 1993. Rate of afterripening in diverse hexaploid genotypes. In: *Pre-Harvest Sprouting in Cereals 1992*, eds. M.K. Walker-Simmons and J.L. Reid, 61-68, Amer. Assoc. Cereal Chem., St. Paul, MN, USA.

Dick, J.W. and J.S. Quick. 1983. A modified screening test for rapid estimation of gluten strength in early generation durum breeding lines. *Cereal Chem.* 60:315-318.

Donelson, J.R. and C.S. Gaines. 1998. Starch-water relationships in the sugar-snap cookie dough system. *Cereal Chem.* 75:660-664.

Donelson, J.R. and Yamazaki, W.T. 1962. Note on a rapid method for the estimation of damaged starch in soft wheat flours. *Cereal Chem.* 39:460-462.

D'Ovidio, R. and O.D. Anderson. 1994. PCR analysis to distinguish between alleles of a member of a multigene family correlated with wheat bread-making quality. *Theor Appl Genet* 88:759-763.

Faridi, H., C. Gaines, and P. Finney. 1994. Soft wheat quality in production of cookies and crackers. In: *Wheat: Production, Properties, and Quality*, eds. W. Bushuk and V.F. Rasper. Chapman & Hall, Glasgow. pp. 154-168.

Finney, P.L. and L.C. Andrews. 1986. A 30 minute conditioning method for micro-, intermediate-, and large-scale experimental milling of soft red winter wheat. *Cereal Chem.* 63:18-21.

Flintham, J. 1993. Grain color and sprout-resistance in wheat. In *Pre-Harvest Sprouting in Cereals 1992*, eds. M.K. Walker-Simmons and J.L. Reid, 30-36, Amer. Assoc. Cereal Chem., St. Paul, MN, USA.

Francis, H.A., A.R. Leitch, and R.M.D. Koebner. 1995. Conversion of a RAPD-generated PCR product, containing a novel dispersed repetitive element, into a fast and robust assay for the presence of rye chromatin in wheat. Theor Appl Genet 90: 636-642.

Fu, B.X., M.I.P. Kovacs, and C. Wang. 1998. A simple wheat flour swelling test. *Cereal Chem.*75:566-567.

Gautier, M.-F., M.-E. Aleman, A. Guirao, D. Marion, and P. Jourdrier. 1994. *Triticum aestivum* puroindolines, two basic cystine-rich seed proteins: cDNA sequence analysis and developmental gene expression. *Plant Mol. Biol.* 25:43-57.

Gaines, C.S. 1985. Associations among soft wheat flour particle size, protein content, chlorine response kernel hardness, milling quality, white layer cake volume, and sugar-snap cookie spread. *Cereal Chem.* 62:290-292.

Gaines, C.S. 2000. Report of the AACC committee on soft wheat flour. Method 56-11, solvent retention capacity profile. *Cereal Foods World* (in press).

Gaines, C.S., Kassuba, A., and Finney, P.L. 1996. Using wire-cut and sugar-snap cookie test baking methods to evaluate distinctive soft wheat flour sets: Implications for quality testing. *Cereal Foods World* 41:155-160.

Gaines, C.S., P.L. Finney, and L.C. Andrews. 2000a. Developing agreement between very short flow and longer flow test wheat mills. *Cereal Sci.* 77:187-192.

Gaines, C.S., P.L. Finney, L.M. Fleege, and L.C. Andrews. 1998. Use of aspiration and the single kernel characterization system to evaluate puffed and shriveled condition of soft wheat grain. *Cereal Chem.* 75:207-211.

Gaines, C.S., Kassuba, A., and Finney, P.L. 1996. Using wire-cut and sugar-snap cookie test baking methods to evaluate distinctive soft wheat flour sets: Implications for quality testing. *Cereal Foods World* 41:155-160.

Gaines, C.S., M.O. Raeker, M. Tilley, P.L. Finney, J.D. Wilson, D.B. Bechtel, R.J. Martin, P.A. Seib, G.L. Lookhart, and T. Donelson. 2000b. Associations of starch gel hardness, granule size, waxy allelic expression, thermal pasting, milling quality and kernel texture of 12 soft wheat cultivars. *Cereal Sci.* 77:163-168.

Giroux, M.J. and C.F. Morris. 1997. A glycine to serine change in puroindoline b is associated with wheat grain hardness and low levels of starch-surface friabilin. *Theor. Appl. Genet.* 95:857-864.

Giroux, M.J. and C.F. Morris. 1998. Wheat grain hardness results from highly conserved mutations in the friabilin components puroindoline a and b. *Proc. Natl. Acad. Sci.* USA. 95:6262-6266.

Glen, G.M. and Saunders, R.M. 1990. Physical and structural properties of wheat endosperm associated with grain texture. *Cereal Chem.* 67:176-182.

Graybosch, R. 1992. The high-molecular-weight glutenin composition of cultivars, germplasm and parents of U.S. red winter wheats. *Crop Sci.* 32: 1151-1155.

Graybosch, R.A., C.J. Peterson, L.E. Hansen, and P.J. Mattern. 1990. Relationships between protein solubility characteristics, 1BL/1RS, high molecular weight glutenin composition and end-use quality in winter wheat germplasm. *Cereal Chem.* 67: 342-349.

Graybosch, R., C.J. Peterson, K. Moore, M. Stearns, and D. Grant. 1993a. Comparative effects of flour protein, lipid and pentosan composition in relation to hard wheat quality characteristics. *Cereal Chem.* 70: 95-101.

Graybosch, R., C.J. Peterson, L.E. Hansen, D. Worrall, D. Shelton, and A. Lukaszewski. 1993b. Comparative flour quality and protein characteristics of 1BL/1RS and 1AL/1RS wheat-rye translocations lines. *J. Cereal Sci.* 17: 95-106.

Graybosch, R.A., C.J. Peterson, D.R. Shelton, and P.S. Baenziger. 1996. Genotypic and environmental modification of wheat flour protein composition in relation to end-use quality. *Crop Science* 36: 296-300.

Graybosch, R.A., C.J. Peterson, G.A. Hareland, D.R. Shelton, M.C. Olewnik, H. He, and M.M. Stearns. 1999. Relationships between small-scale wheat quality assays and commercial test bakes. *Cereal Chem.* 76: 428-433.

Guerinot, M.L. 2000. Enhanced: The green revolution strikes gold. *Science* 287: 241-243.

Guttieri, M.J., J.C. Stark, K. O'Brien, and E. Souza. Relative sensitivity of spring wheat grain yield and quality parameters to drought intensity. *Crop Sci.* (in review).

Hatcher, D.W. and J.E. Kruger. 1993. Distribution of polyphenol oxidase in flour millstreams of Canadian common wheat classes milled to three extraction rates. *Cereal Chem.* 70:51-55.

Hou, L., R.S. Zemetra, and D. Birzer. 1991. Wheat genotype and environment effects on Chinese steamed bread quality. *Crop Sci.* 31:1279-1282.

Howes, N.K., M.I.P Kovacs, A.M. Bernier, R.I.H. McKenzie, and T.F. Townley-Smith. 1995. Unique processing properties of bread wheats having zero polyphenol oxidase. Proceedings: Value Added Cereal Conference, July, 1995, Winnipeg, Manitoba, Canada, p. 55.

Huang, S., K. Quail, R. Moss, and J. Best. 1995. Objective methods for the quality assessment of Northern-style Chinese steamed bread. *J. Cereal Sci.* 21:49-55.

Hoseney, R.C. 1986. *Principles of Cereal Chemistry.* Amer. Assoc. of Cereal Chem., St. Paul, MN.

Interesse, F.S., P. Ruggiero, G. D'Avella, and F. Lamparelli. 1980. Partial purification and some properties of wheat (*Triticum aestivum*) *o*-diphenolase. *J. Sci. Food Agric.* 31:459-466.

Interesse, F.S., P. Ruggiero, G. D'Avella, and F. Lamparelli. 1983. Characterization of wheat *o*-diphenolase isoenzyme. *Phytochem.* 22:1885-1889.

Jeffers, H.C., and G.L. Rubenthaler. 1977. Effect of roll temperature on flour yield with the Brabender Quadrumat experimental mills. *Cereal Chem.* 54:1018-1025.

Jelaca, S.L. and I. Hlynka. 1971. Water-binding capacity of wheat flour crude pentosans and their relation to mixing characteristics of dough. *Cereal Chem.* 48:211.

Kaldy, M.S., G.I. Rubenthaler, G.R. Kereliuk, M.A. Berhow, and C.E. Vandercook. 1991. Relationship of selected flour consitituents to baking quality in soft white wheat. *Cereal Chem.* 68:508-512.

Koebner, R.M.D. 1995. Generation of PCR-based markers for the detection of rye chromatin in a wheat background. *Theor Appl Genet* 90:740-745.

Konik, C.M., D.M. Miskelly, and P.W. Gras. 1993. Starch swelling power, grain hardness and protein: Relationship to sensory properties of Japanese noodles. *Starch* 45: 139-144.

Konik, C.M., L.M. Mikkelsen, R. Moss, and P.J. Gore. 1994. Relationships between physical starch properties and yellow alkaline noodles. *Starch* 93: 292-299.

Kruger, J.E. 1976. Changes in the polyphenol oxidases of wheat during kernel growth and maturation. *Cereal Chem.* 53:201-213.

Kruger, J.E. 1994. Enzymes of sprouted wheat and their possible technological significance. In: *Wheat: Production, Properties, and Quality,* eds. W. Bushuk and V.F. Rasper, 143-153, Chapman and Hall, London, UK.

Kruger, J.E., M.H. Anderson, and J.E. Dexter. 1994. Effect of flour refinement on raw Cantonese noodle color and texture. *Cereal Chem.* 7:177-182.

Kruger, J.E., R.R. Matsuo, and K. Preston. 1992. A comparison of methods for the prediction of Cantonese noodle colour. *Can. J. Plant Sci.* 72: 1021-1029.

Kulp, K. 1968. Pentosans of wheat endosperm. *Cereal Sci. Today* 13:414-417, 426.

Lang, C.E., S.P. Lanning, G.R. Carlson, G.D. Kushnak, P.I. Bruckner, and L.E. Talbert. 1998. Relationship between baking and noodle quality in hard white spring wheat. *Crop Sci.* 38:823-827.

Lillemo, M., and C.F. Morris. 2000. A leucine to proline mutation for puroindoline b is frequently present in hard wheats from Northern Europe. *Theor. Appl. Genet.* (Accepted).

Lee, J.H., R.A. Graybosch, S.M. Kaeppler, and R.G. Sears. 1996. A PCR assay for detection of a 2RL.2BS wheat-rye chromosome translocation. *Genome* 39:605-608.

Lee, J.H., R.A. Graybosch, and C.J. Peterson. 1995. Quality and biochemical effects of a 1BL.1RS wheat-rye chromosomal translocation in wheat. *Theor. Appl. Genet.* 90: 105-112.

Loggini, B., A. Scartazza, E. Brugnoli, and F. Navari-Isso. 1999. Antioxidative defense system, pigment composition, and photosynthetic efficiency in two wheat cultivars subjected to drought. *Plant Physiol.* 119:1091-1100.

Mahoney, R.R. and M. Ramsey. 1992. A rapid tyrosinase test for detecting contamination of durum wheat. *J. Cereal Sci.* 15:267-270.

Mathewson, P. and Y. Pomeranz. 1978. On the relationship between alpha amylase and falling number in wheat. *J. Food Sci.* 43:652-653.

Mares, D.J. 1992. Genetic studies of sprouting tolerance in red and white wheats. In: *Pre-Harvest Sprouting in Cereals 1992*, eds. M.K. Walker-Simmons and J.L. Reid, 21-29, Amer. Assoc. Cereal Chem., St. Paul, MN, USA.

Mares, D.J. and F.W. Ellison. 1989. Dormancy and pre-harvest sprouting tolerance in white-grained and red-grained wheats. In: *Fifth International Symposium on Pre-Harvest Sprouting in Cereals*, eds. K. Ringlund, E. Mosleth and D.J. Mares, 75-84, Westview Press, Boulder, CO, USA.

Miskelly, D.M. 1984. Flour components affecting paste and noodle color. *J. Sci. Food Agric.* 35:463-471.

Miskelly, D.M. 1996. The use of alkali in noodle processing. In: *Pasta and Noodle Technology*, eds. J.E. Kruger, R.B. Matsuo and J.W. Dick, 227-274, Amer. Assoc. Cereal Chem., St. Paul, MN, USA.

Miura, H. and S. Tanii. 1994. Endosperm starch properties in several wheat cultivars preferred for Japanese noodle quality. *Euphytica* 72: 171-175.

Morris, C., R.J. Anderberg, P.J. Goldmark, and M.K. Walker-Simmons. 1991. Molecular cloning and expression of abscisic acid-responsive genes in embryos of dormant wheat seeds. *Plant Physiol.* 95:814-821.

Morris, C.F., H.C. Jeffers, and D.A. Engle. 2000. Effect of processing, formula, and measurement variables on alkaline noodle color toward an optimized laboratory system. *Cereal Chem.* 77:77-85.

Morrison, W.R. 1988. Lipids. In: *Wheat Chemistry and Technology, Vol. I*, ed. Y. Pomeranz, Amer. Assoc. Cereal Chem., St. Paul, MN, USA.

Morrison, W.R., C.N. Law, L.J. Wylie, A.M. Coventry and J. Seekings. 1989. The effect of group 5 chromosomes on the free polar lipids and breadmaking quality of wheat. *J. Cereal Sci.* 9: 41-51.

Nagamine, T., T. Kato, H. Yoshida, and K. Komae. 1998. Genetic variation and developmental changes of the multiple isoforms of starch branching enzyme in wheat endosperm and their influence on the amylopectin structure. In: *Proceedings of the 9th International Wheat Genetics Symposium, Vol. 4*, ed. A.E. Slinkard, 225-228, University Extension Press, University of Saskatchewan, Saskatchewan, Canada.

Nagao, S., S. Imai, T. Sato, Y. Kaneko, and H. Otsubo. 1976. Quality characterization of soft wheats and their use in Japan. I. Methods of assessing wheat suitability for Japanese products. *Cereal Chem.* 53:988-997.

Nakamura, T., M. Yamamori, H. Hirano, and S. Hidaka. 1993. Decrease in waxy (*wx*) protein in two common wheat cultivars with low amylose content. *Plant Breeding* 111: 99-105.

Nakamura, T., M. Yamamori, H. Hirano, S. Hidaka, and T. Nagamine. 1995. Production of waxy (amylose-free) wheats. *Mol. Gen. Genet.* 248: 253-259.

Noll, J.S. and E.M. Czarnecki. 1980. Methods for extending the testing period for harvest time dormancy in wheat. *Cereal Res. Commun.* 8: 233-238.

Panozzo, J. and K. McCormick. 1993. The rapid visco-analyser as a method of testing for noodle quality in a wheat breeding programme. *J. Cereal Sci.* 3: 271-278.

Panozzo, J.F., M.C. Hannah, L. O'Brien, and F. Bekes. 1993. The relationship of free lipids and flour protein to breadmaking quality. *J. Cereal Sci.* 17: 47-62.

Paterson, A.H., and M.E. Sorrells. 1990. Inheritance of grain dormancy in white-kernelled wheat. *Crop Sci.* 30: 25-30.

Payne, P.I. 1987. The genetical basis of breadmaking quality in wheat. *Aspects of Applied Biology* 15: 79-90.

Payne, P.I., L.M. Holt, E.A. Jackson, and C.N. Law. 1984. Wheat storage proteins: Their genetics and their potential for manipulation by plant breeding. *Phil. Trans. R. Soc. Lond. B.* 304: 359-371.

Peterson, C.J., R.A. Graybosch, P.S. Baenziger, and A.W. Grombacher. 1992. Influence of genotype and environment on quality characteristics of hard red winter wheat. *Crop Science*, 32: 98-103.

Peterson, C.J., R.A. Graybosch, P.S. Baenziger, D.R. Shelton, W.D. Worrall, L.A. Nelson, D.V. McVey, and J.H. Hatchett. 1993. Release of N86L177 wheat germplasm. *Crop Sci.* 33: 350.

Peterson, C.J., Graybosch, R.A., D.R. Shelton, and P.S. Baenziger. 1998. Baking quality of hard winter wheat: Response of cultivars to environment in the Great Plains. *Euphytica* 100: 157-162.

Pomeranz, Y. and P.C. Williams. 1990. Wheat hardness: Its genetic, structural, and biochemical background, measurment, and significance. In: *Advances in Cereal Science and Technology*, Vol X, ed. Y. Pomeranz, pp. 471-548, Amer. Assoc. Cereal Chem., St. Paul, MN.

Rayfuse, L.M., M.M. Cadle, P.J. Goldmark, R.J. Anderberg, M.K. Walker-Simmons, and S.S. Jones. 1993. Chromosome location of ABA-inducible genes associated with seed dormancy in wheat. In: *Pre-Harvest Sprouting in Cereals 1992*, eds. M.K. Walker-Simmons and J.L. Reid, 129-135, Amer. Assoc. Cereal Chem., St. Paul, MN, USA.

Rasper, V.F., F. Boucaut, and S. Williams. 1992. Effect of germination on physical properties of Doughs. In: *Pre-Harvest Sprouting in Cereals*, eds. M.K. Walker-Simmons and J.L. Reid, 436-447, Amer. Assoc. Cereal Chem., St. Paul, MN, USA.

Rooke, L., F. Bekes, R. Fido, F. Barro, P. Gras, A.S. Tatham, P. Barcelo, P. Lazzeri, and P.R. Shewry. 1999. Overexpression of a gluten protein in transgenic wheat results in greatly increased dough strength. *J. Cereal Sci.* 30:115-120.

Ross, A.S., K.J. Quail, and G.B. Crosbie. 1996. An insight into structural features leading to desirable alkaline noodle texture. In: *Proceedings of the Eighth Assembly Wheat Breeding Society of Australia*, eds. R.A. Richards, C.W. Wrigley, H.M. Rawson, G.J. Rebetzke, J.L. Davidson, and R.I.S. Brettell, 93-97, Wheat Breeding Society of Australia, The Australian National University, Canberra, Australia.

Schultz, T.F., J. Medina, A. Hill, and R.S. Quatrano. 1998. 14-3-3 Proteins are part of an abscisic acid-VIVPAROUS1 (VP1) response complex in the *Em* promoter and interact with *VP*1 and *EmBP*1. *Plant Cell* 10:837-848.

Shewry, P.R., B.J. Miflin, and D.D. Kasarda. 1984. The structural and evolutionary relationships of the prolamin storage proteins of barley, rye and wheat. *Phil. Trans. R. Soc. Lond. B.* 297-308.

Shoup, N.H., Pell, K.L., Seeborg, E.F., and Barmore, M.A. 1957. A new micromill for preliminary milling-quality tests of wheat. *Cereal Chem.* 34:296-298.

Singh, N.K., R. Donovan, and F. MacRitchie. 1990. Use of sonication and size-exclusion high-performance liquid chromatography in the study of wheat flour proteins. II. Relative quantity of gluten as a measure of breadmaking quality. *Cereal Chem.* 67: 161-170.

Singh, R. and I.S. Sheoran. 1972. Enzymatic browning of whole wheat meal flour. *J. Sci. Food Agric.* 23:121-125.

Slade, L. and Levine, H. Structure-function relationships of cookie and cracker ingredients. In: *The Science of Cookie and Cracker Production*, ed. H. Faridi, Chapman & Hall/AVI, New York, pp. 23-141. 1994.

Sorrells, M.E. and A.H. Paterson. 1986. Registration of NY6432-18 and NY6708-18 wheat germplasm lines. *Crop Sci.* 26: 392-393.

Souza, E., M. Kruk, and D.W. Sunderman. 1994. Association of sugar snap cookie quality with high molecular weight glutenin alleles in soft white spring wheats. *Cereal Chem.* 71:601-605.

Souza, E., J.M. Tyler, K.D. Kephart, and M. Kruk. 1993. Genetic improvement in milling and baking quality of hard red spring wheat cultivars. *Cereal Chem.* 70:280-285.

Souza, E., J.M. Windes, S.O. Guy, L. Robertson, D.W. Sunderman, and K. O'Brien. 1997. Registration of Idaho 377s wheat. *Crop Sci.* 37:1393.

Stenvert, N.L. and K. Kingswood. 1977. The influence of the physical structure of the protein matrix on wheat hardness. *J. Sci. Food Agric.* 28: 11-19.

Thygesen, P.W., I.B. Dry, and S.P. Robinson. 1995. Polyphenol oxidase in potato: A multigene family that exhibits differential expression patterns. *Plant Physiol.* 109: 525-531.

Udall, J.A., E. Souza, J. Anderson, M.E. Sorrells, and R.S. Zemetra. 1999. Quantitative trait loci for flour viscosity in winter wheat. *Crop Sci.* 39:238-242.

Varghese, J.P., D. Struss, and M.E. Kazman. 1996. Rapid screening of selected European winter wheat varieties and segregating populations for the Glu-*D1d* allele using PCR. *Plant Breeding* 115:451-454.

Vasil, V., W.R. Macotte Jr, L. Rosenkrans, S.M. Cocciolone, I.K. Vasil, R.S. Quatrano, and D.R. McCarty. 1995. Overlap of *Viviparous*1 (*VP*1) and abscisic acid response elements in the *Em* promoter: G-box elements are sufficient but not necessary for *VP*1 transactivation. *The Plant Cell* 7:1511-1518.

Watanabe, N. and H. Miura. 1998. Diversity of amylose content in *Triticum durum* and *Aegilops tauschii*. In: *Proceedings of the 9th International Wheat Genetics Symposium, Vol. 4*, ed. A.E. Slinkard, 291-293, University Extension Press, University of Saskatchewan, Saskatchewan, Canada.

Weegels, P.L., R.J. Hamer, and J.D. Schofield. 1996. Functional properties of wheat glutenin. *J. Cereal Sci.* 23: 1-18.

Wu, J. and B.F. Carver. 1999. Sprout damage and preharvest sprout resistance in hard white winter wheat. *Crop Sci.* 39: 441-447.

Windes, J.M. and E. Souza. 1995. Heritability of tyrosinase activity in soft white spring wheats. *Agronomy Abstracts, Amer. Soc. of Agron.*, p. 93-94.

Yamamori, M., T. Nakamura, T. Endo, and T. Nagamine. 1994. Waxy protein deficiency and chromosomal locations of coding genes in common wheat. *Theor. Appl. Genet.* 89: 179-184.

Ye, X., S. Al-Babili, A. Kloti, J. Zhang, P. Lucca, and P. Beyer. 2000. Engineering the provitamin A (B-Carotene) biosynthetic pathway into (carotenoid-free) rice endosperm. *Science* 287: 303-305.

Agronomic Practices
Influence Maize Grain Quality

Stephen C. Mason
Nora E. D'Croz-Mason

SUMMARY. Maize grain yields have increased dramatically over the past fifty years, and concurrently end-uses have proliferated requiring special quality characteristics. Plant breeders have developed many specialty types of maize, all of which are influenced by the agronomic practices used to produce the crop. Grain yield increases have resulted in lower protein concentration except when the yield increase resulted from nitrogen fertilizer application. Irrigation improves the biological value of protein, while higher nitrogen application rates alter the amino acid balance thereby reducing the nutritional value. Kernel breakage susceptibility and kernel density increase with higher nitrogen fertilizer application rates, and are reduced by irrigation. Extractable starch and oil concentration are largely influenced by hybrid choice, but small production practice effects have been documented. Essential mineral nutrient levels are often influenced by soil or foliar fertilizer application. The production of aflatoxin can be reduced by irrigation or other strategies to

Stephen C. Mason is Professor and Nora E. D'Croz-Mason is Adjunct Assistant Professor, Department of Agronomy, 279 Plant Science, University of Nebraska, Lincoln, NE 68583-0915 USA.

Address correspondence to: Stephen C. Mason, Department of Agronomy, 279 Plant Science, University of Nebraska, Lincoln, NE 68583-0915 (E-mail: smason1@ unl.edu).

Contribution of the Department of Agronomy, University of Nebraska, Lincoln, NE 68583-0915 USA. Paper No. 13035 of the Journal Series of the Nebraska Agricultural Research Division.

[Haworth co-indexing entry note]: "Agronomic Practices Influence Maize Grain Quality." Mason, Stephen C., and Nora E. D'Croz-Mason. Co-published simultaneously in *Journal of Crop Production* (Food Products Press, an imprint of The Haworth Press, Inc.) Vol. 5, No. 1/2 (#9/10), 2002, pp. 75-91; and: *Quality Improvement in Field Crops* (ed: A. S. Basra, and L. S. Randhawa) Food Products Press, an imprint of The Haworth Press, Inc., 2002, pp. 75-91. Single or multiple copies of this article are available for a fee from The Haworth Document Delivery Service [1-800-HAWORTH, 9:00 a.m. - 5:00 p.m. (EST). E-mail address: getinfo@haworthpressinc.com].

avoid water stress during grain fill. Although genetics usually exerts the
largest effect on maize grain quality, agronomic practices are also important. *[Article copies available for a fee from The Haworth Document Delivery
Service: 1-800-HAWORTH. E-mail address: <getinfo@haworthpressinc.com>
Website: <http://www.HaworthPress.com> © 2002 by The Haworth Press, Inc. All
rights reserved.]*

KEYWORDS. Protein concentration, lysine, tryptophan, kernel weight,
kernel breakage susceptibility, kernel density, extractable starch, oil
concentration, nutrient levels, aflatoxin

INTRODUCTION

Maize (*Zea mays* L.) is the third most important crop worldwide following rice (*Oryza sativa* L.) and wheat (*Triticum aestivum* L.). Adoption of improved cultivars and production practices have increased
yields by approximately 1.5% per year since 1940 (Tollenaar and Wu,
1999) and estimates indicate that this rate of yield increase will be necessary to continue to meet world demand (Duvic and Cassman, 1999).
Plant breeders have developed specialty maize types with improved
grain quality for specific end-uses for human and livestock nutritional
needs (i.e., opaque-2, high oil), and for food and industrial processing
(i.e., waxy, high amylose). Substantial research on the influence of production practices on maize grain quality has been conducted, especially
fertilizer management and cultural practices. In this paper, we review
the influence of agronomic practices on grain quality factors such as
protein concentration, amino acid balance, kernel size/weight/unifor-
mity, physical quality, wet milling starch extraction, and other factors
of maize.

The maize kernel is composed of approximately 72% starch, 10%
protein, 5% oil, 2% sugar and 1% ash with the remainder being water
(Perry, 1988). Several mutants have been discovered and developed to
alter the starch fractions of the maize endosperm. The maize protein biological value is low due to the low concentration of the essential amino
acids lysine and tryptophan, although opaque-2 maize has been developed with higher levels of these two amino acids (Mertz et al., 1964).
Recent breeding efforts by CYMMYT have improved the hardness of
opaque-2 maize cultivars (Vasal et al., 1980), thus improving the agronomic characteristics of this maize type. The oil in maize is an important energy source for livestock feed, and due to a high degree of

unsaturation, is widely used for human consumption (Perry, 1988). The ash of maize grain contains little Ca, and although the P content is relatively high, only 50% is available to monogastric animals and as a result has become a major environmental concern in manure management (Ertl et al., 1998).

PROTEIN CONCENTRATION

Production factors that increase grain yield also increase the starch concentration of grain while reducinging the grain protein concentration (McDermitt and Loomis, 1981). This inverse relationship has been shown in barley (*Hordeum vulgare* L.), maize, oats (*Avena sativa* L.), rice, sorghum [*Sorghum bicolor* (L.) Moench], and wheat (Frey, 1977). The negative relationship between protein concentration and grain yield is partly associated with the higher glucose costs for synthesis of protein than carbohydrates (Penning de Vries et al., 1974), thus the higher cost of protein synthesis is logically inversely related to the grain yield (Bhatia and Rabson, 1976).

The inverse relationship between maize grain yield and protein concentration has been shown for yield increases due to plant breeding (Kamprath et al., 1982), plant population (Vyn and Tollenaar, 1998; Andrade and Ferreiro, 1996; Ahmadi et al., 1993; Cloninger et al., 1975), planting date (Ahmadi et al., 1993), irrigation (Bullock et al., 1989; Decau and Pujol, 1973; Kniep and Mason, 1991; Sabata and Mason, 1992; Mason and Sabata, 1995), shading (Andrade and Ferreiro, 1996), defoliation (Remison and Omueti, 1982) and growth regulator application (Norberg et al., 1988). In addition, grain yields between 1940 and 1990 have also shown this inverse relationship (Duvick and Cassman, 1999). However, Pollmer et al. (1978) observed no relationship between grain yield and protein concentration for hybrids with a wide range of protein concentrations. Ahmadi et al. (1993) reported that increasing plant population did not always result in increased maize grain yield, but consistently reduced protein concentration.

Benzian and Land (1979) analyzed the relationship between nitrogen supply, grain yield and grain protein concentration for wheat. They found that a greater nitrogen supply increased grain protein concentration linearly while grain yield response to added nitrogen had a diminishing return relationship. They also found that when nitrogen was very limiting, small nitrogen additions resulted in greater grain yield with decreased protein concentration. However, at higher levels of nitrogen, which are far more common, grain and protein yields usually increased while the grain protein concentration decreased. This latter relationship has been confirmed in maize grain in both nitrogen fertilizer application and residual soil nitrate-nitrogen level studies (Anderson et al., 1984; Cromwell et al., 1983; Kniep and Mason, 1991; Oikeh et al; 1998; Olsen et al., 1976; Pierre et al., 1977; Sabata and Mason, 1992; Tsai et al., 1983; Tsai et al., 1992). Research reported by Tsai et al. (1983) suggested that protein concentration of maize grain increases with nitrogen supply due to preferential deposition of zein over the other endosperm proteins. It is apparent that the amount of fertilizer nitrogen required to maximize grain yields is not the same as the amount that will produce maximum grain protein concentrations (Sander et al., 1987).

AMINO ACID BALANCE

As the protein concentration of maize grain increases, zein makes up an increasing proportion of the protein (Frey et al., 1949; Frey, 1951; Tsai et al., 1992). Rendig and Broadbent (1977) reported that concentration of the protein fraction zein in maize grain was closely associated with the level of soil nitrogen, with each added increment of nitrogen increasing the percentage of zein. Tsai et al. (1983) reported that as nitrogen levels increased, zein accumulated preferentially in normal maize grain, but not in grain of opaque-2 hybrids. Tsai et al. (1992) reported that protein yield increase from nitrogen application was accompanied by an increase in the amount of zein present in the endosperm, creating harder, less brittle and more translucent grain. Since zein contains lower amounts of the most limiting essential amino acids lysine and tryptophan, increased grain yields change the amino acid balance by reducing the lysine and typtophan concentrations, thus reducing the biological value of the grain protein. However, this may be compensated for in some cases since nitrogen fertilizer application increases the size of the germ, which has a better amino acid balance than the endosperm (Bhatia and Rabson, 1987). In contrast, studies with opaque-2 maize hybrids indicate that increased nitrogen supply maintains or increases the lysine concentration of grain (Cromwell et al., 1983; Tsai et al., 1983).

Kniep and Mason (1991) found that irrigation increased grain yield, reduced protein concentration, had no effect on percent lysine per sample, and increased percent lysine of protein of normal maize. Nitrogen application increased grain yield, protein concentration and percent lysine of sample, but decreased percent lysine of protein. Irrigation decreased percent lysine per sample for opaque-2 hybrids. The grain from the above study was used in a rat feeding experiment, and found that rats fed grain produced with irrigation had greater and more efficient rates of gain, while those fed grain from plots with nitrogen application had lower and less efficient rates of gain (Hancock et al., 1988). These studies and the one previously reported by Bullock et al. (1989) clearly indicate that irrigation increases grain and protein yields, and lowers protein concentration, but improves the biological value of the protein. In contrast, nitrogen application increases grain and protein yields, and protein concentration, but reduces the biological value of the protein. Similar results for nitrogen fertilizer application, except the adverse effect of nitrogen fertilizer application on percent lysine and tryptophan was less for opaque-2 hybrids than for normal hybrids were reported by Breteler (1976). MacGregor et al. (1961) found that amino acid concentrations of grain did not increase uniformly to nitrogen fertilizer application, and that the concentrations of lysine, methionine and phenyalanine did not increase. Rendig and Broadbent (1979) found that nitrogen fertilizer application decreased the concentrations of tryptophan, lysine, glycine, arginine and threonine in protein, while concentrations of alanine, phenylalanine, tyrosine, glutamic acid, and leucine were increased. Neither choice of normal maize hybrid nor plant population has an effect on lysine or tryptophan concentration of grain (Vyn and Tollenaar, 1998). It is apparent that irrigation has a postive effect on maize grain amino acid balance, while nitrogen fertilizer application has a negative effect.

KERNEL SIZE/WEIGHT/UNIFORMITY

Large uniform kernel size is desired for dry milling (Paulsen and Hill, 1985), wet milling (Watson, 1987), alkaline cooked products (Shumway et al., 1992) and livestock feed when processed by rolling or cracking. Most agronomic studies have focused on kernel weight which often is highly associated with both kernel density and/or size (Watson, 1987). Personal communications from livestock feeders and dry millers indicate that uniformity of kernel size is important, but has not been studied scientifically.

Delayed planting in a temperate climate reduced kernel weight due to decreased interception of solar radiation and lower temperatures during grain fill, resulting in decreased rate and duration of grain filling (Cirilo and Andrade, 1994, 1996). Heat stress (Wilhelm et al., 1999) and water deficit (Claassen and Shaw, 1970) during grain fill reduces kernel weight. Kernel weight has been shown to increase with increasing nitrogen application (Bauer and Carter, 1986; Kniep and Mason, 1989; Cromwell et al., 1983; Rendig and Broadbent, 1975), and to decrease with increasing plant population (Bauer and Carter, 1986; Vyn and Moes, 1988; Vyn and Tollenaar, 1998). Kniep and Mason (1989) found that irrigation with high plant population produced grain with lower kernel weight than nonirrigated maize with low plant population, while Mason and Galusha (unpublished data) showed that irrigation increased kernel weights at plant populations ranging from 24,700 to 84,000 plants ha^{-1}. Ethephon application has also reduced kernel weight (Norberg et al., 1988). Kernel weight is often associated with other quality parameters such as kernel density (Kniep and Mason, 1989; Paulsen et al., 1983), kernel breakage susceptibility (Kniep and Mason, 1989; Johnson and Russell, 1982), protein concentration (Manokarkumar et al., 1978; Oikeh et al., 1998; Arnold et al., 1977), lysine concentration (Arnold et al., 1977), and extractable starch (Zehr et al., 1995). Environmental factors and agronomic practices that influence grain fill usually have an effect on kernel size and weight.

PHYSICAL QUALITY

Physical quality of grain is usually measured by kernel hardness (density), kernel breakage susceptibility (brittleness) and stress cracking. Physical quality is a primary concern of the maize dry milling industry to optimize the production of the highest value end-product of uniform, large flaking grits (Paulsen and Hill, 1985). Design engineers for dry millers are interested in physical quality that optimizes equipment operating conditions, minimizes energy use, and produces an end-product of uniform size (Jindal and Mohsenin, 1978). Dry millers prefer natural air dried grain containing uniformly large, unbroken kernels (Paulsen and Hill, 1985). This also is a concern in export markets since kernel breakage leads to increased percent broken kernels and foreign material, the major quality factor in international markets (Hill, 1981). International purchasers offer lower prices foreseeing decreases in grade due to kernel breakage which in turn reduces the price received by producers at the local elevator (Bauer and Carter, 1986). Paulsen and Hill (1977) documented cases in which the amount of broken corn and foreign material (BCFM) in maize grain exported from the U.S. increased from 3 to 5% at inland terminal elevators to 20 to 22% at point of destination overseas largely due to stress cracked kernels breaking from impacts during handling. International markets are becoming more sophisticated as illustrated by the recent report of Paulsen et al. (1996) that specified desirable quality characteristics by end-use for Japan.

Research shows that the two major means to improve physical quality of maize grain are plant breeding for grain resisting mechanical damage and drying grain at low temperatures (Moes and Vyn, 1988; Vyn and Moes, 1988). Grain with smaller, more dense kernels and with an increased proportion of horny endosperm is the most resistant to mechanical damage (Johnson and Russell, 1982; LeFord and Russell, 1982; Vyn and Moes, 1988). Some hybrids are also better able to withstand high drying temperatures (Moes and Vyn, 1988). Large kernels apparently are more susceptible to kernel breakage than small kernels (Thompson and Foster, 1963; Moes and Vyn, 1988; Miller et al., 1981; Leford and Russell, 1985; Johnson and Russell, 1982) and round kernels are more susceptible than flat kernels (Thompson and Foster, 1963; Moes and Vyn, 1988). However, Vyn and Tollenaar (1998) found the greatest breakage to occur in hybrids producing small kernels. High grain drying temperatures, especially of grain with high moisture content, increases the formation of stress cracks which predisposes the grain to breakage (Paulsen et al., 1983). Little relation between breakage susceptibility and endosperm hardness or kernel density occurs (Dorsey-Redding et al., 1991; Kniep and Mason, 1989) unless stress cracks are present (Weller et al., 1988) when breakage was greater for kernels with greater hardness.

Bauer and Carter (1986) studied kernel breakage susceptibility and kernel density of nine maize hybrids with different maturity classifications using different production practices. They found that grain grown under irrigation and with higher plant populations had higher kernel breakage susceptibility, but that nitrogen fertilizer application decreased kernel breakage susceptibility. Kernel breakage susceptibility was greater for later than for earlier maturity maize, and early planting tended to slightly decrease kernel breakage susceptibility. In a similar study, Kniep and Mason (1989) confirmed that kernel breakage susceptibility increased with irrigation and decreased with nitrogen fertilizer application. They also found that kernel breakage susceptibility increased with increasing grain yield. Moes and Vyn (1988) and Vyn and Moes (1988) reported that kernel breakage susceptibility increased with increasing plant population, but Mason and Galusha (unpublished data) found that plant populations between 24,700 and 84,000 plants ha⁻¹ had no effect. Moes and Vyn (1988) found no relationship between kernel breakage susceptibility and grain yield. Ethephon application influenced kernel breakage susceptibility, but the response varied across years and hybrids (Norberg et al., 1988).

In a study by Bauer and Carter (1986), planting date, plant population, and irrigation had no influence on maize kernel density, while increasing nitrogen rates increased kernel density. Kniep and Mason (1989) observed that increasing nitrogen rates increased kernel density, and also reported that irrigation increased kernel density two years out of three. Shumway et al. (1992) also found that irrigation decreased kernel density. Ahmadi et al. (1993) reported that kernel density was not influenced by planting date or plant population while Vyn and Tollenaar (1998) showed that kernel density decreased with increasing plant population. Shumway et al. (1992) and Cloninger et al. (1975) found that delayed planting decreased kernel density. Bauer and Carter (1986) reported no correlation between kernel density and kernel breakage susceptibility nor grain yield, while Kniep and Mason (1989) found significant correlations. Ethephon application has been shown to slightly decrease kernel density (Norberg et al., 1988). These results indicate that nitrogen application improves while irrigation decreases physical quality of maize grain.

EXTRACTABLE STARCH

Starch extraction by wet-milling is the second largest end-use of maize grain following livestock feed. Desirable grain should have low levels of broken corn and foreign material (BCFM), low breakage susceptibility, and be dried at low temperatures to minimize breakage and maintain high starch yield (Watson, 1987). Weller et al. (1988) and Haros and Suarez (1997) both found that starch recoveries did not differ among maize hybrids with different vitreous-to-floury ratios. Very few studies have been conducted on the influence of agronomic practices on percent extractable starch, at least partially due to the difficulty in measuring starch extraction in the laboratory using small samples. Recently, Wehling et al. (1993) developed a simple, rapid method for estimating extractable starch using near-infrared spectroscopy and Zehr et al. (1995) developed a laboratory method using 100-g samples in order to increase the feasibility of conducting such research. The only study conducted on the influence of production practices on wet milling starch extraction was conducted over four years with very different amounts and distribution of precipitation (Mason and Galusha, unpublished). Experimental treatments included three maize hybrids, water regime (irrigation and nonirrigation), and plant populations ranging from 24,700 to 84,000 plants ha^{-1}. Extractable starch differences were small, but declared significant due to Year × Hybrid, Year × Water Regime, and Year × Plant Population interaction effects. The greatest differences in extractable starch were due to year (Tables 2 and 3), but 1.6% differences occurred among hybrids (Table 1), 0.3% difference between irrigated and nonirrigated (Table 2), and 1.0% greater extractable starch with increasing plant population (Table 3). Fox et al. (1992) reported positive correlation between starch extraction and kernel weight, and negative correlation with grain protein concentration. Merely having higher starch content in the kernel has not been associated with high extractable starch yields.

OIL CONCENTRATION

There are few studies on agronomic practice influencing the oil concentration of maize grain, although the development of the specialty high-oil maize for livestock feed (Alexander, 1988) has spurred much recent interest. Dudley et al. (1977) and Jellum and Marion (1966) indicated that the oil concentration of maize grain was largely due to genetics while environmental factors such as planting dates, years and locations had smaller effects. Bullock et al. (1989) reported that grain oil concentration of 28 maize hybrids ranged from 2.6 to 3.5% on a dry matter ba-

TABLE 1. Summary of maize grain quality parameters response[a] to the agronomic practices of irrigation, nitrogen application and plant population.

Agronomic practice	Maize grain quality parameters[b]					
	Protein concentration	Amino acid balance	Physical quality	Extractable starch	Oil	Aflatoxin
Irrigation	▼	►	▼	►	Small	▼
Nitrogen application	►	▼	►	NA	Small	NA
Plant population increase	▼	NA	Variable	►	Small	NA

[a] Readers should be aware that these responses may be either direct or indirect effects, and that many of them are associated with grain yield.
[b] ► = increase; = ▼ decrease; NA = Not available

TABLE 2. Extractable starch as influenced by Year × Hybrid and Year × Water Regime interaction effects.

Year	Hybrid			Water Regime	
	FR 1128 × LHS1	Pioneer 3168	Crow's 688	Irrigated	Non-irrigated
			%		
1991	62.7	64.4	63.3	63.7	63.5
1992	65.0	66.5	64.5	65.5	65.8
1993	68.6	68.6	67.0	68.1	68.0
1994	63.6	64.2	62.3	63.8	62.9
Mean	65.0	65.9	64.3	65.3	65.0
L.S.D.	Year × Hybrid = 0.65			Year × Water Regime = 0.52	
L.S.D.	Hybrid = 0.33			Water Regime = 0.26	
C.V. (%)	2.3			1.6	

TABLE 3. Extractable starch as influenced by Year × Plant Population interaction effects.

Year	Plant Population (plants ha^{-1})				
	24,700	39,500	54,300	69,100	84,000
			%		
1991	63.6	63.4	63.9	63.3	63.8
1992	65.6	64.8	65.5	66.2	66.2
1993	66.7	67.1	68.5	69.1	68.8
1994	62.9	63.0	63.2	63.9	63.8
Mean	64.7	64.6	65.3	65.6	65.7

Significant Contrasts: (1991 + 1994 vs. 1992 + 1993) × Plant Population Linear
(1992 vs. 1993) × Plant Population Linear
(1992 vs. 1993) × Plant Population Linear

sis. Earle (1977) showed variations in oil concentration by year from 1917 to 1972, but no correlations were found between oil concentration and variations in temperature, rainfall or fertilization. Cloninger et al. (1975) found that oil concentration varied among hybrids and environments, but plant population had no effect. Genter et al. (1956) indicated that protein concentration was more affected than oil, and that fertility level, location and plant population had little effect on oil concentration. Wilhelm et al. (1999) found that heat stress during grain fill had no effect on maize grain oil concentration. Welch (1969) reported that nitrogen, phosphorus and potassium applications increased the oil concentration of maize grain slightly, but more important was that the increased grain yield resulted in greater oil production per unit of land area. In contrast, Jellum et al. (1973) found that increasing nitrogen application rate had no influence on the oil concentration of maize grain. Mason and Galusha (unpublished research) observed that hybrid and year influenced the grain oil concentration, but water regime (irrigation and nonirrigation) and plant population had no effect. Apparently agronomic practices have only a very minor influence on the oil concentration of maize grain.

ESSENTIAL NUTRIENT LEVELS

Humans and animals obtain at least part of their nutritional requirements for phosphorus, potassium, sulfur, calcium, magnesium and several micronutrients from maize grain. In general, fertilizer application of these nutrients either to the soil or to plant foliage in excess of that required for maximum grain yield increases the concentration in maize grain (Sander et al., 1987; Usherwood, 1985). Phosphorus availability in maize grain is low since most phosphorus is stored as phytic acid, which also chelates copper, zinc, manganese, iron, and calcium reducing their bioavailability (Ravindran et al., 1995). Recent breeding efforts with low phytic acid mutants may significantly increase maize grain phosphorus availability in the future (Ertl et al., 1998). Maize grain has very low calcium concentration, while magnesium is adequate for human and livestock diets (Sander et al., 1987). Sulfur is a structural component of the amino acids methionine and cystine, and sulfur fertilizer application may increase concentrations of protein and sulfur containing amino acids. Improving micronutrient concentrations of grain crops has become a new paradigm for meeting some human nutritional needs (Welch and Graham, 1999). Recently plant breeding (Graham et al., 1999), molecular (Schachtman and Barker, 1999) and agronomic (Rengel et al., 1999) approaches to improve micronutrient concentrations of cereal grains has been reviewed. Agronomic practices emphasized in this review were soil nutrient applications with crop and variety selection, and crop rotation indicated as other means to improve micronutrient concentration of grain.

TOXINS

Several fungal infections can occur in the field during grain maturation and drying that reduce grain quality. The fungi *Aspergillus flavus* and *Aspergillus parasiticus*, that cause ear and kernel rot, are most important due to the production of aflatoxin, which is highly toxic and carcinogenic (Watson, 1987). Infection is most serious when high temperatures and high moisture conditions during the late-grain fill growth stage follow a period of water stress. Apparently temperature is the most important factor in the infection process, but water stress increases the fungal spore production (Anderson et al., 1975) as illustrated by the serious infection present in the northern U.S. maize belt during the hot, dry summer of 1983 (Cote et al., 1984; Tuite et al., 1984). Irrigation or other means to minimize water stress of the maize crop reduce the potential for aflatoxin production. Other toxins can be produced in maize grain by *Fusarium* species, but infection is largely due to environmental conditions and agronomic practices have little effect (Watson, 1987).

CONCLUDING REMARKS

Maize grain quality varies greatly depending on the specific end-use. High quality for one end-use sometimes is low quality for another end-use, thus a generic set of quality goals is impossible to develop. This review shows that agronomic practices influence maize grain quality, especially protein concentration and the amino acid balance (Table 1). Environmental and genetic factors appear to be more important than agronomic practices for physical quality and extractable starch parameters. Most agronomic research has focused on normal field and opaque-2 maize, but there are major gaps in knowledge about agronomic practice influence on wet-milling extractable starch and nutrient levels other than nitrogen. Much future research for specialty maize types such as high oil, waxy, amylose, and low-phytate is merited and should be directed to the quality parameters specifically associated with the intended end-use.

REFERENCES

Ahmadi, M., W.J. Wiebold, J.E. Beuerlein, D.S. Eckert, and J. Schoper. (1993). Agronomic practices that affect kernel characteristics. *Agronomy Journal* 85:615-619.
Alexander, D.E. (1988). Breeding special nutritional and industrial types. In: *Corn and Corn Improvement*, 3rd Edition. ed. C.F. Sprague and J.W Dudley, Madison, WI: American Society of Agronomy, pp. 869-880.

Anderson, H.W., E.W. Nehring, and W.R. Wichser. (1975). Aflatoxin contamination of corn in the field. *Journal of Agriculture, Food and Chemistry* 23:775-782.

Anderson, E.L., E.J. Kamprath, and R.H. Moll. (1984). Nitrogen fertility effects on accumulation, remobilization, and partitioning of N and dry matter in corn genotypes differing in prolificacy. *Agronomy Journal* 76:397-404.

Andrade, F.H. and M.A. Ferreiro. (1996) Reproductive growth of maize, sunflower and soybean at different source levels during grain fill. *Field Crops Research* 48:155-165.

Arnold, J.M., L.F. Bauman, and H.S. Aycock. (1977). Interrelations among protein, lysine, oil, certain mineral element concentrations, and physical kernel characteristics in two maize populations. *Crop Science* 17:421-425.

Bauer, P.J. and P.R. Carter. (1986). Effect of seeding date, plant density, moisture availability, and soil nitrogen fertility on maize kernel breakage susceptibility. *Crop Science* 17:362-366.

Benzian, B. and P. Lane. (1979). Some relationships between grain yield and grain protein of wheat experiments in south-east England and comparisons with such relationships elsewhere. *Journal of the Science of Food and Agriculture* 30:59-70.

Bhatia, C.R. and R. Rabson. (1976). Bioenergetic considerations in cereal breeding for protein improvement. *Science* 194:1418-1421.

Bhatia, C.R. and R. Rabson. (1987). Relationship of grain yield and nutritional quality. In: *Nutritional Quality of Cereal Grains: Genetics and Agronomic Management.* eds. R.A. Olson and K.J. Frey. Madison, WI: American Society of Agronomy, pp. 11-44.

Breteler, H. (1976). Nitrogen fertilisation, yield and protein quality of a normal and high-lysine maize variety. *Journal of the Science of Food and Agriculture* 27: 978-982.

Bullock, D.G., P.L. Raymer, and S. Savage. (1989). Variation of protein and fat concentration among commercial corn hybrids grown in the southeastern USA. *Journal of Production Agriculture* 12:157-160.

Cirilo, A.G. and F.H. Andrade. (1994). Sowing date and maize productivity: I. Crop growth and dry matter partitioning. *Crop Science* 34:1039-1043.

Cirilo, A.G. and F.H. Andrade. (1996). Sowing date and kernel weight in maize. *Crop Science* 36:325-331.

Claassen, M.M. and R.H. Shaw. 1970. Water deficit effects on corn. II. Grain components. *Agronomy Journal* 62:652-655.

Cloninger, F.D., R.D. Horrocks, and M.S. Zuber. (1975). Effects of harvest date, plant density, and hybrid on corn grain quality. *Agronomy Journal* 67:693-695.

Cote, L.M., J.D. Reynolds, R.F. Vesonder, and W.B. Bulk. (1984). Survey of vomitoxin-contaminated feed grains in midwestern United States, and associated health problems with swine. *Journal of the American Veterinary Medical Association* 184:189-192.

Cromwell, G.L., M.J. Bitzer, T.S. Stahly, and T.H. Johnson. (1983). Effects of soil nitrogen fertility on the protein and lysine content and nutritional value of normal and opaque-2 corn. *Journal of Animal Science* 57:1345-1351.

Decau, J. and B. Pujol. (1973). Effets compares de l'irrigation et de la fumure azotee sur les productions qualitatives et quantitatives de mais de varietes differentes. II. Production de proteines. *Annual Agronomique* 24:359-373.

Dorsey-Redding, C., C.R. Hurburgh, Jr., L.A. Johnson, and S.R. Fox. (1991). Relationships among maize quality factors. *Cereal Chemistry* 68:602-605.

Bullock, D.G., P.L. Raymer, and S. Savage. (1989). Variation of protein and fat concentration among commercial corn hybrids grown in the southeastern U.S.A. *Journal of Production Agriculture* 2:157-160.

Dudley, J.W., R.J. Lambert, and I.A. de la Roche. (1977). Genetic analysis of crosses among corn strains divergently selected for percent oil and protein. *Crop Science* 17:114-117.

Duvick, D.N. and K.G. Cassman. (1999). Post-green revolution trends in yield potential of temperate maize in the north-central United States. *Crop Science* 39: 1622-1630.

Earle, F.R. (1977). Protein and oil in corn: Variation by crop years from 1907 to 1972. *Cereal Chemistry* 54:70-79.

Ertl, D.S., K.A. Young, and V. Raboy. (1998). Plant genetic approaches to phosphorus management in agricultural production. *Journal of Environmental Quality* 27: 299-304.

Fox, S.R., L.A. Johnson, C.R. Hurburgh, Jr., C. Dorsey-Redding, and T.B. Bailey. (1992). Relations of grain proximate composition and physical properties to wet-milling characteristics of maize. *Cereal Chemistry* 69:191-197.

Frey, K.J. (1951). The inter-relationships of protein and amino acids in corn. *Cereal Chemistry* 28:123-132.

Frey, K.J. (1977). Proteins of oats. *Zeitschrift fur Pflanzenzuchtung* 78:185-215.

Frey, K.J., B. Brimhall, and G.F. Sprague. (1949). The effects of selection upon protein quality in the corn kernel. *Agronomy Journal* 41:399.

Genter, C.G., J.F. Eheart, and W.N. Linkous. (1956). Effect of location, hybrid, fertilizer, and rate of planting on the oil and protein content of corn grain. *Agronomy Journal* 48:63-67.

Graham, R., D. Senadhira, S. Beebe, C. Iglesias, and I. Monasterio. (1999). Breeding for micronutrient density in edible portions of staple food crops: Conventional approaches. *Field Crops Research* 60:57-80.

Haros, M. and C. Suarez. (1998). Effect of drying, initial moisture and variety in corn wet milling. *Journal of Food Engineering* 34:473-481.

Hancock, J.D., E.R. Peo Jr., A.J. Lewis, K.R. Kniep, and S.C. Mason. (1988). Effects of irrigation and nitrogen fertilization of normal and high lysine corn on protein utilization by the growing rat. *Nutritional Reports International* 38:413-422.

Hill, L.D. (1981). Quality problems in exporting U.S. corn. *Proceedings of the Annual Corn and Sorghum Research Conference* 36:191-198.

Jellum, M.D. and J.E. Marion. (1966). Factors affecting oil content and oil composition of corn (*Zea mays* L.) grain. *Crop Science* 6:41-42.

Jellum, M.D., F.C. Boswell, and C.T. Young. (1973). Nitrogen and boron effects on protein and oil of corn grain. *Agronomy Journal* 65:330-331.

Jindal, V.K. and N.N. Mohsenin. (1978). Dynamic hardness determination of corn kernels from impact tests. *Journal of Agricultural Engineering Research* 23:77-84.

Johnson, D.Q. and W.A. Russell. (1982). Genetic variability and relationships of physical grain quality traits in the BSSS population of maize. *Crop Science* 22:805-809.

Kamprath, E.J., R.H. Moll, and N. Rodriguez. (1982). Effects of nitrogen fertilization and recurrent selection on performance of hybrid populations of corn. *Agronomy Journal* 74:955-958.

Kniep, K.R. and S.C. Mason. (1991). Lysine and protein content of normal and opaque-2 maize grain as influenced by irrigation and nitrogen. *Crop Science* 31:177-181.

Kniep, K.R. and S.C. Mason. (1989). Kernel breakage and density of normal and opaque-2 maize grain as influenced by irrigation and nitrogen. *Crop Science* 29:158-163.

LeFord, D.R. and W.A. Russell. (1985). Evaluation of physical quality in BS17 and BS1(HS)C1. *Crop Science* 25:421-425.

MacGregor, J.M., L.T. Taskovitch, and W.P. Martin. (1961). Effect of nitrogen fertilizer and soil type on the amino acid content of corn grain. *Agronomy Journal* 53:211-214.

Manoharkumar, B., P. Gerstenkorn, H. Zwingelberg, and H. Bolling. (1978). On some correlations between grain composition and physical characteristics to the dry milling performance for maize. *Journal of Food Science and Technology* 15:1-6.

Mason, S.C. and R.J. Sabata. (1995). Water regime and nitrogen impact on maize grain yield, grain quality, and soil nitrate. *Agronomy (Trends in Agricultural Science)* 3:5-10.

McDermitt, D.K. and R.S. Loomis. (1981). Elemental composition of biomass and its relation to energy content, growth efficiency, and growth yield. *Annals of Botany* 48:275-290.

Mertz, E.T., L.S. Bates, and O.E. Nelson. (1964). Mutant gene that changes protein composition and increases lysine content of maize endosperm. *Science* 145: 279.

Miller, B.S., J.W. Hughes, R. Rousser, and Y. Poneranz. (1981). Measuring the breakage susceptibility of shelled corn. *Cereal Foods World* 26:75-80.

Moes, J. and T.J. Vyn. (1988). Management effects on kernel breakage susceptibility of early maturing corn hybrids. *Agronomy Journal* 80:699-704.

Norberg, O.S., S.C. Mason, and S.R. Lowry. (1988). Ethephon influence on harvestable yield, grain quality, and lodging of corn. *Agronomy Journal* 80:768-772.

Oikeh, S.O., J.G. Kling, and A.E. Okoruwa. (1998). Nitrogen management effects on maize grain quality in the West Africa moist savanna. *Crop Science* 38:1056-1061.

Olsen, R.A., K.D. Frank, E.J. Deibert, A.F. Dreier, D.H. Sander, and V.A. Johnson. (1976). Impact of residual mineral N in soil on grain protein yields of winter wheat and corn. *Agronomy Journal* 68:769-772.

Paulsen, M.R. and L.D. Hill. (1977). Corn breakage in overseas shipments–Two case studies. *Transactions of the American Society of Agricultural Engineers* 20:550-552.

Paulsen, M.R. and L.D. Hill. (1985). Corn quality factors affecting dry milling performance. *Journal of Agricultural Engineering Research* 31:255-263.

Paulsen, M.R., L.D. Hill, D.G. White, and G.F. Sprague. (1983). Breakage susceptibility of corn-belt genotypes. *Transactions of the American Society of Agricultural Engineers* 32:1007-1014.

Paulsen, M.R., S.L. Hofing, L.D. Hill, and S.R. Eckhoff. (1996). Corn quality characteristics for Japan markets. *Applied Engineering in Agriculture* 12:731-738.

Penning de Vries, F.W.T., A.H.M. Brunsting, and H.H. van Laar. (1974). Products, requirements and efficiency of biosynthesis: A quantitative approach. *Journal of Theoretical Biology* 45:339-377.

Perry, W.P. (1988). Corn as a livestock feed. In: *Corn and Corn Improvement*, 3rd Edition. eds. C.F. Sprague and J.W Dudley, Madison, WI: American Society of Agronomy, pp. 941-963.

Pierre, W.H., L Dumenil, V.D. Jolley, J.R. Webb, and W.D. Shrader. (1977). Relationship of corn yield, expressed as a percentage of maximum and the N percentage in grain: I. Various N-rate experiments. *Agronomy Journal* 69:215-220.

Pollmer, W.G., D. Eberhard, and D. Klein. (1978). Inheritance of protein and yield of grain and stover in maize. *Crop Science* 18:757-759.

Ravindran, V., W.L. Bryden, and E.T. Kornegay. (1995). Phytates: Occurrence, bioavailability and implications in poultry nutrition. *Poultry and Avian Biological Review* 6:125-143.

Rendig, V.V. and F.E. Broadbent. (1979). Proteins and amino acids in grain of maize grown with various levels of N. *Agronomy Journal* 71:509-512.

Rengel, Z., G.D. Batten, and D.E. Crowley. (1999). Agronomic approaches for improving the micronutrient density in edible portions of field crops. *Field Crops Research* 60:27-40.

Remison, S.U. and O. Omueti. (1982). Effects of nitrogen application and leaf clipping after mid-silk on yield and protein content of maize. *Canadian Journal of Plant Science* 62:777-779.

Sabata, R.J. and S.C. Mason. (1992). Corn hybrid interactions with soil nitrogen level and water regime. *Journal of Production Agriculture* 5:137-142.

Sander, D.H., W.H. Allaway, and R.A. Olson. (1987). Modification of nutritional quality by environment and production practices. In: *Nutritional Quality of Cereal Grains: Genetics and Agronomic Management*, eds. R.A. Olson and K.J. Frey, Madison, WI: American Society of Agronomy, pp. 45-82.

Schachtman, D.P. and S.J. Barker. (1999). Molecular approaches for increasing the micronutrient density in edible portions of food crops. *Field Crops Research* 60:81-92.

Shumway, C.R., J.T. Cothren, S.O. Serna-Saldivar, and L.W. Rooney. (1992). Planting date and moisture effects on yield, quality and alkaline-processing characteristics of food-grade maize. *Agronomy Journal* 32:1265-1269.

Thompson, R.A. and G.H. Foster. (1963). Stress cracks and breakage in artificially dried corn. USDA Agricultural Marketing Service, Marketing Research Report 631. Washington, DC.

Tollenaar, M. and J. Wu. (1999). Yield improvement of temperate maize is attributable to greater stress tolerance. *Crop Science* 39:1597-1604.

Tsai, C.Y., H.L. Warren, D.M. Huber, and R.A. Bressan. (1983). Interactions between the kernel N sink, grain yield and protein nutritional quality of maize. *Journal of the Science of Food and Agriculture* 34:255-263.

Tsai, C.Y., I. Dweikat, D.M. Huber, and H.L. Warren. (1992). Interrelationship of nitrogen nutrition with maize (*Zea mays*) grain yield, nitrogen use efficiency and grain quality. *Journal of the Science of Food and Agriculture* 58:1-8.

Tuite, J.C., Koh-Knox, R. Strosine, F.A. Cantone, and L.F. Bauman. (1984). Preharvest aflatoxin contamination of dent corn in Indiana in 1983. *Plant Disease* 68:893-895.

Usherwood, N.R. (1985). The role of potassium in crop quality. In: *Potassium in Agriculture*, ed. R.D. Munson, Madison, WI: American Society of Agronomy, pp. 489-513.

Vasal, S.K., E. Villegas, M. Bjarnason, B. Gelaw, and P. Goertz. (1980). Genetic modifers and breeding strategies in developing hardendosperm opaque-2 materials. In: *Improvement of Quality Traits for Grain and Silage Use*, eds. W.G. Pollmer and R.H. Phipps., The Hague, Netherlands: Martinus Nijhoff Publishers, pp. 37-71.

Vyn, T.J. and J. Moes. (1988). Breakage susceptibility of corn kernels in relation to crop management under long growing season conditions. *Agronomy Journal* 80:915-920.

Vyn, T.J. and M. Tollenaar. (1998). Changes in chemical and physical quality parameters of maize grain during three decades of yield improvement. *Field Crops Research* 59:135-140.

Watson, S.A. (1987). Measurement and maintenance of quality. In: *Corn: Chemistry and Technology*, eds. S.A. Watson and P.E. Ramstad, St. Paul, MN: American Association of Cereal Chemists, pp.125-183.

Wehling, R.L., D.S. Jackson, D.G. Hooper, and A.R. Ghaedian. (1993). Prediction of wet-milling starch yield from corn by near-infrared spectroscopy. *Cereal Chemistry* 70:720-723.

Welch, L.F. (1969). Effect of N, P, and K on the percent and yield of oil in corn. *Agronomy Journal* 61:890-891.

Welch, R.M. and B.D. Graham. (1999). A new paradigm for world agriculture: Meeting human needs: Productive, sustainable, nutritious. *Field Crops Research* 60:1-10.

Weller, C.L., M.R. Paulsen, and M.P. Steinberg. (1988). Correlation of starch recovery with assorted quality factors of four corn hybrids. *Cereal Chemistry* 65:392-397.

Wilhelm, E.P., R.E. Mullen, P.L. Keeling, and G.W. Singletary. (1999). Heat stress during grain filling in maize: Effects on kernel growth and metabolism. *Crop Science* 39:1733-1741.

Zehr, B.E., S.R. Eckhoff, S.K. Singh, and P.L. Keeling. (1995). Comparison of wet-milling properties among maize inbred lines and their hybrids. *Cereal Chemistry* 72:491-497.

Making Rice a Perfect Food:
Turning Dreams into Reality

Ju-Kon Kim

Hari B. Krishnan

SUMMARY. Despite a significant increase in food production during the last century, world production will need to be doubled or tripled by the year 2050 to meet the needs of an expected 10 billion global population. Fortunately, a second revolution in agriculture appears to be taking place from advances in biotechnology. Worldwide in 1999, about 40 million hectares of transgenic plants were grown, and this area is expected to increase significantly for years to come. Rice (*Oryza sativa* L.) is the staple food of the majority of 3.5 billion people in Asia. Increases in population in the rice growing regions of Asia will require 70% more rice in 2025 than is consumed today. In addition, rice, in its milled form, is poor in essential amino acids and a range of vitamins and micronutrients, which creates malnutrition. With the increasing number of genes discovered in plants and other organisms, transgenic research is being utilized to improve agronomic traits of rice, such as resistance to biotic and abiotic stress, and to increase photosynthetic efficiency which collectively increase yield. Research is also moving toward improvement of

Ju-Kon Kim is affiliated with the Department of Biological Science, Myongji University, Yongin 449-728, Korea.

Hari B. Krishnan is affiliated with Plant Genetics Research Unit, USDA-ARS, and Department of Agronomy, University of Missouri, Columbia, MO 65211 USA.

Address correspondence to: Dr. Hari B. Krishnan, Plant Genetics Research Unit, USDA-ARS, 108W Curtis Hall, Department of Agronomy, University of Missouri, Columbia, MO 65211 (E-mail: KrishnanH@missouri.edu).

The authors would like to thank Drs. Mike McMullen, Robert Sharp, Larry Darrah, and Jerry White for critical reading of the manuscript.

[Haworth co-indexing entry note]: "Making Rice a Perfect Food: Turning Dreams into Reality." Kim, Ju-Kon, and Hari B. Krishnan. Co-published simultaneously in *Journal of Crop Production* (Food Products Press, an imprint of The Haworth Press, Inc.) Vol. 5, No. 1/2 (#9/10), 2002, pp. 93-130; and: *Quality Improvement in Field Crops* (ed: A. S. Basra, and L. S. Randhawa) Food Products Press, an imprint of The Haworth Press, Inc., 2002, pp. 93-130. Single or multiple copies of this article are available for a fee from The Haworth Document Delivery Service [1-800-HAWORTH, 9:00 a.m. - 5:00 p.m. (EST). E-mail address: getinfo@ haworthpressinc.com].

grain quality traits, including amino acids, micronutrients, and vitamins. Genome sequencing and the techniques for rice transformation have been developed. Thus, collaborative efforts in genomics, transformation, and molecular breeding of rice are expected to lead to a significant contribution to global food security. In this article, we review the current status of genetic improvement of rice. Improved methods for transgene expression in rice and potential modifications that will significantly improve yield and grain quality of rice are also discussed. *[Article copies available for a fee from The Haworth Document Delivery Service: 1-800-HAWORTH. E-mail address: <getinfo@haworthpressinc.com> Website: <http://www. HaworthPress.com>]*

KEYWORDS. Rice, biotechnology, grain quality traits, genetic improvement

INTRODUCTION

In the past 12 years, world population has grown from 5 billion to 6 billion and is expected to be 10 billion by 2050. Despite significant achievements in increasing food production during the last 20 to 30 years, more than 800 million people consume less than 2,000 calories a day and are chronically undernourished. Large numbers of people in developing countries exist on simple diets composed primarily of a few staple foods (rice, cassava, wheat, and maize) that are poor sources of some macronutrients and many essential micronutrients. For example, milled rice is poor in fat, protein, and a whole range of vitamins and micronutrients which results in widespread malnutrition in developing countries (FAO, 1998). Over 250 million children are at risk for vitamin A deficiency, and each year 500,000 children suffer irreversible blindness from vitamin A deficiency (UNICEF, 1998). Two billion people are at risk for iron deficiency and 1.5 billion people are at risk for iodine deficiency (DellaPenna 1999).

Rice (*Oryza sativa* L.) is the staple food of the majority of the Asian population. In 1997, 572 million metric tons of rice were produced worldwide (FAO, 1998). Yields ranged from as little as 1 t/ha in many countries of Africa to more than 6 t/ha in China, Japan, and South Korea (FAO, 1998). Genetic improvements in rice and the development of modern rice varieties, along with improved cultivation practices, account for the impressive growth in production over the past 20 to 30 years. Total rice production has doubled since 1965. Scientists have incorporated many yield-enhancing traits into modern rice, including greater pest resistance, shorter crop duration, and better quality grain. There is, however, increasing evidence of decline in the rate of growth of crop yields (Figure 1; Conway and Toenniessen, 1999).

Rice is the smallest (430 Mb) in genome size among the important cereal crops, allowing results from rice research to be used to understand of other crops. The genome size of maize and wheat is in the range 10^9 to 10^{10} bp per haploid genome making their genomics difficult to work with. Comparative mapping conducted among rice, wheat, millet, maize, sorghum, and barley revealed a synteny or co-linearity of genes along the chromosomes (Moore et al., 1995; Gale and Devos, 1998a, 1998b). Furthermore, it is becoming evident that co-linearity of genes exists even between rice and Arabidopsis, a model plant for dicots. Two of the five Arabidopsis chromosomes have been sequenced and sequences of the other three are expected to come by the end of year 2000 (Gura, 2000). The International Rice Genome Sequencing Project (IRGSP) has been organized and some of their results are available on publicly accessible databases (http://www.staff.or.jp). More recently, Monsanto (St. Louis, MO, USA) and collaborators announced that they had almost completed a rough draft of the entire rice genome and decided to turn their data over to IRGSP (Pennisi, 2000). As a result, IRGSP, with Monsanto's data integrated, could complete its work in just two to three years.

Rice has been focused on for improvement by breeding practices from the beginning of the last century and much genetic knowledge has been accumulated. Available DNA sequence and genetic information will not only prompt us to understand what all the new genes do, but

FIGURE 1. Average annual increase in yields of three major crops in developing countries by periods [Reprinted by permission from *Nature* (Conway G. and G. Toenniessen. Feeding the world in the twenty-first century. *Nature* 402: C55-C58), copyright (1999) Macmillan Magazines Ltd.]

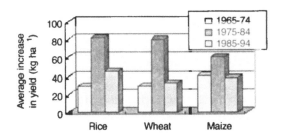

also will open up many new opportunities for making transgenic plants for use in cultivar development. Once useful genes are available from any source, transgenic approaches can support the breeders by transferring genes that are not available in the rice gene pool. These can also support the breeders by transferring desired genes from one cultivar to another without linked undesirable genes.

Recent reviews have summarized studies about transgenic rice plants, including transformation methods and applications (Hiei et al., 1999), genetically engineered virus-resistant rice (Waterhouse and Upadhyaya, 1999), and nutritional improvement of rice (Klöti and Potrykus, 1999). In this article, we describe improved methods for transgene expression in rice and will consider the types of modifications that can significantly improve yield and grain quality of rice.

IMPROVEMENTS IN TRANSGENE EXPRESSION

Improvement in plant transformation technology has allowed novel approaches for studying gene expression and opened new avenues for the genetic modification of crop plants. During the last four years, the area planted to transgenic crops increased dramatically, from 1.7 million hectares in 1996 to 39.9 million hectares in 1999 (James, 1999). Rice is the first cereal crop from which fertile transgenic plants were obtained (Zhang and Wu, 1988; Toriyama et al., 1988; Shimamoto et al., 1989). Since then, many fertile transgenic rice lines have been produced by protoplast-mediated DNA transformation (Datta et al., 1990; Peng et al., 1992), by microprojectile bombardment (Christou et al., 1991; Cao et al., 1992; Duan et al., 1996; Xu et al., 1996; Christou, 1997), or by *Agrobacterium*-mediated DNA transfer (Hiei et al., 1994, 1997; Komari et al., 1996). Thus, generation of transgenic plants is now becoming increasingly routine for many rice varieties. Despite many attempts at using transgenesis to improve important agronomic traits of rice (Christou, 1997; Hiei et al., 1997), only few cases are documented where transgenic plants had acceptable agronomic performance (Christou, 1997).

High-Level Expression System in Transgenic Rice: Targeting of Gene Products to Plastids

For efficient and stable gene expression in transgenic plants with desired phenotypes, there are several factors that need to be considered before construction of expression cassettes for transformation. These include transcriptional (such as promoters and enhancers), post-transcriptional (introns and 3'-untranslated regions), translational (5'-untranslated regions and codons) and post-translational (modification and intracellular targeting) factors. There is a range of well-characterized promoters that permit high-level expression in the seed (Russell and Fromm, 1997). In particular, an improved expression system is desirable for genes that are expressed in cytoplasm at low levels or those that are deleterious to cells when expressed in cytoplasm. One strategy is to make gene products localized to other intracellular compartments such as chloroplasts and thus alleviate cytoplasmic toxicity of the gene products.

The small subunit gene (*rbcS*) of ribulose bisphosphate carboxylase/oxygenase (Rubisco) is nuclear-encoded and its expression is regulated by both light and an endogenous, leaf-specific pattern of activation (Dean et al., 1989; Gilmartin et al., 1990). The *rbcS* mRNA is translated in the cytoplasm to produce a large precursor protein which is subsequently processed to its mature size by cleavage of transit peptide during transport into the chloroplast (von Heijne et al., 1989; Keegstra and Cline, 1999). There are some examples of *rbcS* transit peptides that increased expression levels of fused genes as well as targeting of gene products (De Almeida et al., 1989; Herminghaus et al., 1991, 1996; Jang et al., 1999). In addition, chloroplast targeting of gene products can alleviate the toxic effects of the products on cells. Targeting of Cry1A(c) (Wong et al., 1992) and the polyhydroxybutyrate synthesizing enzymes (Nawrath et al., 1994) to chloroplasts of transgenic tobacco and Arabidopsis, respectively, gave rise to normal plants, which otherwise would have been stunted.

Jang et al. (1999) have designed the expression vector *rbcS-Tp* for chloroplast targeting (Figure 2), which consists of rice *rbcS* promoter and transit peptide (*Tp*) sequence translationally fused to the green fluorescent protein gene *sgfp*. In transgenic rice plants, addition of the transit peptide to the *rbcS* promoter not only targeted sGFP to chloroplasts, but also elevated accumulation of the protein 20-fold, yielding up to 10% of total soluble protein. Accumulation of sGFP was determined by immunoblot and SDS-PAGE analysis (Figure 2, middle and lower panels, respectively). Thus, the *Tp* sequence of the *rbcS* gene significantly enhanced the activity of the *rbcS* promoter in mature leaf tissues of transgenic rice plants.

It is possible that the rice *rbcS* transit peptide sequence may act as a transcriptional enhancer, increasing the expression levels of the fused sGFP gene. Alternatively, continuous removal of sGFP from the cyto-

FIGURE 2. A high-level transgene expression system in rice. Expression vectors used for rice transformation (top panel). The *rbcS-Tp* consists of the rice *rbcS* promoter and its transit peptide sequence (*Tp*) fused to *sgfp* coding region, and 3' region of the potato proteinase inhibitor II gene (*3' PinII*) plus the *bar* gene expression cassette that contains *35S* promoter/*bar* coding region/3' region of the nopaline synthase gene (*3' NOS*). The *Act1* consists of the rice *Act1* 5' region including the promoter, first non-coding exon, and first intron linked to the *sgfp* coding region of the *rbcS-Tp*. Expression analysis of sGFP in transgenic rice plants by Western blot (middle panel) and by SDS-PAGE gel stained with Coomassie blue (bottom panel). The amounts of protein extracts loaded per lane are also indicated. Proteins extracted from the Act1-transformed (Act1) and rbcS-Tp-transformed (rbcS-Tp) rice plants are marked. First and second lanes are the purified GFP and protein extracts from an untransformed plant, respectively.

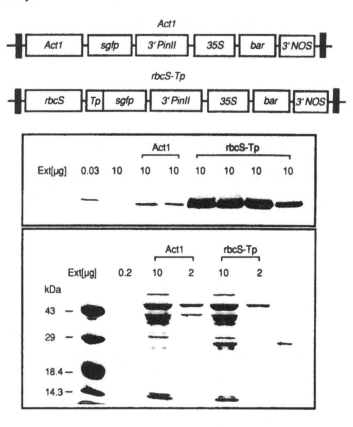

plasm may prevent the protein from reaching a toxic threshold level resulting in efficient translation. The latter is supported by the observation that in the leaf mesophyll cells, intracellular space shared by cytoplasm and nuclei, where the untargeted GFP is present, is much smaller than that shared by chloroplasts, where the targeted GFP is present (Jang et al., 1999). Consequently, the amount of GFP per cell producing the targeted GFP could be larger than that per cell producing the untargeted GFP. This expression system will be useful in increasing the expression of other genes which are agronomically important. For example, expression of *Bacillus thuringiensis* toxin genes that have reached levels of 0.02-0.10% of the total soluble protein so far, can be significantly increased. This system can also be applied for production of foreign proteins or peptides that are medically important, for example, vaccines or monoclonal antibodies.

Use of Matrix Attachment Region (MAR) Sequences to Reduce Variation in Transgene Expression Levels

There are important genetic factors that affect transgene expression and that have to be carefully integrated before transformation. These include transgene silencing and influence of chromosomal microenvironment on transgene expression. One method to escape from transgene silencing in rice is to use genes from genetically distant organisms, such as dicotyledonous plants or microorganisms, and to select transgenic lines with a single copy of the transgene after transformation. This is because silencing is caused mostly by sequence homology between the genes being introduced and the endogenous counterparts or between the newly introduced transgenes (Matzke and Matzke, 1995; Meyer and Saedler, 1996). In most eukaryotic organisms, injection of dsRNA longer than 500 bp specifically suppresses the expression of a gene with a corresponding DNA sequence, but has no effect on genes unrelated in sequence (Sharp and Zamore, 2000; Grishok et al., 2000). The dsRNA suppresses gene expression by a post-transcriptional process, although there is convincing evidence, at least in plants, that dsRNA can also regulate DNA methylation (another mechanism for silencing genes) (Wassenegger et al., 1994). The surprising nature of dsRNA is highlighted by the recent discovery of small RNAs, about 25 nucleotides long, in plants displaying post-transcriptional gene silencing (Hamilton and Baulcombe, 1999). These small RNAs are complementary to both the sense and antisense strands of the silenced gene, suggesting that they are derived from a dsRNA signal.

In addition to transgene silencing, genetic transformation of plants generally results in a large and random variation in the expression of the newly introduced transgene between individual transformants. This variability is attributed to different integration sites of the transgene, reflecting the influence of the surrounding chromatin known as position effects. Matrix attachment region (MAR) sequences have been repeatedly shown to insulate transgenes from surrounding chromatin, reducing variability in the transgene expression in transgenic tobacco (Allen et al., 1996; Mlynarova et al., 1994, 1995, 1996). In some cases, addition of MAR sequences to expression vectors resulted in an increase in expression levels of the linked reporters in transgenic plants (Allen et al., 1996). Interestingly, in a study of genomic context around transgenic loci (Iglesias et al., 1997), the authors found that the stably expressed loci were flanked by a MAR-like sequence, whereas the unstably expressed loci were not, suggesting that MAR sequences also play a role in decreasing risks of transgene silencing. In recent studies, tobacco MAR (Rb7) significantly improved the stability of transgene expression levels in transgenic rice (Vain et al., 1999) and tobacco (Ulker et al., 1999) plants. Thus, use of MAR sequences in construction of expression vectors can minimize position effects, increase expression levels, and possibly stabilize transgene expression.

To investigate whether the MAR sequence functions in transgenic rice plants, we inserted the chicken lysozyme MAR sequences into borders of the T-DNA of the plasmid pSBG700 containing *Act1* promoter linked to the *sgfp* coding sequence, generating the plasmid pSBG-MAR (Figure 3). A number of fertile transgenic rice plants for each construct were produced by *Agrobacterium*-mediated transformation. Southern analysis of 60 primary transformants resulted in distinct band patterns, indicating that all the transformants had been generated by independent integration events. Interestingly, many more single-copy transgenic plants were produced (60%) in the case of the MAR-containing vector than for the control vector (20%). Northern analysis demonstrated that in the absence of MAR, expression levels of the *sgfp* transgene varied markedly with the integration sites (Figure 3). In contrast, all the lines in which the transgene was flanked by the MAR sequence at both ends showed position-independent expression, resulting in similar levels in expression among different lines. Thus, the MAR sequence significantly reduced variation of transgene expression between independent transformants. Moreover, the sequence appeared to confer copy number-dependence in gene expression (Figure 3). Since the MAR sequence turned out to be effective in rice chromosomes, this system

FIGURE 3. A MAR sequence reduces variation in expression levels of transgene in rice. Expression vectors for transformation with the MAR sequence (pSBG-MAR) and without it (pSBG700). Other abbreviations are the same as those in Figure 2. Total RNAs extracted from the pSBG-MAR-transformed (M1-M12), pSBG700-transformed (G1-G12) and from an untransformed (NC) rice plants were hybridized with the *sgfp* coding region. RNA gel blot analysis showing levels of *sgfp* transcripts in pSBG-MAR (upper panel) or pSBG700 (lower panel)-transformed rice lines. EtBr-staining of total RNA was used as a control for equal RNA loading.

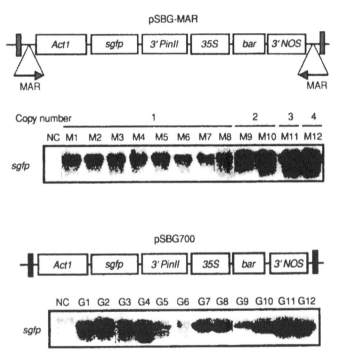

should provide two significant advantages in transformation practices. One is to increase the frequency of transformation because transgenes that are flanked by MAR sequences will be expressed regardless of chromosomal regions they are integrated into. Transferred genes without MAR sequences can be expected to be expressed only when they land on transcriptionally active regions of chromosomes. The genes integrated into transcriptionally inactive regions are not expressed and are thus excluded during selection in tissue culture. The other advantage is to reduce the number of transgenic plants that one needs to generate in the initial stage of transformation. Generally, for a few fertile transgenic plants with desired levels of transgene expression, a large number of primary transformants have to be generated and analyzed. Because the MAR sequence can direct evenly high levels of expression of transgenes in different transgenic lines by suppressing position effects, a large number of transgenic lines are not needed. These advantages will synergistically help us to perform efficient transformations by reducing the amount of work for one transformation and by obtaining transgenic plants stably expressing genes of interest.

TARGETS FOR MODIFICATION: INCREASE IN YIELD

Dehydration, Cold, and Salt Tolerance

Plants respond to a wide range of internal and external stimuli by modulating transcription of diverse genes. Crop productivity is greatly affected by environmental stresses such as drought, freezing, and salinity. Such a stress causes a battery of genes to be activated as part of the plant stress response (Shinozaki and Yamaguchi-Shinozaki, 1996, 1997; Thomashow, 1998). Although conventional plant breeding programs have improved yields for crops grown in stressful environments, there is a growing belief that further gains can only be achieved through targeted manipulation of genes involved in stress resistance. Many stress-inducible genes have been identified over the past decade (Ingram and Bartels, 1996; Shinozaki and Yamaguchi-Shinozaki, 1996, 1997; Thomashow, 1998), but only recently have functional roles in stress tolerance been identified. The products of these genes can be classified into two groups: those that directly protect against environmental stresses and those that regulate gene expression and signal transduction in the stress response. To improve the stress tolerance of plants, several genes that belong to the first group have been employed for transformation (Holmberg and Bulow, 1998). These genes include those encoding enzymes for biosynthesis of stress protective molecules such as proline (Kavi-Kishor et al., 1995), betaine (Hayashi et al., 1997), mannitol (Tarczynski and Bohnert, 1993), and trehalose (Holmström et al., 1996). Also included are genes for LEA protein (Xu et al., 1996), desaturase (Kodama et al., 1994; Ishizaki-Hishizawa et al., 1996), and SOD (McKersie et al., 1996). These transgenic studies suggest that these proteins play an important role in stress tolerance, but the mechanism of protection remains poorly characterized. The second group of gene products includes transcription activators and protein kinases involved in signal transduction pathways. To make stress-tolerant plants, one can transform plants with genes encoding stress-inducible transcription factors that activate many genes involved in stress tolerance. Recently, this was illustrated by studies of the molecular basis of plant tolerance to freezing (Jaglo-Ottosen et al., 1998) and to dehydration (Kasuga et al., 1999). Low temperature and dehydration stress induce physiological changes in plants that make them better able to tolerate subsequent severe stresses (Figure 4). Jaglo-Ottosen et al. (1998) have shown that overexpression of a single transcription factor, CBF1, in Arabidopsis can induce the expression of cold-response genes even in the absence of the normal stimulus of low temperatures. Usually, the cold-response genes are induced gradually when the plant is exposed to slowly declining temperatures. In transgenic Arabidopsis plants the genes are active all the time, enabling them to survive sudden temperature drops to as low as −8°C. However, overexpression of these genes resulted in severe growth retardation under normal growth conditions, presumably due to the 35S CaMV promoter driving expression of *CBF1* in a constitutive manner. Similarly, Kasuga et al. (1999) transformed Arabidopsis with a gene encoding dehydration-inducible transcription factor, DREB1A, that also activates many genes involved in dehydration tolerance. By using the dehydration-inducible *rd29A* promoter to drive expression of *DREB1A*, the authors minimized the negative effects on plant growth experienced with use of the 35S CaMV promoter. By overexpressing the *DREB1A* gene, they succeeded in inducing the expression of several stress-related genes involving *rd29A(cor78)*, *rd17(cor47)*, *kin2(cor6.6)*, *cor15a*, *erd10*, and *kin1*, leading to striking improvements in plant tolerance to freezing, salt loading, and dehydration. Liu et al. (1998) had previously shown DREB1A also mediated transcription of several genes in response to cold and water stress. Stress-related gene expression is induced by the binding of DREB1A, which is itself induced by cold and water stress, to a *cis*-acting DNA element (DRE, drought responsive element) in the promoters of genes such as *rd29A*, *rd17*, *cor6.6*, *cor15a*, *erd10*, and *kin1*. This binding initiated synthesis of gene products implicated in plant acclimation responses to low temperature and water stress (Gilmour et al., 1998; Stockinger et al., 1997). The signal transduction pathway mediating these inductive events is independent of abscisic acid (ABA), a hormone involved in expression of similar genes in response to water stress (Shinozaki and Yamaguchi-Shinozaki, 1997; Thomashow, 1998). Presumably, low temperature and water stress induce the expression of endogenous *DREB1A* in the transgenic plants, which then interact with the DRE on the *rd29A-DREB1A* construct, rap-

FIGURE 4. A strategy to improve tolerance to dehydration, low temperature, and salt stress by a regulated expression of transcription factor genes from *Arabidopsis*. DREB/CBF and ABF genes, that are induced by dehydration and cold stress in *Arabidopsis,* encode transcription factors that bind to the DRE and ABRE promoter element of stress-responsive genes, and turn on their expression (left panel). By expressing of a fusion of DREB/CBF and ABF or individual gene in rice under a stress-inducible (DRE and ABRE-containing) promoter, one can amplify the stress signaling pathway(s), hence, improving tolerance to such a stress without causing growth retardation. The fusion protein of DREB/CBF and ABF is designed to be post-translationally cleaved by inserting an intra-molecular protease (P), such that two proteins are produced at a similar level.

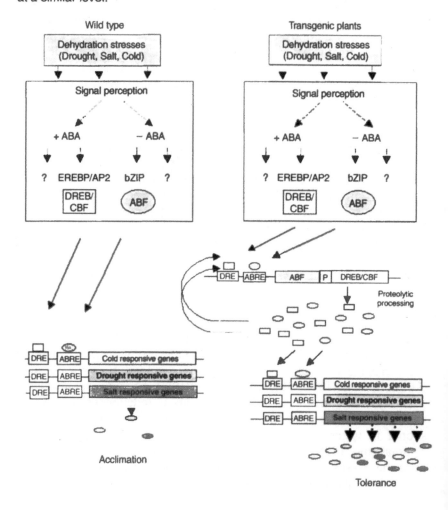

idly amplifying the initial signal. The final outcome is overexpression of *DREB1A* and its target genes, including *rd29A*, *rd17*, *cor6.6*, *cor15a*, *erd10*, and *kin1*, and enhanced stress tolerance for the plant. Other reports have also shown that ectopic expression of transcriptional activators resulted in changes in plant responses to salinity (Winicov and Bastola, 1999), disease resistance (Cao et al., 1998), and metabolic products in plants (Tamagnone et al., 1998) and cultured cells (Grotewold et al., 1998) by affecting the levels of expression of downstream endogenous genes.

It is known in Arabidopsis that at least four independent signal pathways function in the induction of genes in response to dehydration stresses; two are ABA-independent and two are ABA-dependent (Shinozaki and Yamaguchi-Shinozaki, 1997). As described above, DREB1A and CBF1 are involved in one of the ABA-independent pathways, activating genes that contain DRE in their promoters. The ABA-dependent responses are the induced expression of a large number of genes, which are mediated by *cis*-regulatory elements known as ABA-responsive elements (ABREs). More recently, a family of ABRE binding factors (ABFs) were isolated from Arabidopsis and they appear to mediate ABA-dependent stress responses during vegetative growth (Choi et al., 2000). Assuming that similar mechanisms are used for freezing, high salt, and drought tolerance in an important crop, such as rice, the use of the genes encoding CBF1, DREB1A, and ABFs may help make it more tolerant to such stresses. A strategy as depicted in Figure 4 (right panel) could be employed to develop a stress-tolerant transgenic rice. To activate a battery of genes that are responsible for abiotic stress tolerance in a coordinate manner, one can transform rice with a construct(s) in which genes for DREB/CBF, ABFs or DREB/CBF-ABFs fusion proteins are expressed under the control of a stress-inducible promoter like *rd29A* or *cor15a* (Figure 4). The stress-induced expression of the transcription factors that respond to both ABA-independent and ABA-dependent signaling will turn on a broad spectrum of downstream stress-responsive genes in addition to their own expression in transgenic plants. This autoregulated expression of the factors will significantly amplify the stress response of the plants, making them highly stress-tolerant without causing growth inhibition in the absence of the stress.

Salt Tolerance

About one-third of the world's irrigated land is unsuitable for growing crops because of contamination with high levels of salt (Serrano and

Gaxiola, 1994). Most crop plants are highly sensitive to salty conditions, experiencing a water deficit because of osmotic stress and biochemical perturbations due to the influx of sodium ions. The Arabidopsis Ser/Thr kinase gene (*At-DBF2*) enhances salt, drought, cold, and heat tolerance upon overexpression in transgenic tobacco cells (Lee et al., 1999). Given the polygenic character of the stress response, the engineering of transcriptional regulators looks promising. Like CBF1 and DREB1A, the overproduction of At-DBF2 in rice may enhance multiple stress tolerance, which constitutes a clear advantage in agriculture. *Alfin1* cDNA encodes a putative transcription factor associated with NaCl tolerance in alfalfa. Transgenic alfalfa calli and plant roots overexpressing *Alfin1* showed enhanced levels of salt-induced *MsPRP2* mRNA accumulation (Winicov and Bastola, 1999), resulting in salinity tolerance comparable to NaCl-tolerant mutant plants.

Recently, Apse et al. (1999) have engineered salt-tolerant Arabidopsis by overexpressing a single endogenous gene (*AtNHX1*) encoding a Na^+/H^+ antiport protein. They were able to identify and characterize the Na^+/H^+ antiport gene in Arabidopsis because of its similarity to bacterial, fungal, and mammalian homologs. They showed by cell fractionation that the antiport protein was accumulated primarily in the membrane of the large intracellular vacuole. Thus, the Na^+/H^+ antiport protein is the principal protein responsible for sequestering Na^+ in the large vacuole of the plant cell. Engineered Arabidopsis expressed greater levels of AtNHX1 and showed increased vacuolar uptake of Na^+ compared with wild-type plants. Transgenic plants were also significantly more salt-tolerant, thriving in soil irrigated with 200 mM NaCl. With this strategy, it should be possible to engineer a whole spectrum of salt-tolerant crop plants including rice, enabling them to be irrigated with seawater or water of marginal quality. Plants living in a saline environment must contend not only with the toxicity of Na^+ ions, but also with water loss caused by osmotic stress. It has been suggested that the vacuolar salt gradients in salt-tolerant transgenic Arabidopsis should help to drive water into plant cells, resulting in plants that are not only salt-resistant, but that use water more efficiently.

Plant cells accumulate potassium and exclude sodium. The resulting high potassium/sodium ratios in the cells enable potassium to perform essential functions. This selectivity in favor of potassium is especially important in the arid regions of the world, where excess sodium salts in the soil cause severe problems for crop production. It is known that calcium is required to maintain or enhance the selective absorption of potassium by plants at high concentrations of sodium. Liu et al. (1998,

2000) provide a molecular basis for the calcium sensitivity of potassium/sodium discrimination. In Arabidopsis plants carrying a mutation in the *SOS3* gene, the potassium/sodium ratio is changed to lower values than in wild-type plants. The mutant phenotype is abolished by high external calcium concentrations, suggesting that in the mutants, there is an impairment in the signaling pathway essential for the normal function of calcium in mitigating salt stress. The *Arabidopsis thaliana SOS2* and *SOS3* genes are required for intracellular Na^+ and K^+ homeostasis and plant tolerance to high Na^+ and low K^+ environments (Halfter et al., 2000). SOS2 was reported to interact with SOS3 by the C-terminal regulatory domain of SOS2, activating SOS2 protein kinase in a Ca^{2+}-dependent manner (Halfter et al., 2000).

Salinity has resulted in a decline in crop production over large areas. Yields should rise by several percent worldwide if crops can be engineered to thrive in saline environments that are currently unsuitable for agriculture. Salt-tolerant crops will be particularly important for developing countries. Increasing yields from marginal soils by growing transgenic crops is one step toward solving the problem of feeding the world's rapidly growing population.

High-Temperature Tolerance

Photosynthesis is inhibited by high temperature, but the causes of this inhibition are not clear. An increase in temperature is thought to be detrimental to the membranes of chloroplasts where photosynthesis takes place. Murakami et al. (2000) have shown that the number of unsaturated lipids in the thylakoid membrane of chloroplasts is important in determining a plant's ability for growth and photosynthesis at high temperatures. They silenced the gene (*FAD7*) encoding the chloroplast version of the w-3 fatty acid desaturase in transgenic tobacco plants. This enzyme converts 16:2 fatty acids to 16:3 molecules, or 18:2 fatty acids to 18:3 molecules. The silencing resulted in a decrease in unsaturated lipids with three double bonds and a corresponding increase in lipids with two double bonds in thylakoid membranes. The reduced level of lipid unsaturation improved the rate of photosynthesis at 40°C and markedly improved plant growth at 36°C. This provides an unambiguous demonstration of how reducing the unsaturation of thylakoid membrane lipids improves photosynthesis and growth at high temperatures. Murakami et al. (2000) showed that their transgenic plants grew much better than controls at higher temperatures. Differences in growth rate were noted at 36°C, and transgenic plants survived for 2 hr at 47°C, a treatment that killed their wild-type counterparts. With increasing concentrations of greenhouse gases in the atmosphere, the effect of high temperature on growth of tropical crops including rice may cause a decline in crop production particularly in tropical regions. This report, therefore, provides valuable information about the approach to engineering of rice plants that can carry out photosynthesis in the presence of heat stress.

Acid-Soil or Aluminum Tolerance

The most common metal in soils, aluminum, is a problem on 30% to 40% of the world's arable lands where acid soil releases aluminum ions into the ground water. For some important crops, such as maize, it is second only after drought as an obstacle to crop yields, reducing production by up to 80%. Acid soil is also one of the major problems that prevent rice from growing in Southeast Asian countries like Indonesia. In nonacid soils, aluminum is present in insoluble forms. In acid soils, which are most common in the tropics (where heavy rains leach alkaline materials from the land), it becomes soluble. It can be taken up into the cells of plant roots, where it affects cell metabolism and prevents healthy root growth. de la Fuente et al. (1997) found that some naturally aluminum-resistant plant strains have roots that secrete citric or malic acid, which binds to aluminum and prevents it from entering the roots. They introduced the bacterial gene for citrate synthase from *Pseudomonas aeruginosa*, the enzyme that makes citric acid, into tobacco and papaya. The plants carrying the citrate synthase gene secreted five to six times more citrate from their roots than control plants did, resulting in aluminum tolerance. The citrate-producing plants could grow well in aluminum concentrations 10-fold higher than those tolerated by control plants. The fact that the transgenic plants synthesize more citrate may have been a trade-off regarding the plants' productivity. Nevertheless, this provides a strategy for making acid-soil or aluminum tolerant crops that are already bred for high yield and pest resistance, hence securing more arable lands.

Aiming at C_4 Type Photosynthesis in Rice

Rice is a C_3 type photoassimilator because the first product of atmospheric CO_2 fixation is the C_3 acid, 3-phosphoglycerate. Some plants, such as maize possess a different photosynthetic pathway in which C_4 dicarboxylic acids (oxaloacetate, malate, or aspartate) are the first product of atmospheric CO_2 fixation. It has been known for more than 30 years that C_4 plants are more efficient assimilators of CO_2 than C_3 plants under high light intensity, high temperature, and drought conditions. By using the two chloroplast-containing cell types, mesophyll (MC) and bundle sheath (BSC) cells, C_4 plants can achieve high photosynthetic efficiency. Oxygen is a competitive inhibitor with respect to CO_2 fixation by Rubisco. Under current atmospheric conditions (21% O_2, 0.035% CO_2), O_2 reduces the photosynthetic efficiency of C_3 plants by as much as 40% (Ku et al., 1999). Although there was some interest in engineering Rubisco in favor of carboxylase over oxygenase, it is unlikely that Rubisco can be engineered for increased carboxylation efficiency (Somerville, 1990). C_4 plants have evolved a CO_2-pumping mechanism that increases CO_2 concentration in BSC, preventing CO_2 from limiting the rate of photosynthesis. The C_4 plants first fix CO_2 to oxaloacetate (OAA) by phosphoenolpyruvate carboxylase (PEPC) in MC (Figure 5). This process is highly efficient and insensitive to O_2. The CO_2 acceptor PEP is generated from pyruvate by pyruvate orthophosphatedikinase (PPDK). OAA is then converted to malate by NADP malate dehydrogenase (MDH) or to aspartate by aspartate aminotransferase. Either malate or aspartate is transferred to BSC where CO_2 is released by NADP- or NAD-malic enzyme (NME) or PEP carboxykinase (PEPCK), and assimilated by Rubisco through the Calvin cycle, thus suppressing Rubisco's oxygenase activity and avoiding photorespiration (Edwards and Walker, 1983). In addition to enriching CO_2, the C_4 plants also minimize O_2 concentration in BSC by keeping it away from intracellular air space that is connected to stomata and also by having small amounts of photosystem II in BSC that generates O_2 from H_2O in a light-dependent manner (Heldt, 1997).

Evaluation of the potential to transfer C_4 traits into C_3 plants to improve their photosynthetic efficiency has been an active area of research (Ku et al., 1999; Furbank and Taylor, 1995). Many nuclear genes encoding the enzymes in the C_4 pathways have been isolated (Sheen, 1999) and some of them have been studied in transgenic rice plants (Matsuoka et al., 1993, 1994; Ku et al., 1999). In addition, considerable progress has been made in characterization of the enzymes of the C_4 pathway and the genes responsible for light-dependent and cell-specific expression in C_4 plants (Sheen, 1999). Based on these studies, various strategies could be considered to improve CO_2 assimilation in C_3 crops. Ku et al. (1999) reported transformation of rice with an intact maize PEPC gene, showing a very high level of expression in transgenic rice.

FIGURE 5. Differences in pathways of photosynthetic carbon assimilation in C$_4$ and C$_3$ plants and one strategy that may be useful for making C$_4$ transgenic rice plants. OAA, oxaloacetate; PEPC, phosphoeno*l*pyruvate; PEPC, PEP carboxylase; PPDK, pyruvate orthophosphatedikinase; MDH, NADP malate dehydrogenase; NME, NADP- or NAD-malic enzyme; PEPCK, PEP carboxykinase; RuBP, ribulose 1,5-bisphosphate; Rubisco, RuBP carboxylase/oxygenase.

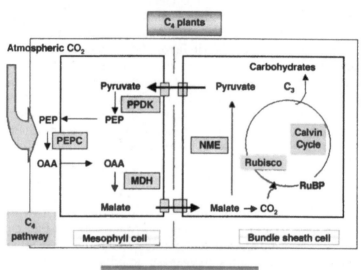

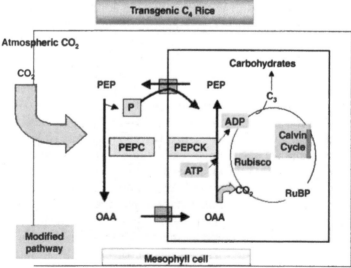

Interestingly, the O_2 sensitivity of photosynthesis in rice was decreased in the transgenic plants, suggesting substantial CO_2 fixation by the introduced O_2-insensitive PEPC. In the absence of other components necessary for a functional C_4 pathway, however, single gene transfer may not be sufficient to increase the supply of CO_2 to Rubisco, as described by Edwards (1999). He pointed out that PEPC in the transformants may increase flux of fixed carbon into C_4 acids, with decreased synthesis of carbohydrates. This is because PEPC activity is likely to be dependent on Rubisco to generate three carbon precursors for synthesis of PEP. Ku et al.'s (1999) results, nevertheless, provide a step toward developing C_4 transgenic rice. Now, the question is whether C_3 plants can be manipulated to perform C_4 photosynthesis. With current transformation methods and knowledge of C_4 gene structure and regulation, genetic engineering of C_4 components into C_3 plants may not be very far from reality. Since several genes can be transferred to rice at the same time, one could transform rice with four to five genes that are responsible for the C_4 cycle such as in NADP-malic enzyme species, with specific intracellular targeting of the introduced gene products. It seems difficult to engineer a new cell type like BSC of maize in rice because the genes required for development of Kranz anatomy (two different types of cell, MC and BSC) are not well characterized yet. It is, however, possible to engineer MC of rice in which genes that are working in MC of C_4 plants are designed to be expressed in cytosol of transgenic rice MC and genes in BSC of maize are in chloroplasts of MC. Alternatively, as illustrated in Figure 5, one could start with a much simplified version of the C_4 cycle that uses only two C_4 genes, *PEPC* and *PEPCK*, localized in the cytosol and chloroplast of transgenic rice, respectively. It appears informative and logical to consider a stepwise engineering of C_3 plants with analysis of corresponding changes in carbon flux, to ultimately improve their capacity of carbon assimilation.

TARGETS FOR MODIFICATION: GRAIN QUALITY IMPROVEMENT

Humans require an energy supply that can be provided by a mixture of carbohydrates, lipids, and protein, as well as several mineral nutrients and vitamins (Grusak and DellaPenna, 1999). Among the lipids, linoleic and linolenic acid cannot be synthesized by humans, and thus must be obtained from plant dietary sources. Nine essential amino acids, including lysine, must be obtained in various amounts from ingested protein, which also provides dietary sulfur and nitrogen. The remaining essential mineral nutrients and vitamins are required in varying amounts. Milled rice is relatively poor in fat, protein, and a number of vitamins and micronutrients (FAO, 1993). In particular, as mentioned earlier, deficiencies in lysine, vitamin A, iron, and iodine cause malnutrition in developing countries in Asia. Considering that half of the world's population depends on rice as their staple food, improving the grain quality of rice should be one of the primary goals of the current agricultural biotechnology.

Proteins

About 5-8% of rice seed is made up of protein, which is much lower than that of legume seeds (30-40%). Such storage proteins are utilized as an important source of amino acids for human diets. However, the amino acid composition of rice storage proteins is not well balanced, having low levels of lysine, threonine, and methionine (Takaiwa, 1999). In addition, significant parts of the rice storage proteins are indigestible and some albumins are known to cause allergic reactions in hypersensitive patients (Limas et al., 1990; Matsuda et al., 1991).

Rice tansformation methods have been refined and also structures and regulation of many rice storage proteins are known in detail (Takaiwa, 1999). It is therefore possible to design specific modifications to protein quality, either to alter the content of a particular amino acid (Saalbach et al., 1994; Coleman et al., 1997; Molvig et al., 1997; Falco et al., 1998) or to induce functional changes (Katsube et al., 1999). To increase lysine content in rice endosperm, Zheng et al. (1995) transformed rice with a gene for a seed storage protein, β-phaseolin, from *Phaseolus vulgaris* L. under control of the rice glutelin 1 promoter. The transgenic plants accumulated β-phaseolin to about 4% of total endosperm protein. Katsube et al. (1999) transformed rice with a wild-type and a methionine-modified soybean glycinin coding sequence under the control of the rice glutelin (*GluB-1*) promoter. Interestingly, they found that glycinins can assemble with glutelins to form soluble heteropolymers and insoluble complexes in rice endosperm, suggesting that the two proteins are compatible for formation of higher-order protein structures. The transgenic rice seeds were expected to provide a better balance of essential amino acids and also to lower the cholesterol levels in human serum. The latter was mentioned based on previous studies that soybean proteins exhibit a hypocholesterolemic effect in human diets (Kito et al., 1993).

Some lessons can be learned from the development of high-lysine maize and soybean for use in improved animal feeds. Two of the key enzymes in the lysine biosynthetic pathway are aspartokinase (AK) and dihydrodipicolinic acid synthase (DHDPS), which are both feedback inhibited by lysine (Brinch-Pederson, 1996; Tzchori et al., 1996). Falco et al. (1995) isolated bacterial genes encoding lysine-insensitive forms of AK and DHDPS from *Escherichia coli* and *Corynebacterium*, respectively. A deregulated form of the plant DHDPS was created by site-specific mutagenesis (Shaver et al., 1996). Expression of these genes in tobacco leaves produced high concentrations of free lysine (Shaul and Galili, 1993), but no accumulation was observed in tobacco seed with the use of either constitutive or seed-specific promoters. In soybean and canola seeds, however, lysine accumulated sufficiently to more than double the total seed lysine content. In maize, expression with an endosperm-specific promoter did not lead to lysine accumulation, whereas expression with an embryo-specific promoter gave high concentrations of lysine, sufficient to raise overall lysine concentrations in seed by 50 to 100% (Mazur et al., 1999). These results indicate that both bacterial or plant enzymes can be successfully expressed in specific plant tissues to alter metabolic pathways. However, different tissues and different species may react differently to metabolic engineering. Application of a strategy similar to that which was used in maize to transgenic rice may not be of great value because substantial parts of the embryo of rice seeds are removed during the milling process.

There are reports of reducing the content of allergenic proteins by antisense technology (Nakamura and Matsuda, 1996; Tada et al., 1996). The rice gene *RA17* codes for a seed protein of about 16 kDa with reactivity for IgE of rice-allergic patients. An antisense sequence from *RA17* cDNA under various seed-specific promoters was introduced into rice, reducing accumulation of RA17 to one-fifth of wild-type protein level (Tada et al., 1996).

In addition to increasing the nutritional value of seeds, the production of recombinant proteins in plant systems has many advantages for generating biopharmaceuticals such as proteins, vaccines, and industrial enzymes. These advantages include the following: (1) plant systems are more economical than industrial facilities using mammalian cell expression systems; (2) it is easy to harvest and process plant products on a large scale; (3) proteins are post-translationally modified in plants similarly to what occurs in mammalian systems; (4) plants can be directed to localize proteins into seed endopserm, which can easily be collected and stored for a long time; and (5) health risks due to contamination with pathogens and/or toxins are minimized. To date, genes for several antibodies (Owen et al., 1992; Fiedler and Conrad, 1995; Ma et al., 1995), enkephallins (Vandekerckhove et al., 1989), and human serum albumin (Sijmons et al., 1990) have all been successfully expressed in plants. Plants have also been used to express viral proteins which may serve as candidates for human vaccines (Mason et al., 1992; Haq et al., 1995; McGarvey et al., 1995; Turpen et al., 1995). Recently, Horvath et al. (2000) reported a method for production of recombinant proteins in transgenic barley grains. They made transgenic barley grains with a codon-optimized β-glucanase gene under the barley hordein gene *Hor3-1* promoter, obtaining seeds producing 54 μg/mg recombinant soluble protein. They found that high-level expression of the enzyme in the endosperm of the mature grain required codon optimization to a C+G content of 63% and synthesis as a precursor with a signal peptide for transport through the endoplasmic reticulum and targeting into the storage vacuoles. Advances in elucidation of mechanisms by which storage proteins are synthesized, targeted, and sequestered for long-term storage in the cereal endosperm provide the basis for the production of recombinant proteins in cereal grains.

Carbohydrates

Milled rice contains about 80% starch in its endosperm. The endosperm starch consists of two glucose polymers, amylose and amylopectin. Cooking quality of rice is determined by the percentage of amylose. If it is low (10-18%), the rice will be soft and sticky. If it is high (25 to 30%), the rice will be hard and fluffy. Amylose content is generally lower in *Japonica* rice varieties (17-22%) than in *Indica* varieties (20-30%). Three enzymes have been considered to catalyze key regulatory steps in starch biosynthesis: ADP-glucose pyrophospharylase (AGPase), granule-bound starch synthase (GBSS or *waxy*) and starch branching enzyme (SBE).

There are several approaches to modification of seed carbohydrates of crops. These include alteration of an existing compound in a qualitative or quantitative manner, usually sucrose (Klann et al., 1996) or starch (Bruinenberg et al., 1995; Sivak and Preiss, 1995). In transgenic rice plants (cv. Nipponbare), the amylose content was decreased from about 19% to 6% by expression of an antisense sequence of the waxy gene (Shimada et al. 1993). The initial reaction in starch biosynthesis is catalyzed by AGPase, and the observed allosteric properties of AGPase identify it as a potential site for controlling starch biosynthesis (Nelson and Pan, 1995; Hannah, 1997). This was supported by the increase in starch by 30% in transgenic potato tubers that express an allosterically altered AGPase (*glg-C16*) from *E. coli* (Stark et al., 1992), which indicated that the endogenous AGPase of potato was sub-optimal in the production of AGP-glucose for starch biosynthesis. Various carbon sources and intermediates activate bacterial AGPase. Plant AGPase, in contrast, is activated by 3-phosphoglyceric acid and inhibited by inorganic phosphate. In maize, the genes *shrunken 2* (*sh2*) and *brittle 2* (*bt2*) encode the two subunits of the maize endosperm tetrameric AGPase (Hannah and Greene, 1998). A mutation in the 3′ terminus of the *sh2* gene in maize resulted in increase in seed weight from 15 to 18% (Giroux et al., 1996). More recently, Greene et al. (1998) isolated up-regulated mutant forms of the large subunit of the potato AGPase. The mutant enzymes are more sensitive to 3-phosphoglycerate activation and more resistant to phosphate inhibition. In considering the general public perception of genetically engineered food products, such a modified plant gene may be more acceptable than a similar version of a bacterial gene in increasing starch biosynthesis in transgenic rice plants.

Amylopectin is synthesized by the combined actions of the enzymes soluble starch synthase and SBE. Branching degree of tuber starch increased up to 25% when the *E. coli* branching enzyme gene (*glgB*) was expressed under the control of the potato *GBSS* promoter in tubers of amylose-free transgenic potato (Kortstee et al., 1996). Experiments are under way to express the *glgB* gene in rice under control of the rice *SBE* promoter (W. Kim, 1999, unpublished results). The effect of the increased branching degree on physical properties of starch, such as swelling and visco-elastic behavior, remains to be determined.

Another approach to modify carbohydrates involves the introduction of a novel product. For example, genes for fructan synthesis were introduced into chicory (Vijn et al., 1997) and tobacco (Ebskamp et al., 1994). Fructans are of great importance to many food companies interested in relatively high-value, non-calorific carbohydrates. A similar example of high-value carbohydrate production in plants includes bacterial cyclodextrins that have been produced in transgenic potato tubers (Oakes et al., 1991). Thus, it is likely that new carbohydrates will be available with any desired chain length and degree of branching in near future.

Increase in Vitamin Content

Each year, about 500,000 children suffer irreversible blindness from vitamin A deficiency (UNICEF, 1998). Rice, in its milled form, lacks β-carotene and all of its immediate precursors. Until recently, however, it was not possible to selectively modify the vitamin A content of plants by genetic means. There is now sufficient biochemical understanding of some of the critical pathways (Hirschberg, 1999), and modified plants have been produced. For example, phytoene synthase, the enzyme that condenses two molecules of geranyl geranyl diphosphate (GGPP), is necessary for β-carotene biosynthesis in plants (Figure 6; Guerinot, 2000). Immature rice endosperm synthesizes the carotenoid precursor GGPP. In order to convert GGPP to β-carotene in rice endosperm, two molecules of GGPP must be condensed to produce phytoene, which is then desaturated to lycopene and finally cyclized to form β-carotene (Figure 6). The gene encoding phytoene synthase in daffodil (*Narcissus pseudonarcissus*) has been expressed in rice endosperm, producing a high-level of phytoene (Burkhardt et al., 1997). The same research group reported the engineering of rice endosperm to produce β-carotene (Ye et al., 2000), representing the culmination of progress made both in rice transformation and in dissecting the carotenoid biosynthetic pathway. The authors made transgenic plants by introducing three genes that are involved in the β-carotene biosynthetic pathway; the gene for a plant phytoene synthase (*psy*) from daffodil (Schledz, 1996), the gene for a bacterial phytoene desaturase (*crtI*) from *Erwinia uredovora*, and the gene for a lycopene β-cyclase from doffodil (Al-Babili et al., 1996). The three genes, with their own or the pea *rbcS* plastid targeting signal and under the endosperm-specific glutelin (*Gt1*) or the constitutive CaMV 35S promoter, have been transferred into rice by *Agrobacterium*-mediated transformation. β-Carotene production in rice endosperm appears to be sufficient for children; 2 mg β-carotene/g dry seed weight, corresponding to 100 mg retinol equivalent at a daily intake of 300 g of rice per day. This would cover the daily pro-vitamin A needs of children. Humans can make vitamin A from β-carotene and excess dietary β-carotene, as opposed to excess vitamin A, has no harmful effect, so plants with enhanced β-carotene content should be a safe and effective means of vitamin A delivery. In Southeast Asia, 70% of children under the age of five suffer from vitamin A deficiency, leading to vision impairment and increased susceptibility to disease (www.who.int/nut/).

FIGURE 6. The carotenoid biosynthetic pathway of plants (Reprinted with permission from M. L. Guerinot. 2000. The green revolution strikes gold. *Science* 287:241-243. Copyright 2000 American Association for the Advancement of Science) and three genes that were introduced into rice by the *Agrobacterium*-mediated method, making the golden (β-carotine or pro-vitamin A producing) rice (Reprinted with permission from Ye, X., S. Al-Babili, A. Kloti, J. Zhang, P. Lucca, P. Beyer, and I. Potrykus. 2000. Engineering the provitamin A (β-carotene) biosynthetic pathway into (carotenoid-free) rice endosperm. *Science* 287:303-305. Copyright 2000 American Association for the Advancement of Science). Isopentenyl diphosphate (IPP) is a common precursor for all isoprenoids. Since GGPP is abundant in rice endosperm, Ye et al. transformed rice with three genes encoding phytoene synthase (*psy*), phytoene desaturase (*crtl*), and lycopene β-cyclase (*lcy*) to convert GGPP to β-carotene.

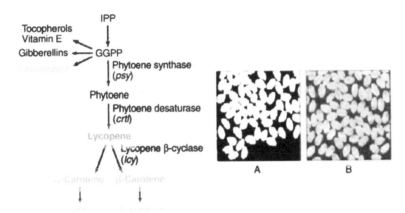

The United Nations Children's Fund predicts that improved vitamin A nutrition could prevent 1 to 2 million deaths each year among children aged one to four years (www.unicef.org/vitamina/). Thus, in considering the importance of vitamin A daily delivery, Ye et al.'s (2000) work exemplifies the best that agricultural biotechnology has to offer the world.

α-Tocopherol (vitamin E), a lipid-soluble antioxidant synthesized only in plant chloroplasts or photosynthetic organisms, is important to human health. The recommended daily allowance (RDA) in United States is 7 to 9 mg. In addition, daily intake of vitamin E in excess of the RDA is associated with reduced risk of cardiovascular disease and some cancers, and improved immune function (Grusak and DellaPenna, 1999). Plant oils, the main dietary source of α-tocopherols, typically contain α-tocopherol as a minor component and high levels of its biosynthetic precursor, γ-tocopherol. In soybean oil, for example, α-tocopherol and its immediate biosynthetic precursor, γ-tocopherol, account for 7 and 70%, respectively, of the total tocopherol pool (Shintani and DellaPenna, 1998). Therefore, substantial increases in the α-tocopherol content of major food crops are needed to provide the public with dietary sources of vitamin E that can approach the desired therapeutic levels. Shintani and DellaPenna (1998) used a genomics-based approach to clone the final enzyme in α-tocopherol synthesis, γ-tocopherol methyltransferase (γ-TMT). Overexpression of γ-TMT in Arabidopsis seeds with a seed-specific promoter shifted oil compositions in favor of α-tocopherol, resulting in a nine-fold increase in vitamin E activity. We may assume that γ-TMT activity is also limiting in commercially important crops that have low α- to γ-tocopherol ratios, such as soybean, corn, canola, and rice. Targeted expression of the γ-TMT gene together with genes involved in γ-tocopherol synthesis in chloroplasts of rice using an endosperm-specific promoter should similarly elevate α-tocopherol levels of seeds and thereby increase the nutritional value of the crop.

Increase in Iron Content

Iron deficiency affects more than 2 billion people worldwide, and plants are the principal source of iron in most diets. However, as with vitamins, many of the world's staple foods are poor in iron. Rice, in its milled form, is characterized by a very low content of iron, 0.2 to 2.8 mg/100 g. In addition, absorption of iron in human intestines from a meal is in the range of 1 to 20% (Klöti and Potrykus, 1999). Thus, to improve the iron availability for human use, one needs to engineer rice to increase the iron content in rice endosperm and the absorption of iron in human intestine. According to Klöti and Potrykus (1999), these could be achieved by expression of ferritin genes to increase iron content and phytase genes to reduce the level of phytic acid that inhibits iron absorption. Goto et al. (1999) engineered rice to have higher levels of the iron storage protein ferritin in the grain. The authors have transformed rice with the coding sequence of the soybean ferritin gene under the control of the rice glutelin promoter, increasing iron content of the rice endosperm as much as threefold greater than that of the non-transformed control. Unlike vitamins, however, which are synthesized by the plants themselves, plants must take up iron from the soil. To increase iron content of rice endosperm further, therefore, understanding of how plants take up and store iron at molecular levels is essential. Higher plants utilize one of two strategies for iron acquisition (Marschner and Römheld, 1994; Grusak and DellaPenna, 1999). In all dicots and the non-grass monocots, a ferric iron first needs to be reduced to Fe^{2+} prior to membrane influx. Grasses, including rice, employ ferric chelators, called phytosiderophores, that are released by roots and chelate ferric iron in the rhizosphere. The Fe(III)-phytosiderophore is absorbed via a transport protein. Increasing the iron content of either type usually necessitates increases in total iron input to the plant, and may require modifications to whole-plant partitioning (Grusak and Pezeshgi, 1996; Grusak et al., 1999). A major focus is, therefore, the identification and isolation of genes required for the uptake, partitioning, and accumulation of the target compound such that its levels can be modified in staple crops to effect the desired dietary change (Samuelsen et al., 1998). Recently, the *FRO2* gene encoding the plant Fe(III) reductase has been isolated and sequenced from Arabidopsis (Robinson et al., 1999). The gene is expressed in iron-deficient roots of Arabidopsis and is allelic to *frd1* mutations that impair the activity of ferric-chelate reductase. Introduction of functional *FRO2* complements the *frd1* phenotype in transgenic plants. The Arabidopsis *IRT1* gene (for *I*ron *R*egulated *T*ransporter) was also identified by functional complementation of a yeast mutant defective in iron uptake (Eide et al., 1996). It would be of interest to see how much iron accumulated in rice if the Arabidopsis iron uptake system was adapted together with the iron storage protein ferritin (Goto et al., 1999) by transforming rice with both *FRO2* and *IRT1*. Alternatively, in the case of seeds, whose predominant supply of nutrients is provided via the phloem pathway, improvements in iron content will require modifications to the phloem loading system. For example, overexpression of a phloem-mobile iron-chelator increased phloem iron transport (Marentes and Grusak, 1998). In rice seeds, which import only about 4% of total shoot iron (Grusak and DellaPenna, 1999), targeting increased phloem-mobile chelate expression to source regions that contain available iron could help to increase the seed iron content. This is supported by studies with the *brz* and *dgl* iron-hyper-accumulating pea mutants demonstrating that iron needs to be chelated prior to phloem loading (Grusak, 1994). The iron transported to seeds must then be sequestered in a non-reactive form by overexpression of a phytoferritin gene in seeds to store the excess iron chelated to the ferritin, further enhancing the seed's iron nutritional quality.

CONCLUSION

Functional genomics and bioinformatics will greatly accelerate the acquisition of knowledge about the function of plant genes (Somerville and Somerville, 1999). The genomes of Arabidopsis and rice will soon be completely sequenced. The genomes of wheat, maize, sorghum, millet, and other cereals can be partially deciphered on the basis of their similarity to the rice genome because of extensive synteny among the cereal genomes (Gale and Devos, 1998a, 1998b). Extensive partial cDNA sequence information will become available for a majority of the genes of many plant species. As genes become associated with functions or traits in one plant, it will likely be possible to use a database search to identify orthologs responsible for the trait in other plant species. Thus, unraveling the complex relationship between genes and phenotypes and applying this information to developing better crops will be dependent on the cooperative work of genomics, transformation, and plant breeding, which will eventually lead to significant contributions to global food security. The initial phase of a second revolution in agriculture appears to be taking place. Large acreages of genetically modified crops of soybeans, corn, cotton, canola and others have been grown. Worldwide in 1999, about 40 million hectares of transgenic plants were grown, and this is expected to increase significantly for years to come (James, 1999).

The International Program on Rice Biotechnology of the Rockefeller Foundation has devoted much effort to developing and improving techniques for rice transformation over the last decade or so. Because of these efforts, enormous progress has been made in the field of transgenic rice as well as in the molecular breeding of the crop including marker development, mapping genes on chromosomes, and marker-assisted selection for breeding populations. Most transgenic rice research has been concentrated on the improvement of agronomic traits, such as resistance to biotic and abiotic stresses and herbicides. Research is now moving on to the improvement of grain quality traits including micronutrients and vitamins. With the increasing number of genes discovered in plants and other organisms and the increasing knowledge of their function(s), transgenic research can contribute more and more to the introduction of traits for rice grain yield and grain quality. As such, making rice a perfect food by the use of agricultural biotechnology is not very far from reality.

REFERENCES

Al-Babili, S., J. von Lintig, H. Haubruck, and P. Beyer (1996). A novel soluble form of phytoene desaturase from *Narcissus pseudonarcissus* chromoplasts is Hsp70-complexed and component for flavinylation, membrane association and enzymatic activation. *The Plant Journal* 9: 601-612.

Allen, G. C., G. Hall, S. Michalowski, W. Newman, S. Spiker, A. K. Weissinger, and W. F. Thompson (1996). High-level transgene expression in plant cells: effects of a strong scaffold attachment region from tobacco. *Plant Cell* 8: 899-913.

Apse, M. P., G. S. Aharon, W. A. Snedden, and E. Blumwald (1999). Salt tolerance conferred by overexpression of a vacuolar Na^+/H^+ antiport in Arabidopsis. *Science* 285: 1256-1258.

Brinch-Pedersen, H., G. Galili, S. Knudsen, and P. B. Holm (1996). Engineering of the aspartate family biosynthetic pathway in barley (*Hordeum vulgare* L.) by transformation with heterologous genes encoding feed-back-insensitive aspartate kinase and dihydrodipicolinate synthase. *Plant Molecular Biology* 32:611-620.

Bruinenberg, P. M., E. J. Jacobsen, and R. G. F. Visser (1995). Starch from genetically engineered crops. *Chemistry and Industry* 6: 881-884.

Burkhardt, P. K., P. Beyer, J. Wunn, A. Kloti, G. A. Armstrong, M. Schledz, J. von Lintig, and I. Potrykus (1997). Transgenic rice (*Oryza sativa*) endosperm expressing daffodil (*Narcissus pseudonarcissus*) phytoene synthase accumulates phytoene, a key intermediate of provitamin A biosynthesis. *The Plant Journal* 11:1071-1078.

Cao, H., X. Li, and X. Dong (1998). Generation of broad-spectrum disease resistance by overexpression of an essential regulatory gene in systemic acquired resistance. *Proceedings of the National Academy of Sciences, USA* 95: 6531-6536.

Cao, J., X. Duan, D. McElroy, and R. Wu (1992). Regeneration of herbicide resistant transgenic rice plants following microprojectile-mediated transformation of suspension cells. *Plant Cell Reports* 11: 586-591.

Choi, H.-I., J.-H. Hong, J.-O. Ha, J.-Y. Kang, and S. Y. Kim (2000). ABFs, a family of ABA-responsive element binding factors. *The Journal of Biological Chemistry* 275: 1723-1730.

Christou, P. (1997). Rice transformation: bombardment. *Plant Molecular Biology* 35: 197-203.

Christou, P., T. L. Ford, and M. Kofron (1991). Production of transgenic rice (*Oryza sativa* L.) plants from agronomically important indica and japonica varieties via electric discharge particle acceleration of exogenous DNA into immature zygotic embryos. *Bio/Technology* 9: 957-962.

Coleman, C. E., A. M. Clore, J. P. Ranch, R. Higgins, A. M. Lopes, and B. A. Larkins (1997). Expression of a mutant alpha-zein creates the floury 2 phenotype in transgenic maize. *Proceedings of the National Academy of Sciences, USA* 94: 7094-7097.

Conway G. and G. Toenniessen (1999). Feeding the world in the twenty-first century. *Nature* 402: C55-C58.

Datta, S. K., A. Peterhans, K. Datta, and I. Potrykus (1990). Genetically engineered fertile indica-rice recovered from protoplasts. *Bio/Technology* 8: 736-740.

DeAlmeida, E. R. P., V. Gossele, C. G. Muller, J. Dock, A. Reynaerts, J. Botterman, E. Krebbers, and M. P. Timko (1989). Transgenic expression of two marker genes under the control of an *Arabidopsis rbcS* promoter: sequences encoding the rubisco transit peptide increase expression levels. *Molecular and General Genetics* 218: 78-86.

Dean, C., E. Pichersky, and P. Dunsmuir (1989). Structure, evolution, and regulation of *RbcS* genes in higher plants. *Ann. Rev. Plant Physiol. and Plant Molecular Biology* 40: 415-439.

de la Puente, J. M., V. Ramírez-Rodríguez, and J. L. Cabrera-Ponce (1997). Aluminum tolerance in transgenic plants by alteration of citrate synthesis. *Science* 276: 1566-1568.

DellaPenna, D. (1999). Nutritional genomics: manipulating plant micronutrients to improve human health. *Science* 285:375-379.

Duan, X., X. Li, Q. Xue, M. Abo-El-Saad, D. Xu, and R. Wu (1996). Transgenic rice plants harboring an introduced potato proteinase inhibitor II gene are insect resistant. *Nature Biotechnology* 14: 494-498.

Ebskamp, M. J., I. M. van der Meer, B. A. Spronk, P. J. Weisbeek, and S. C. Smeekens (1994). Accumulation of fructose polymers in transgenic tobacco. *Bio/Technology* 12: 272-275.

Edwards, G. (1999). Tuning up crop photosynthesis. *Nature Biotechnology* 17: 22-23.

Edwards, G. E. and D. A. Walker (1983). C_3, C_4: *Mechanisms and Cellular and Environmental Regulation of Photosynthesis*. Blackwell Science Publishers, London, UK.

Eide, D., M. Broderius, J. Fett, and M. L. Guerinot (1996). A novel iron-regulated metal transporter from plants identified by functional expression in yeast. *Proceedings of the National Academy of Sciences, USA* 93:5624-5628.

Falco, S. C., T. Guida, M. Locke, J. Mauvais, C. Sanders, R. T. Ward, and P. Webber (1995). Transgenic canola and soybean seeds with increased lysine. *Bio/Technology* 13: 577-582.

Falco, S. C., S. J. Keeler, and J. A. Rice (1998). Chimeric genes and methods for increasing the lysine and threonine content of the seeds of plants. U.S. Patent 5773691.

FAO (1993). Rice in human nutrition. FAO, Rome, pp. 17-84.

FAO (1998). FAOSTAT Agriculture Statistics Database; see <http://www.fao.org/WAICENT/Agricul.htm>.

Fiedler, U. and U. Conrad (1995). High-level production and long-term storage of engineered antibodies in transgenic tobacco seeds. *Bio/Technology* 13: 1090-1093.

Furbank, R. T. and W. C. Taylor (1995). Regulation of photosynthesis in C_3 and C_4 plants: a molecular approach. *Plant Cell* 7: 797-807.

Gale, M. D. and K. M. Devos (1998a). Plant comparative genetics after 10 years. *Science* 282:656-659.

Gale, M. D. and K. M. Devos (1998b). Comparative genetics in the grasses. *Proceedings of the National Academy of Sciences, USA* 95:1971-1974.

Gilmartin, P. M., L. Sarokin, J. Memelink, and N.-H. Chua (1990). Molecular light switches for plant genes. *Plant Cell* 2: 369-378.

Gilmour, S. J., D. G. Zarka, E. J. Stockinger, M. P. Salazar, J. M. Houghton, and M. F. Thomashow (1998). Low temperature regulation of the Arabidopsis CBF family of AP2 transcription activators as an early step in cold-induced COR gene expression. *The Plant Journal* 16: 433-442.

Giroux, M. J., J. Shaw, G. Barry, B. G. Cobb, T.W. Greene, T.W. Okita, and L. C. Hannah (1996). A single gene mutation that increase maize seed weight. *Proceedings of the National Academy of Sciences, USA* 93: 5824-5829.

Goto, F., T. Yoshihara, N. Shigemoto, S. Toki, and F. Takaiwa (1999). Iron fortification of rice seed by the soybean ferritin gene. *Nature Biotechnology* 17:282-286.

Greene, T. W., I. H. Kavakli, M. L. Kahn, and T. W. Okita (1998). Generation of up-regulated allosteric variants of potato ADP-glucose pyrophosphorylase by reverse genetics. *Proceedings of the National Academy of Sciences, USA* 95: 10322-10327.

Grishok, A., H. Tabara, and C. C. Mello (2000). Genetic requirements for inheritance of RNAi in *C. elegans. Science* 287: 2494-2497.

Grotewold, E., M. Chamberlin, M. Snook, B. Siame, L. Butler, J. Swenson, S. Maddock, G. St. Clair, and B. Bowen (1998). Engineering secondary metabolism in maize cells by ectopic expression of transcription factors. *Plant Cell* 10: 721-740.

Grusak, M. A. (1994). Iron transport to developing ovules of *Pisum sativum*. I. Seed import characteristics and phloem iron-loading capacity of source regions. *Plant Physiology* 104:649–655.

Grusak, M. A. and D. DellaPenna (1999). Improving the nutrient composition of plants to enhance human nutrition. *Annual Review of Plant Physiology Plant Molecular Biology* 50: 133-161.

Grusak, M. A. and S. Pezeshgi (1996). Shoot-to-root signal transmission regulates root Fe(III) reductase activity in the *dgl* mutant of pea. *Plant Physiology* 110:329–334.

Grusak, M. A., N. Pearson, and E. Marentes (1999). The physiology of micronutrient homeostasis in field crops. *Field Crops Research* 60:41–56.

Gura, T. (2000). Reaping the plant gene harvest. *Science* 287:412-414.

Guerinot, M. L. (2000). The green revolution strikes gold. *Science* 287: 241-243.

Halfter, U., M. Ishitani, and J.-K. Zhu (2000). The *Arabidopsis* SOS2 protein kinase physically interacts with and is activated by the calcium-binding protein SOS3. *Proceedings of the National Academy of Sciences, USA* 97: 3735-3740.

Hamilton, A. J. and D. C. Baulcombe (1999). A species of small antisense RNA in posttranscriptional gene silencing in plants. *Science* 286: 950-953.

Hannah, L. C. (1997). Starch synthesis of maize endosperm. In: *Cellular and Molecular Biology of Plant Seed Development*, ed. B. A. Larkins and I. K. Vasil, Dordrecht, The Netherlands: Kluwer Academic Publishers, pp. 375-405.

Hannah, L. C. and T. W. Greene (1998). Maize seed weight is dependent on the amount of endosperm ADP-glucose pyrophosphorylase. *Journal of Plant Physiology* 152: 649-652.

Haq, T. A., H. S. Mason, J. D. Clements, and C. J. Arntzen (1995). Oral immunization with a recombinant bacterial antigen produced in transgenic plants. *Science* 268: 714-716.

Hayashi, H., L. Mustardy, P. Deshnium, M. Ida, and N. Murata (1997). Transformation of *Arabidopsis thaliana* with the *codA* gene for choline oxidase: accumulation of glycine-betaine and enhanced tolerance to salt and cold stress. *The Plant Journal* 12:133-142.

Heldt, H.-W. (1997). *Plant Biochemistry and Molecular Biology*. Oxford, UK: Oxford University Press. Inc.

Herminghaus, S., D. Tholl, C. Rugenhagen, L. F. Fecker, C. Leuschner, and J. Berlin (1996). Improved metabolic action of a bacterial lysine decarboxylase gene in tobacco hairy root cultures by its fusion to a *rbcS* transit peptide coding sequence. *Transgenic Research* 5: 193-201.

Herminghaus, S., P. H. Schreier, J. E. G. McCarthy, J. Landsmann, J. Botterman, and J. Berlin (1991). Expression of a bacterial lysine decarboxylase gene and transport of the protein into chloroplasts of transgenic tobacco. *Plant Molecular Biology* 17: 475-486.

Hiei, Y., S. Ohta, T. Komari, and T. Kumashiro (1994). Efficient transformation of rice (*Oryza sativa* L.) mediated by *Agrobacterium* and sequence analysis of the boundaries of the T-DNA. *The Plant Journal* 6: 271-282.

Hiei, Y., T. Komari, and T. Kubo (1997). Transformation of rice mediated by *Agrobacterium tumefaciens*. *Plant Molecular Biology* 35: 205-218.

Hiei, Y., T. Komari, and T. Kubo (1999). *Agrobacterium*-mediated transformation. In: *Molecular Biology of Rice*, ed. K. Shimamoto, Tokyo, Japan: Springer-Verlag, pp. 235-256.

Hirschberg, J. (1999). Production of high-value compounds: carotenoids and vitamin E. *Current Opinion in Biotechnology* 10:186-191.

Holmberg, N. and L. Bulow (1998). Improving stress tolerance in plants by gene transfer. *Trends in Plant Science* 3: 61-66.

Holmström, K.-O., E. Mäntylä, B. Welln, A. Mandal, and E. T. Palva (1996). Drought tolerance in tobacco. *Nature* 379: 683-684.

Horvath, H., J. Huang, O. Wong, E. Kohl, T. Okida, C. G. Kannangara, and D. von Wettstein (2000). The production of recombinant proteins in transgenic barley grains. *Proceedings of the National Academy of Sciences, USA* 97: 1914-1919.

Iglesias, V. A., E. A. Moscone, I. Papp, F. Neuhuber, S. Michalowski, T. Phelan, S. Spiker, M. Matzke, and A. J. M. Matzke (1997). Molecular and cytogenetic analysis of stably and unstably expressed transgene loci in tobacco. *Plant Cell* 9: 1251-1264.

Ingram, J. and D. Bartels (1996). The molecular basis of dehydration tolerance in plants. *Annual Review of Plant Physiology and Plant Molecular Biology* 47: 377-403.

Ishizaki-Nishizawa, O., T. Fujii, M. Azuma, K. Sekiguchi, N. Murata, T. Ohtani, and T. Toguri (1996). Low-temperature resistance of higher plants is significantly enhanced by a nonspecific cyanobacterial desaturase. *Nature Biotechnology* 14: 1003-1006.

Jaglo-Ottosen, K. R., S. J. Gilmour, D. G. Zarka, O. Schabenberger, and M. F. Thomashow (1998). Arabidopsis *CBF1* overexpression induces *COR* genes and enhances freezing tolerance. *Science* 280: 104-106.

James, C. (1999). Global review of commercialized transgenic crops: 1999. The Internation Service for the Acquisition of Agri-biotech Applications (ISAAA) Briefs No. 1, ISAAA, Ithaca, NY.

Jang, I.-C., B. H. Nahm, and J.-K. Kim (1999). Subcellular targeting of green fluorescent protein to plastids in transgenic rice plants provides a high-level expression system. *Molecular Breeding* 5: 453-461.

Kasuga, M., Q. Liu, S. Miura, K. Yamaguchi-Shinozaki, and K. Shinozaki (1999). Improving plant drought, salt, and freezing tolerance by gene transfer of a single stress-inducible transcription factor. *Nature Biotechnology* 17:287-291.

Katsube, T., N. Kurisaka, M. Ogawa, N. Maruyama, R. Ohtsuka, S. Utsumi, and F. Takaiwa (1999). Accumulation of soybean glycinin and its assembly with the glutelins in rice. *Plant Physiology* 120: 1063-1073.

KaviKishor, P. B., Z. Hong, G.-U. Miao, C.-A. H. Hu, and D. P. S. Verma (1995). Overexpression of (Δ'-pyrroline-5-carboxylate synthetase increases proline production and confers osmotolerance in transgenic plants. *Plant Physiology* 108: 1387-1394.

Keegstra, K. and K. Cline (1999). Protein import and routing systems of chloroplasts. *Plant Cell* 11:557-570.

Kito, M., T. Moriyama, Y. Kimura, and H. Kambara (1993). Changes in plasma lipid levels in young healthy volunteers by adding an extruder-cooked soy protein to conventional meals. *Bioscience, Biotechnology and Biochemistry* 57: 354-355.

Klann, E. M., B. Hall, and A. B. Bennett (1996). Antisense acid invertase (TIV1) gene alters soluble sugar composition and size in transgenic tomato fruit. *Plant Physiology* 112: 1321-1330.

Klöti, A. and I. Potrykus (1999). Rice improvement by genetic transformation. In: *Molecular Biology of Rice*, ed. K. Shimamoto, Tokyo, Japan: Springer-Verlag, pp. 283-301.

Kodama, H., T. Hamada, G. Horiguchi, M. Nishimura, and K. Iba (1994). Genetic enhancement of cold tolerance by expression of a gene for chloroplast ω-3 fatty acid desaturase in transgenic tobacco. *Plant Physiology* 105: 601-605.

Komari, T., Y. Hiei, Y. Saito, N. Murai, and T. Kumashiro (1996). Vectors carrying two separate T-DNAs for co-transformation of higher plants mediated by *Agrobacterium tumefaciens* and segregation of transformants free from selection markers. *The Plant Journal* 10: 165-174.

Kortstee, A. J., A. M. S. Vermeesch, B. J. de Vries, E. Jacobsen, and R. G. F. Visser (1996). Expression of *Escherichia coli* branching enzyme in tubers of amylose-free transgenic potato leads to an increased branching degree of the amylopectin. *The Plant Journal* 10: 83-90.

Ku, M. S. B., S. Agarie, M. Nomura, H. Fukayama, H. Tsuchida, K. Ono, S. Hirose, S. Toki, M. Miyao, and M. Matsuoka (1999). High-level expression of maize phosphoenolpyruvate carboxylase in transgenic rice plants. *Nature Biotechnology* 17: 76-80.

Lee, J. H., M. Van Montagu, and N. Verbruggen (1999). A highly conserved kinase is an essential component for stress tolerance in yeast and plant cells. *Proceedings of the National Academy of Sciences, USA* 96: 5873-5877.

Limas, G. G., M. Salinas, I. Moneo, S. Fisher, B. Wittman-Liebold, and E. Hendez (1990). Purification and characterization of ten new rice NaCl-soluble proteins: identification of four protein-synthesis inhibitors and two immunoglobulin-binding proteins. *Planta* 181: 1-9.

Liu, J., M. Ishitani, U. Halfter, C.-S. Kim, and J.-K. Zhu (2000). The *Arabidopsis thaliana SOS2* gene encodes a protein kinase that is required for salt tolerance. *Proceedings of the National Academy of Sciences, USA* 97: 3730-3734.

Liu, Q., M. Kasuga, Y. Sakuma, H. Abe, S. Miura, K. Yamaguchi-Shinozaki, and K. Shinozaki (1998). Two transcription factors, DREB1 and DREB2, with an EREBP/AP2 DNA binding domain, separate two cellular signal transduction pathways in drought- and low temperature-responsive gene expression, respectively, in Arabidopsis. *Plant Cell* 10: 1391-1406.

Ma, J. K.-C., A. Hiatt, M. Hein, N. D. Vine, F. Wang, P. Stabila, C. van Dolleweerd, K. Mostov, and T. Lehner (1995). Generation and assembly of secretory antibodies in plants. *Science* 268: 716-719.

Marentes, E. and M. A. Grusak (1998). Iron transport and storage within the seed coat and embryo of developing seeds of pea (*Pisum sativum* L.). *Seed Science Research* 8: 367-375.

Marschner, H. and V. Römheld (1994). Strategies of plants for acquisition of iron. *Plant Soil* 165: 261-274.

Mason, H. S., D. M.-K. Lam, and C. J. Arntzen (1992). Expression of hepatitis-B surface antigen in transgenic plants. *Proceedings of the National Academy of Sciences, USA* 89: 11745-11749.

Matsuda, T., R. Nomura, M. Sugiyama, and R. Nakamura (1991). Immunochemical studies on rice allergenic proteins. *Agricultural Biology and Chemistry* 55: 509-513.

Matsuoka, M., J. Kyozuka, K. Shimamoto, and Y. Kano-Murakami (1994). The promoters of two carboxylases in a C4 plant (maize) direct cell-specific, light regulated expression in a C3 plant (rice). *The Plant Journal* 6: 311-319.

Matsuoka, M., Y. Tada, T. Fujimura, and Y. Kano-Murakami (1993). Tissue-specific light-regulated expression directed by the promoter of a C_4 gene, maize pyruvate orthophosphate dikinase, in a C_3 plant, rice. *Proceedings of the National Academy of Sciences, USA* 90: 9586-9590.

Matzke, M. A. and A. J. M. Matzke (1995). How and why do plants inactivate homologous (trans)genes. *Plant Physiology* 107: 679-685.

Mazur, B., E. Krebbers, and S. Tingey (1999). Gene discovery and product development for grain quality traits. *Science* 285: 372-375.

McGarvey, P. B., J. Hammond, M. M. Dienelt, D. C. Hooper, Z. F. Fu, B. Dietzschold, H. Koprowski, and F. H. Michaels (1995). Expression of the rabies virus glycoprotein in transgenic tomatoes. *Bio/Technology* 13: 1484-1487.

McKersie, B. D., S. R. Bowley, E. Harjanto, and O. Leprince (1996). Water-deficit tolerance and field performance of transgenic alfalfa overexpressing superoxide dismutase. *Plant Physiology* 111: 1177-1181.

Meyer, P. and H. Saedler (1996). Homology-dependent gene silencing plants. *Annual Review Plant Physiol and Plant Molecular Biolology* 47: 23-48.

Mlynarova, L., A. Loonen, A. Heldens, R. C. Jansen, L. C. P. Keizer, W. J. Stiekema, and J.-P. Nap (1994). Reduced position effect in mature transgenic plants conferred by the chicken lysozyme matrix-associated region. *Plant Cell* 6: 417-426.

Mlynarova, L., L. C. P. Keizer, W. J. Stiekema, and J.-P. Nap (1996). Approaching the lower limits of transgene variability. *Plant Cell* 8: 1589-1599.

Mlynarova, L., R. C. Jansen, A. J. Conner, W. J. Stiekema, and J.-P. Nap (1995). The MAR-mediated reduction in position effect can be uncoupled from copy-dependent expression in transgenic plants. *Plant Cell* 7: 599-609.

Molvig, L., L. M. Tabe, B. O. Eggum, A. E. Moore, S. Craig, D. Spencer, and T. J. Higgins (1997). Enhanced methionine levels and increased nutritive value of seeds of transgenic lupins (*Lupinus angustifolius* L.) expression a sunflower seed albumin gene. *Proceedings of the National Academy of Sciences, USA* 94: 8393-8398.

Moore, G., K. M. Devos, Z. Wang, and M. D. Gale (1995). Cereal genome evolution, grasses, line up and form a circle. *Current Biology* 5:737-739.

Murakami, Y., M. Tsuyama, Y. Kobayashi, H. Kodama, and K. Iba (2000). Trienoic fatty acids and plant tolerance of high temperature. *Science* 287: 476-479.

Nakamura, R. and T. Matsuda (1996). Rice allergenic protein and molecular-genetic approach for hypoallergenic rice. *Bioscience, Biotechnology and Biochemistry* 60: 1215-1221.

Nawrath, C., Y. Poirier, and C. Somerville (1994). Targeting of the polyhydroxybutyrate biosynthetic pathway to the plastids of *Arabidopsis thaliana* results in high levels of polymer accumulation. *Proceedings of the National Academy of Sciences, USA* 91: 12760-12764.

Nelson, O. E., Jr. and D. Pan (1995). Starch synthesis in maize endosperms. *Annual Review of Plant Physiology and Plant Molecular Biology* 46: 475-496.

Oakes, J. V., C. K. Shewmaker, and D. M. Stalker (1991). Production of cyclodextrins, a novel carbohydrate, in the tubers of transgenic potato plants. *Bio/Technology* 9: 982-986.

Owen, M., A. Gandecha, W. Cockburn, and G. Whitelam (1992). Synthesis of a functional anti-phytochrome single-chain Fv protein in transgenic tobacco. *Bio/Technology* 10: 790-794.

Peng, J., H. Kononowicz, and T. K. Hodges (1992). Transgenic indica rice plants. *Theoretical and Applied Genetics* 83: 855-863.

Pennisi, E. (2000). Stealth genome rocks rice researchers. *Science* 288: 239-241.

Robinson, N. J., C. M. Procter, E. L. Connolly, and M. L. Guerinot (1999). A ferric-chelate reductase for iron uptake from soils. *Nature* 397:694-697.

Russell, D. A. and M. E. Fromm (1997). Tissue-specific expression in transgenic maize of four endosperm promoters from maize and rice. *Transgenic Research* 6: 157-168.

Saalbach, I., T. Pickardt, F. Machemehl, G. Saalbach, O. Schieder, and K. Muntz (1994). A chimeric gene encoding the methionine-rich 2S albumin of the Brazil nut (*Bertholletia excelsa* H. B. K.) is stably expressed and inherited in transgenic legumes. *Molecular and General Genetics* 242: 226-236.

Samuelsen, A. I., R. C. Martin, D. W. S. Mok, and M. C. Mok (1998). Expression of the yeast FRE genes in transgenic tobacco. *Plant Physiology* 118: 51-58.

Schledz, M., S. al-Babili, J. von Lintig, H. Haubruck, S. Rabbani, H. Kleinig, and P. Beyer (1996). Phytoene synthase from *Narcissus pseudonarcissus*: functional expression, galactolipid requirement, topological distribution in chromoplasts and induction during flowering. *The Plant Journal* 10:781-792.

Serrano, R. and R. Gaxiola (1994). Microbial models and salt stress tolerance in plants. *Critical Reviews in Plant Sciences* 13: 121-138.

Sharp, P. A. and P. D. Zamore (2000). RNA interference. *Science* 287: 2431-2433.

Shaul, O. and G. Galili (1993). Concerted regulation of lysine and threonine synthesis in tobacco plants expressing bacterial feedback-insensitive aspartate kinase and dihydrodipicolinate synthase. *Plant Molecular Biology* 23:759-768.

Shaver, J. M., D. C. Bittel, J. M. Sellner, D. A. Frisch, D. A. Somers, and B. G. Gengenbach (1996). Single-amino acid substitutions eliminate lysine inhibition of maize dihydrodipicolinate synthase. *Proceedings of the National Academy of Sciences, USA* 93: 1962-1966.

Sheen, J. (1999). C_4 gene expression. *Annual Review of Plant Physiology and Plant Molecular Biology* 50: 187-217.

Shimada, H., Y. Tada, T. Kawasaki, and T. Fujimura (1993). Antisense regulation of the rice waxy gene expression using a PCR-amplified fragment of the rice genome reduces the amylose content in grain starch. *Theoretical and Applied Genetics* 86: 665-672.

Shimamoto, K., R. Terada, T. Izawa, and H. Fujimoto (1989). Fertile transgenic rice plants regenerated from transformed protoplasts. *Nature* 338: 274-276.

Shinozaki, K. and K. Yamaguchi-Shinozaki (1996). Molecular responses to drought and cold stress. *Current Opinion in Biotechnology* 7: 161-167.

Shinozaki, K. and K. Yamaguchi-Shinozaki (1997). Gene expression and signal transduction in water-stress response. *Plant Physiology* 115: 327-334.

Shintani, D. and D. DellaPenna (1998). Elevating the vitamin E content of plants through metabolic engineering. *Science* 282: 2098-2100.

Sijmons, P. C., B. M. M. Dekker, B. Schrammeijer, T.C. Verwoerd, P.J. van den Elzen, and A. Hoekema (1990). Production of correctly processed human serum albumin in transgenic plants. *Bio/Technology* 8: 217-221.

Sivak, M. N. and J. Preiss (1995). Progress in the genetic manipulation of crops aimed at changing starch structure and increasing starch accumulation. *Journal of Environmental Polymer Degradation* 3: 145-152.

Somerville, C. R. (1990). The biochemical basis for plant improvement. In: *Plant Physiology and Plant Molecular Biology*, ed. D. T. Dennis and D. H. Turpin, Longman group, pp. 490-501.

Somerville, C. and S. Somerville (1999). Plant functional genomics. *Science* 285: 380-383.

Stark, D. M., K. P. Timmerman, G. Barry, J. Preiss, and G. M. Kishore (1992). Regulation of the amount of starch in plant tissues by ADP-glucose pyrophosphorylase. *Science* 258: 287-292.

Stockinger, E. J., S. J. Glimour, and M. F. Thomashow (1997). *Arabidopsis thaliana* CBF1 encodes an AP2 domain-containing transcription activator that binds to the C-repeat/DRE, a cis-acting DNA regulatory element that stimulates transcription in response to low temperature and water deficit. *Proceedings of the National Academy of Sciences, USA* 94: 1035-1040.

Tada, Y., M. Nakase, and T. Adachi (1996). Reduction of 14-16 kDa allergenic proteins in transgenic rice plants by antisense genes. *FEBS Letters* 391: 341-345.

Takaiwa, F. (1999). Structure and expression of rice seed storage protein genes. In: *Molecular Biology of Rice*, ed. K. Shimamoto, Tokyo, Japan: Springer-Verlag, pp. 179-199.

Tamagnone, L., A. Merida, A. Parr, S. Mackay, F. A. Culianez-Macia, K. Roberts, and C. Martin (1998). The AmMYB308 and AmMYB330 transcription factors from *Antirrhinum* regulate phenylpropanoid and lignin biosynthesis in transgenic tobacco. *Plant Cell* 10: 135-154.

Tarcznski, M. and H. Bohnert (1993). Stress protection of transgenic tobacco by production of the osmolyte mannitol. *Science* 259: 508-510.

Thomashow, M. F. (1998). Role of cold-responsive genes in plant freezing tolerance. *Plant Physiology* 118: 1-7.

Toriyama, K., Y. Arimoto, H. Uchimiya, and K. Hinata (1988). Transgenic rice plants after direct gene transfer into protoplasts. *Bio/Technology* 6: 1072-1074.

Turpen, T. H., S. J. Reini, Y. Charoenvit, S. L. Hoffman, V. Fallarme, and L.K. Grill (1995). Malarial epitopes expressed on the surfaces of recombinant tobacco mosaic virus. *Bio/Technology* 13: 53-57.

Tzchori B.-T., I. A. Perl, and G. Galili (1996). Lysine and threonine metabolism are subject to complex patterns of regulation in *Arabidopsis*. *Plant Molecular Biology* 32: 727-734.

Ulker, B., G. C. Allen, W. F. Thompson, S. Spiker, and A. K. Weissinger (1999). A tobacco matrix attachment region reduces the loss of transgene expression in the progeny of transgenic tobacco plants. *The Plant Journal* 18: 253-263.

UNICEF (1998). *The State of the World's Children 1998*. Oxford Univ. Press: Oxford/New York.

Vain, P., B. Worland, A. Kohli, J. W. Snape, P. Christou, G. C. Allen, and W. F. Thompson (1999). Matrix attachment regions increase transgene expression levels and stability in transgenic rice plants and their progeny. *The Plant Journal* 18: 233-242.

Vandekerckhove, J., J. Van Damme, and M. Van Lijsabettens (1989). Enkephalins produced in transgenic plants using modified 2S seed storage proteins. *Bio/Technology* 7: 929-932.

Vijn, I., A. van Dijken, N. Sprenger, K. van Dun, P. Weisbeek, A. Wiemken, and S. Smeekens (1997). Fructan of the inulin neoseries is synthesized in transgenic chicory plants (*Cichorium intybus* L.) harboring onion (*Allium cepa* L.) fructan:fructan 6G-fructosyltransferase. *The Plant Journal* 11: 387-398.

von Heijne, G., J. Steppuhn, and R. G. Herrmann (1989). Domain structure of mitochondrial and chloroplast targeting peptides. *European Journal of Biochemistry* 180: 535-545.

Wassenegger, M., S. Heimes, L. Riedel, and H. L. Sanger (1994). RNA-directed de novo methylation of genomic sequences in plants. *Cell* 76, 567-576.

Waterhouse, P. M. and N. M. Upadhyaya (1999). Genetic engineering of virus resistance. In: *Molecular Biology of Rice*, ed. K. Shimamoto, Tokyo, Japan: Springer-Verlag, pp. 257-281.

Winicov, I. and D. R. Bastola (1999). Transgenic overexpression of the transcription factor Alfin1 enhances expression of the endogenous MsPRP2 gene in alfalfa and improves salinity tolerance of the plants. *Plant Physiology* 120: 473-480.

Wong, E. Y., C. M. Hironaka, and D. A. Fischoff (1992). *Arabidopsis thaliana* small subunit leader and transit peptide enhance the expression of *Bacillus thuringiensis* proteins in transgenic plants. *Plant Molecular Biology* 20: 81-93.

Xu, D., X. Duan, B. Wang, B. Hong, T.-H. D. Ho, and R. Wu (1996). Expression of a late embryogenesis abundant protein gene, *HVA1*, from barley confers tolerance to water deficit and salt stress in transgenic rice. *Plant Physiology* 110: 249-257.

Ye, X., S. Al-Babili, A. Kloti, J. Zhang, P. Lucca, P. Beyer, and I. Potrykus (2000). Engineering the provitamin A (β-carotene) biosynthetic pathway into (carotenoid-free) rice endosperm. *Science* 287: 303-305.

Zhang, W. and R. Wu (1988). Efficient regeneration of transgenic plants from rice protoplasts and correctly regulated expression of the foreign gene in the plants. *Theoretical and Applied Genetics* 76: 835-840.

Zheng, Z., K. Sumi, K. Tanaka, and N. Murai (1995). The bean seed storage protein β-phaseolin is synthesized, processed and accumulated in the vacuolar type-II protein bodies of transgenic rice endosperm. *Plant Physiology* 109: 777-786.

Genetic Control of Quantitative Grain and Malt Quality Traits in Barley

E. Igartua
P. M. Hayes
W. T. B. Thomas
R. Meyer
D. E. Mather

SUMMARY. Many of the important determinants of barley grain and malt quality exhibit quantitative variation and are affected by both genetic and environmental factors. In recent years, the application of molecular marker techniques has permitted the detection and mapping of quantitative trait loci (QTL) for grain and malt quality traits in many populations. Here, the quantitative traits affecting the grain and malt quality of barley grain are reviewed, and results from two analyses of

E. Igartua and D. E. Mather are affiliated with the Department of Plant Science, McGill University, 21111 Lakeshore Road, Ste-Anne-de-Bellevue, Quebec H9X 3V9, Canada.

P. M. Hayes is affiliated with the Department of Crop and Soil Science, Oregon State University, Corvallis, OR 97331 USA.

W. T. B. Thomas and R. Meyer are affiliated with the Scottish Crop Research Institute, Invergowrie, Dundee DD2 5DA, Scotland, UK.

Address correspondence to: D. E. Mather, Department of Plant Science, McGill University, 21111 Lakeshore Road, Ste-Anne-de-Bellevue, Quebec H9X 3V9, Canada.

Financial support for the analyses presented here was provided by the Natural Sciences and Engineering Research Council of Canada. Our analyses and interpretation relied upon information from numerous previous experiments in which barley grain and malt quality QTL were mapped. The authors are grateful to everyone who contributed to the success of those experiments.

[Haworth co-indexing entry note]: "Genetic Control of Quantitative Grain and Malt Quality Traits in Barley." Igartua, E. et al. Co-published simultaneously in *Journal of Crop Production* (Food Products Press, an imprint of The Haworth Press, Inc.) Vol. 5, No. 1/2 (#9/10), 2002, pp. 131-164; and: *Quality Improvement in Field Crops* (ed: A. S. Basra, and L. S. Randhawa) Food Products Press, an imprint of The Haworth Press, Inc., 2002, pp. 131-164. Single or multiple copies of this article are available for a fee from The Haworth Document Delivery Service [1-800-HAWORTH, 9:00 a.m. - 5:00 p.m. (EST). E-mail address: getinfo@haworthpressinc.com].

grain and malt quality data from barley mapping populations (a multivariate analysis of interrelationships among traits and a comparative analysis of QTL positions and effects among five populations) are presented and discussed. *[Article copies available for a fee from The Haworth Document Delivery Service: 1-800-HAWORTH. E-mail address: <getinfo@ haworthpressinc.com> Website: <http://www.HaworthPress.com> © 2002 by The Haworth Press, Inc. All rights reserved.]*

KEYWORDS. Barley, grain quality, malt quality, quantitative trait loci

INTRODUCTION

Malting of barley and the use of barley malt to produce fermented beverages are among the earliest biotechnological processes used by humans, with records dating back to the beginnings of agriculture (Dickson 1979). These processes have been profusely studied and are known to be influenced by over 30 physical, chemical and biochemical properties (Burger and LaBerge 1985), most of which are affected by both the genetic makeup of the barley and the environmental conditions under which it is grown, stored and malted. Quality specifications for barley grain and malt vary between Europe and North America, reflecting different brewing practices. In general, European maltsters and brewers favor low grain protein content and moderate enzymatic activity, whereas the North American industry prefers moderate levels of protein and higher enzymatic activity.

Decades of research have generated an enormous amount of information on the biochemistry and physiology of the malting and brewing processes (Burger and LaBerge 1985; Fincher and Stone 1993). Until recently, the genetic factors affecting variation in these processes were poorly understood. With the application of molecular genetic techniques, major advances have now been made towards understanding of the genetic control of barley grain and malt quality traits. Several genes already known from classical genetic studies have been located and cloned. Many quantitative trait loci (QTL) have been mapped, some coincident with previously known major genes and others at new locations in the genome.

Here, we present a detailed review of barley grain and malt quality traits and the results of two analyses of grain and malt quality data from barley mapping populations: (1) a multivariate analysis of interrelationships among traits, and (2) a comparative analysis of QTL positions and effects among five populations. We focus mainly on results generated by the North American Barley Genome Mapping Project using three populations: Steptoe × Morex, Harrington × TR306, and Harrington × Morex. The marker maps for these three crosses contain many common markers, and this facilitates the alignment of the three maps for the purpose of comparing QTL positions. Morex (a six-rowed cultivar) and Harrington (a two-rowed cultivar) were standard-setting malting barley cultivars in the USA and Canada, respectively. For development of the first two mapping populations, these cultivars were crossed to feed-type lines of similar spike morphology, in order to detect and map grain and malt quality QTL from Morex and Harrington. This approach proved successful (Hayes et al. 1993; Mather et al. 1997; Ullrich et al. 1997 and other references in Table 1). Many QTL were mapped. At most (but not all) of these, the favorable allele came from the malting-quality parent. For the third population, Harrington was crossed with Morex, in order to detect and map QTL that could explain differences between the two-rowed and six-rowed malting barley, and to open up the possibility of combining favorable QTL alleles from both types (Marquez-Cedillo et al. 2000). In addition, we compare the results obtained with these North American populations to results from other populations (see Table 1) and we make use of data from two populations developed and mapped at the Scottish Crop Research Institute: Blenheim × E224/3 and Derkado × B83-12/21/5. One of the parents of Blenheim, Triumph, has had a major impact in Europe, both as a standard-setting malting barley cultivar and as a parent of many other malting barley cultivars.

BARLEY GRAIN AND MALT QUALITY TRAITS

Barley must be malted before it can be used for brewing. Malt is produced by germinating the barley grains and growing the young seedlings for four to six days. Then germination is arrested and the malt is dried. During germination, the endosperm carbohydrate and protein reserves are mobilized by the enzymatic action. This process is known as 'modification.' Malt is then mashed (mixed) with water in prescribed proportions. This mixture is stirred and subjected to a time/temperature program that facilitates enzymatic reactions. After cooling, the mash is filtered and the filtrate (wort) is recovered.

Burger and LaBerge (1985) and Bamforth and Barclay (1993) have discussed the roles of grain and malt characters in malting and brewing. Bamforth and Barclay (1993) also provided a thorough description of

TABLE 1. Barley populations in which QTL for grain and malt quality traits have been reported.

Population	References
Blenheim × E224/3	Powell et al. 1992; Chalmers et al. 1993; Thomas et al. 1996; Powell et al. 1997
Blenheim × Kym	Bezant et al. 1997
Calicuchima-sib × Bowman	Hayes et al. 1996
Chebec × Harrington	Langridge et al. 1996
Clipper × Sahara	Langridge et al. 1996
Derkado × B83-12/21/5	W.T.B. Thomas et al. (unpublished.)
Dicktoo × Morex	Oziel et al. 1996
Galleon × Haruna Nijo	Langridge et al. 1996
Harrington × Morex	Márquez-Cedillo et al. 2000
Harrington × TR306	Mather et al. 1997, among others
Steptoe × Morex	Hayes et al. 1993; Han et al. 1995; Oberthur et al. 1995; Zwickert-Menteur et al. 1996; Ullrich et al. 1997, among others

the entire malting process. Much of the information in this section is drawn from those two works, and from the review by MacGregor (1999).

In purchasing barley, maltsters seek specific malting cultivars, kernel homogeneity, adequate grain moisture, and a series of characteristics that can be summarized under three headings:

- *Size.* Uniform kernels are desirable because they germinate at a uniform rate. Plump kernels may malt more slowly than thin kernels, but are desirable because they usually produce more malt extract, as do larger kernels in general.
- *Protein content.* Proteins play a multifaceted role in malting and brewing, so their influence on malt and beer quality is complex. For a given kernel size, the more protein that is present, the less starch (and therefore potential fermentable sugar). Excessive grain

protein is undesirable because it is associated with lower malt extract levels and because it can cause problems with beer stability and viscosity. However, some grain protein is needed to provide amino acids for yeast nutrition during brewing, and also to provide starch-degrading enzymes.

- *Modification potential.* The main components of the modification process are the degradation of starch, cell walls, and protein.

The conversion of starch to fermentable carbohydrates is carried out by several enzymes: α-amylase, β-amylase, limit dextranase, and α-glucosidase. α-Amylase is of prime importance for rapidly hydrolyzing the starch, once it has been solubilized, and producing substrate for the other enzymes. β-Amylase is most active on the products formed by the action of α-amylase on starch, and is a major contributor to the 'diastatic power' (overall starch-degrading capacity) of malt. Complete conversion of starch to fermentable products requires the hydrolysis of the α-(1$\rightarrow$6) branch linkages in starch by the debranching enzyme, limit dextranase. It is known that barley contains proteins that complex with, and strongly inhibit, the limit dextranase (MacGregor et al. 1994). Sufficient levels of these proteins remain in the finished malt to inhibit a high proportion of limit dextranase during malting. Finally, α-glucosidase has the least well-characterized activities of the amylolytic complex.

Cell walls must be degraded to allow enzymes to access the cellular contents. β-Glucan is a major component of the endosperm cell walls of barley. It is a glucose-containing polysaccharide, which forms very viscous solutions and causes filtration problems during brewing, and stability problems in beer during storage if it is not adequately degraded during malting and mashing.

Within the barley endosperm, starch granules are embedded in a protein matrix. This matrix must be broken down during malting to release the starch. This task is performed by proteases. Insufficient breakdown prevents complete release of starch during malting with concomitant reduction in starch conversion during mashing, beer storage problems and insufficient production of amino acids for yeast metabolism during brewing, but excessive degradation of protein adversely affects the foam stability of beer. In practice, it is difficult to lower levels of β-glucan without generating excessive amounts of soluble nitrogen-containing compound (usually referred to as 'soluble protein'). Therefore, it might be advantageous to de-link the synthesis of proteases from that of β-glucanase so that rapid synthesis of β-glucanase need not be accompanied by high levels of soluble protein in extracts.

In the development of new malting barley cultivars, parents (usually cultivars or lines with good quality profiles) are crossed and progeny are subjected to detailed evaluations of grain and malt quality. Kernel size, kernel plumpness and concentration of protein are evaluated on the grain. Solutions of ground malt are filtered to produce malt extracts. These extracts are analyzed to determine the amount of soluble material (malt extract), the concentration of soluble protein and β-glucan, the viscosity, and the activity of starch degrading enzymes. In some cases, preliminary assessments of one or more of these characteristics are based on indirect tests (Wainwright and Buckee 1977; Burger and LaBerge 1985).

The traits measured to estimate malt quality differ, especially between North American and European laboratories. However, with information on the methodologies used and with knowledge of the malting process, similarities among traits can be seen. In the following discussion, certain trait names are underlined. These are the traits for which we will later present analytical results from North American and European mapping populations.

Kernel plumpness is usually recorded as the percentage in weight of kernels that are retained by a sieve of specific hole diameter. *Kernel weight* is recorded as the weight of 1000 kernels. *Grain protein content* is usually expressed as the percentage by weight in a grain sample. *Grain nitrogen content* can be considered similar to grain protein content, as nitrogen content is often multiplied by a constant to give a protein content estimate.

Milling energy, commonly measured in British studies, is a surrogate measure for malt extract (Allison et al. 1976). It is related to the overall carbohydrate and protein structure of the endosperm, since the ease with which the structure of the endosperm can be disrupted mechanically appears to relate to the ease with which it is broken down by enzymes during malting. *Milling energy loss* is measured as the difference in energy required to grind grain and malt samples (Swanston 1990), and reflects the extent of endosperm breakdown during the malting process. The more energy loss, the better the endosperm was modified.

Malt extract is a key quality indicator because it reflects the amount of beer that can be produced from a given quantity of malt. It is a measure of the percentage of the malt rendered soluble upon mashing in the laboratory, i.e., the solids content of wort. Though there are different protocols to measure malt extract in different countries, the results should be comparable. Malt extract is usually measured on two malt samples, one ground to a finer consistency than the other. A small difference between the two measures (*fine-coarse difference*) is one indication that the endosperm has been well modified during malting. Other indicators of thorough endosperm modification are high levels of soluble protein and low levels of *extract* β-*glucan*. The activities of some enzymes, such as β-glucanase, α-*amylase*, β-amylase, and limit dextranase can be directly assayed. *Diastatic power* is a widely used measure of the combined hydrolysing capacity of all starch degrading enzymes. Proteolytic activity is expressed as the *ratio of soluble to total protein* (Kolbach index). *Wort protein* is the soluble protein measured in the wort after mashing. β-Glucan can be measured at several stages of the malting process. When measured in the grain, it is an indicator of the expected resistance of the endosperm to modification. It should be related to milling energy. When measured in the extract, it indicates the extent of modification, since only the residual un-degraded β-glucan are left at that stage. Extract *viscosity* is another related measure. High viscosities slow filtering in the brewhouse, and are unacceptable. The primary contributor to viscosity is β-glucan. *Wort color* can also be determined, as it can be related to final beer color.

Objectives for the breeding of new malting barley cultivars for Canada were clearly stated by Edney (1999), and the most important of these can be considered worldwide objectives. They include: improving hull adherence, increasing carbohydrate extract, increasing the amount of final product; increasing modification capacity; and addressing the problem of favoring β-glucan breakdown while not producing an excess of soluble protein.

INTERRELATIONSHIPS AMONG BARLEY GRAIN AND MALT QUALITY TRAITS

Many studies have reported interrelationships among barley grain and malting quality traits. The causes of such relationships may be genetic and/or environmental, with the possible genetic causes being pleiotropy (with some traits acting as links in chains of events leading to the expression of other traits) and genetic linkage. Relationships among traits can be studied by looking at the correlation among traits. Here, we examined correlation among traits for populations of doubled haploid

lines derived from the barley crosses Steptoe × Morex, Harrington ×
TR306, Harrington × Morex, Blenheim × E224/3, and Derkado ×
B83-12/21/5. Within each population, correlation coefficients were
computed using mean phenotypic values for each line over all environments. Principal component analysis was used to summarize the information conveyed by the correlation coefficients.

For Steptoe × Morex, the first principal component was related to
modification and the only traits with large loads on the second component were kernel weight and plumpness (Figure 1a). This pattern indicates a certain independence of grain and malt (mostly modification)
quality traits, possibly influenced by different loci. The negative effect
of increased protein content on malt extract was evident on the third
component (Figure 1b), for which these two traits presented the only
large loads, with opposite signs. In this population, grain protein content was not tightly associated with kernel weight and plumpness.

For Harrington × TR306, the first three principal components accounted for 41, 19, and 15% of the combined variance for all traits (Figures
1c and 1d). The first component was clearly related to modification,
with the three traits for which large values denote poor modification
(malt β-glucan, fine-coarse difference, and wort viscosity) opposed to
traits whose higher values are related to better modification (diastatic
power, α-amylase, soluble/total protein ratio, with negative loads on
first axis). High malt extract had a large negative load on the first axis,
and thus was associated with good modification in this population.
Grain quality traits (kernel weight, protein content, and plumpness) had
minor loads on this component, meaning that they were quite independent of modification. These three traits were positively correlated with
each other, as indicated by their proximity in Figure 1c. They had large
loads on the second component, mostly opposed to malt extract (the
larger, plumper, and more protein-rich the kernels, the less extract produced) and fine-coarse difference. The pattern of relationships among
these three traits suggests the difficulty of selecting concurrently for
low protein content and large kernel weight and plumpness in this population. The third component, though still large, had no simple explanation.

In Harrington × Morex (Figure 1e), tight associations among kernel
weight, kernel plumpness and grain protein content dominated the first
principal component. These three traits (and malt β-glucan) were opposed to malt extract, the main quality indicator, and soluble/total protein ratio. The most noticeable feature in the second component was the

FIGURE 1. Principal component analysis for grain and malt quality traits for five barley doubled haploid populations, based on the matrix of correlation coefficients. KWT, kernel weight; PLM, kernel plumpness; PRO, grain protein content; NIT, grain nitrogen content; WPR, wort protein; STR, soluble/total protein ratio; MEX, fine-grind malt extract; CME, coarse-grind malt extract; DIF, fine-coarse grind extract difference; MEN, milling energy; MEL, milling energy loss; DST, diastatic power; AMY, α-amylase activity; GBG, grain β-glucan; MBG, malt β-glucan; VSC, wort viscosity; WCO, wort color.

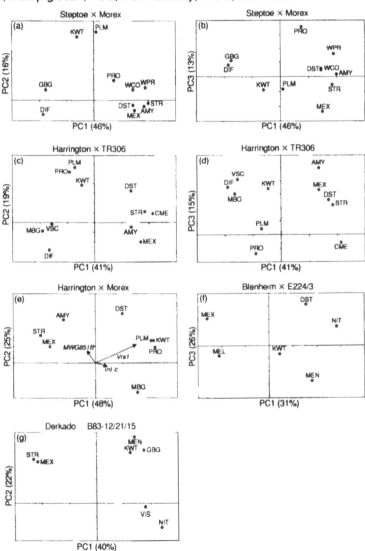

contrast between malt β-glucan and the carbohydrate-hydrolyzing enzymes, with the largest loads on this component.

In Blenheim × E224/3, two principal components accounted for 57% of the total variance (Figure 1f). Most of the malt extract variance was accounted for by the first principal component. Loads on the remaining axes for this variable were negligible. Kernel weight contributed almost nothing to these first two axes. Kernel plumpness, usually in close association with kernel weight, was not determined for this population. The first axis was predominantly a contrast between grain protein content and, to a lesser degree, milling energy and diastatic power, versus milling energy loss and malt extract. It can be interpreted as a combination of modification potential and grain protein content. Grain protein was positively correlated (r = 0.56) with diastatic power, giving these two traits relatively close positions on the graph. This association has been consistently found across many studies (Peterson and Foster 1973; Eagles et al. 1995; Goblirsch et al. 1996). The third axis for this population, not represented, was merely a contrast of kernel weight versus all other traits. This emphasizes the independence, for this population, of kernel size and malting quality characteristics. This pattern of trait relationships (as in Steptoe × Morex) is more favorable for breeding than the one displayed in Harrington × TR306, where antagonistic correlations between kernel size and protein content must be broken through extensive recombination (if due to linkage), to achieve concurrent progress for both traits.

For Derkado × B83-12/21/5, malt extract and soluble/total protein ratio are very close on the graph of the first two principal components (Figure 1g). They formed a distinct group, separated from the remaining variables, which had the opposite sign on the first axis. Soluble/total protein ratio is the only modification trait on the graph. Its position suggests a close relationship between extent of modification and malt extract production in Derkado × B83-12/21/5. Malt extract was plotted completely opposite to grain protein content (r = −0.55), a much stronger relationship than in Blenheim × E224/3 (r = −0.25). Grain protein and kernel weight had a moderate correlation coefficient (r = 0.33). They had positive loads on the first principal component but were split apart in the second, suggesting the presence of some common loci and some independent loci. In this population, grain and malt quality could apparently be improved by selecting for lower grain protein content.

QTL POSITIONS AND EFFECTS FOR GRAIN
AND MALT QUALITY TRAITS

The list of published works on QTL for these traits covers many of the germplasm groups available for barley breeding: two-rowed and six-rowed types and winter and spring types of diverse geographic origin (Table 1). Summaries of grain and malt quality QTL results have been posted on the internet *(http://www.css.orst.edu/barley/nabgmp/ QTL71400.htm)* and printed in Barley Genetics Newsletter (Zale et al. 2000).

Here, we examine QTL positions and effects for grain and malt quality traits in five mapping populations: Steptoe × Morex, Harrington × TR306, Harrington × Morex, Blenheim × E224/3, and Derkado × B83-12/21/5. A brief description of the number of markers, progenies, and field trials for each population is presented in Table 2. For further details on materials and methods regarding types of markers; locations, and dates of field trials; and the methods of determination of grain and malt quality traits, we refer the reader to the original publications (Table 1). The original studies on these populations had employed different analytical tools and approaches. To facilitate comparison of the results among crosses, we reanalyzed all data by simple interval mapping using the software MQTL (Tinker and Mather 1995). Results of these analyses are presented in Figure 2 (Steptoe × Morex, Harrington × TR306, and Harrington × Morex) and Figure 3 (Blenheim × E224/3 and Derkado × B83-12/21/5). In Figure 2, the QTL scans for the three populations are superimposed, using the Harrington × Morex marker map as a common axis. It was possible to display the scans for the other two populations on this axis by aligning common markers and slightly 'stretching' or 'compressing' intervals between common markers. The maps for these three crosses contain many common markers (mostly restriction fragment length polymorphisms), arranged in identical orders, with similar map distances between them.

This permitted a good alignment with only slight distortion of map distances for Steptoe × Morex and Harrington × TR306. In contrast, the maps for the two European crosses had few markers in common and could not be aligned with each other or with the North American maps. Thus, comparisons of QTL positions involving the European populations are less precise than within the North American set.

Here, we indicate QTL positions according to the closest marker or marker interval, and when possible we assign them to 'BINs,' following Kleinhofs (1999). In the two-rowed by six-rowed Harrington × Morex

TABLE 2. Numbers of doubled haploid progenies, field trials, and genetic markers for five barley mapping populations for which data were analyzed in this study.

Population	Number of progenies	Number of trials	Number of markers
Blenheim × E224/3	59	5	222
Derkado × B83-12/21/5	156	6	127
Harrington × Morex	140	8	107
Harrington × TR306	145	6	127
Steptoe × Morex	150	9	223

population, the expression of several traits was clearly dominated by the main locus determining inflorescence type, *Vrs1*. Thus, for that population, separate analyses were conducted for the two-rowed and six-rowed subsets of doubled-haploids (as in Marquez-Cedillo et al. 2000).

Here, we present the grain and malt quality QTL regions that have been detected on each chromosome. Where no literature citations are given, the results referred to are presented in Figures 2 (North American populations) and 3 (European populations).

Chromosome 1 (7H)

Several studies have reported QTL for malting quality traits in the central part of chromosome 1 (7H). An important genetic feature of this region seems to be the presence of *Amy2*, one of the two α-amylase coding loci known for barley (Mitsui and Itoh 1997). The region around this locus, interval *ABC308-Amy2* (BIN 7), is associated with several enzymatic activity traits in Steptoe × Morex and Harrington × Morex, and also possibly in Blenheim × E224/3. An effect on the activity of α-amylase itself, however, was detected only in Steptoe × Morex. This suggests that *Amy2* is not the only locus in this genomic region contributing to diastatic power. Thomas et al. (1996) suggested that the diastatic power they measured in Blenheim × E224/3 was in effect β-amylase activity. This point could not be confirmed for the North American populations, in which β-amylase activity was not specifically assayed. It appears that the Harrington and Morex, carry alleles with similar effects at the Amy2 locus, but that they differ for another linked QTL that af-

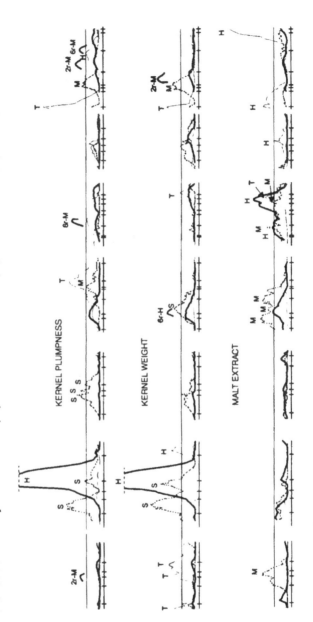

FIGURE 2. Scans of main-effect test statistics from simple interval mapping for Harrington × TR306, Steptoe × Morex, and Harrington × Morex barley populations. The chromosomes are displayed left to right, starting with chromosome 1 (7H), and ending with chromosome 7 (5H). Each chromosome is oriented with the 'plus' arm on the right. The maps for the three populations were aligned and the scans for Harrington × TR306 and Steptoe × Morex are represented using the Harrington × Morex marker map as a template. Relevant markers mentioned in the text are represented as tickmarks on chromosomes. Test statistics were normalized by dividing each one by its corresponding threshold, for each trait and population. Thresholds are represented as continuous horizontal lines. A letter close to each peak indicates the parent contributing the higher-value allele. Partial scans showing only the significant peaks are included to show the results of analyses within sub-populations of two-rowed (2r) and six-rowed (6r) Harrington × Morex lines.

FIGURE 2 (continued)

GRAIN PROTEIN

SOLUBLE/TOTAL PROTEIN

α-AMYLASE ACTIVITY

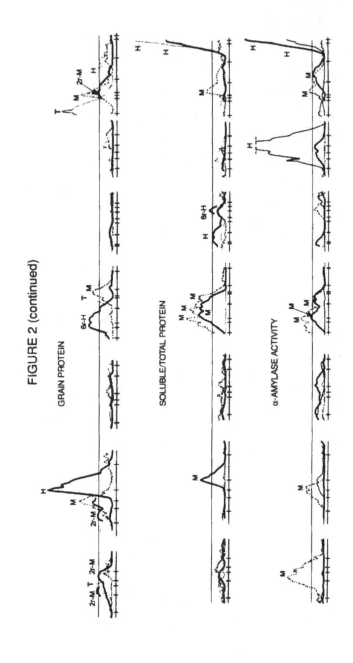

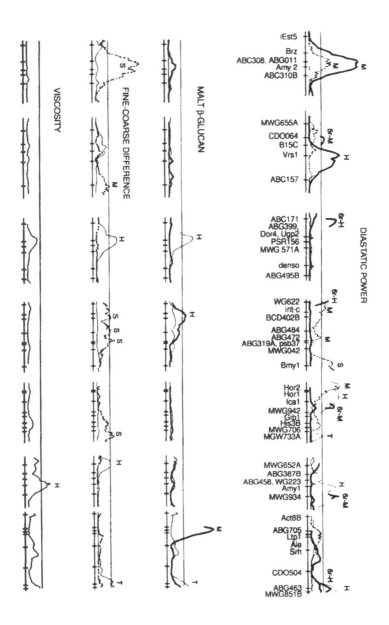

VISCOSITY

FINE-COARSE DIFFERENCE

MALT β-GLUCAN

DIASTATIC POWER

iEst5
Brz
ABC308, ABG011
Amy 2
ABC310B

MWG655A
CDO064
B15C
Vrs1
ABC157

ABC171
ABG399
Dor4, Ugp2
PSR156
MWG 571A
denso
ABG495B

WG622
int-c
BCD402B
ABG484
ABG472
ABG319A, psb37
MWG042
Bmy1

Hor2
Hor1
lca1
MWG942
Gib1
His3B
MWG706
MGW733A

MWG652A
ABG387B
ABG458, WG223
Amy1
MWG934

Act8B
ABG705
Ltp1
Ale
Srh
CDO504
ABG463
MWG851B

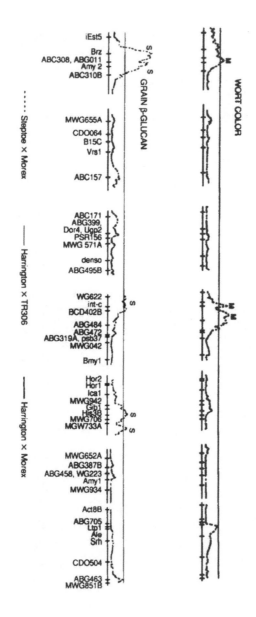

FIGURE 2 (continued)

FIGURE 3. Scans of main-effect test statistics from simple interval mapping for Blenheim × E224/3 and Derkado × B83-12/21/5. The chromosomes are displayed left to right, starting with chromosome 1 (7H), and ending with chromosome 7 (5H). Each chromosome is oriented with the 'plus' arm on the right. Tickmarks represent background markers used in the composite interval mapping in the original studies; they are different for the two populations. A letter close to each peak represents the parent contributing the higher-value allele. Letters within parentheses represent QTL identified in the original studies that were not significant in the present analysis.

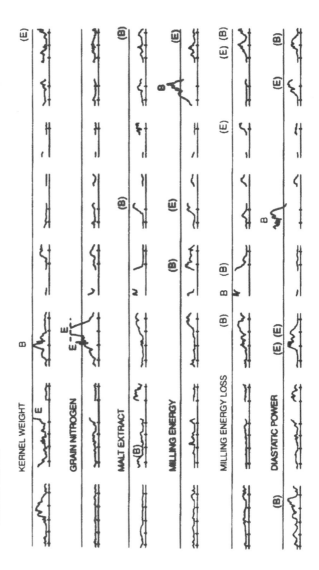

Blenheim × E224/3

147

FIGURE 3 (continued)

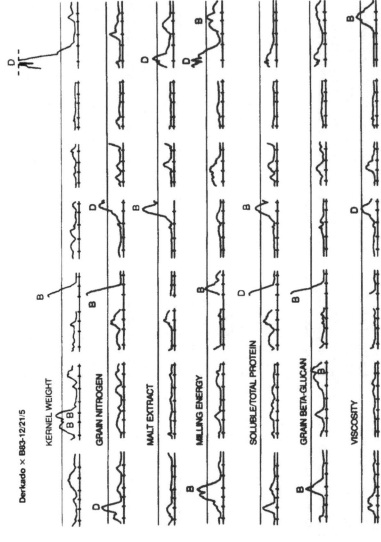

fects diastatic power (with Morex contributing the higher diastatic power allele). Also in BIN 7, several studies found QTL for malt β-glucan and β-glucanase activity (Zwicker-Menteur et al. 1996; Ullrich et al. 1997) in Steptoe × Morex. This region of the genome did not affect β-glucan in Harrington × TR306, nor in Harrington × Morex.

Ullrich et al. (1997) detected another QTL affecting the same traits at a nearby region, interval *Brz-ABG011* in BIN 5, which they declared to be different from the QTL in BIN 7. Li et al. (1999) later found a QTL for limit dextranase activity in the same region, co-segregating with RFLP marker *ABC154A* (8.5 cM distal to *Brz*) in the Steptoe × Morex population. No polymorphism was found in the Harrington × TR306 population, consistent with the fact that no enzymatic effect was found for this population on this region. Langridge et al. (1996) reported a QTL for β-glucanase in exactly the same region, for the Galleonx-Haruna Nijo population. Oziel et al. (1996) reported QTL for diastatic power, α-amylase activity, and fine-coarse difference distal to *Brz* in DicktooxMorex. From these reports, it is not clear whether there are one or two QTL regions in the vicinity of *Brz*.

The central region of this chromosome for Derkado × B83-12/21/5 also has QTL for grain β-glucan and milling energy, but not for malt extract. This suggests that whereas the structure of the endosperm of the barley kernel was affected by QTL in this region, the modification process was powerful enough to override these differences. Evidence from Blenheim × Kym (Bezant et al. 1997) also suggests the presence of QTL for malt extract, and grain protein content on the central part of this chromosome.

In Steptoe × Morex, the scan for malt extract has two peaks coincident with the regions described above. However, in Harrington × Morex, the QTL with a large effect on diastatic power had no apparent effect on malt extract. Perhaps in the progeny of this cross between two malting-quality parents, diastatic power is high enough so as not to limit malt extract.

Finally, there are two (or possibly three) peaks for kernel weight in Harrington × TR306, one of them coincident with another grain protein peak, and very close to *Amy2*.

Chromosome 2 (2H)

On chromosome 2 (2H), there are few QTL in common among Steptoe × Morex, Harrington × Morex, and Harrington × TR306, although kernel weight QTL were detected for the three populations. One of them, shared by Steptoe × Morex and Harrington × Morex, is in the vicinity of the *Vrs1* locus (BIN 10). This locus is the main one controlling the fertility of lateral spikelets. It determines the spike type of barley: two-rowed or six-rowed. Kernel plumpness, grain protein content, diastatic power, and soluble/total protein ratio also have large peaks coincident with the *Vrs1* locus in Harrington × Morex. The Harrington allele contributed higher kernel weight, diastatic power and, surprisingly, grain protein content. Harrington itself had much lower grain protein content than Morex (11.5% vs. 13.2%) and the mean grain protein content of the doubled haploid progeny was above that of the higher parent (13.9%). This suggests that there are epistatic interactions for this trait. The very large QTL at *Vrs1* may be involved in these interactions (Marquez-Cedillo et al. 2000). The other two-rowed by six-rowed cross that has been studied for grain and malt quality traits is Calicuchima-sib × Bowman. In that cross, Hayes et al. (1996) observed a similar phenomenon, with both high kernel weight and high grain protein content associated with the two-rowed parent (Bowman) at the *Vrs1* locus. Grain protein content for the two parents was very similar, between 12 and 12.5%, and again, the mean grain protein content for the doubled haploid population was higher, around 13.2% (see Figure 2 in Hayes et al. 1996), suggesting epistasis. The similarity of results for kernel weight and especially grain protein content between from Calicuchima-sib × Bowman and Harrington × Morex reinforces observations made by Marquez-Cedillo et al. (2000), about the importance of inflorescence architecture in determining grain size and allocation of protein and enzymes to kernels. Whether this association is due to linkage or pleiotropy remains unsolved. Marquez-Cedillo et al. (2000) speculated that because *Vrs1* is near the centromere, recombination might be suppressed, leading to conservation of linkage blocks of contrasting alleles within two-rowed and six-rowed germplasm groups.

Examination of pedigrees reveals a possible reason for the similar effects of the *Vrs1* region in Harrington × Morex and Calicuchima-sib × Bowman. Both two-rowed parents (Harrington and Bowman) have the older malting barley cultivars Klages, Betzes, and Isaria in their ancestry. Thus, Bowman and Harrington may carry alleles that are identical-by-descent and these may underlie the trait associations. However, it should be noted that associations between the two-rowed allele at *Vrs1* and high grain protein content have been reported often and for cultivars with diverse pedigrees (Day et al. 1955; McGuire and Hockett

1983; Swanston 1983; Takahashi et al. 1976). Therefore, it seems to be a rather general phenomenon.

The Harrington × Morex and Calicuchima-sib × Bowman populations also present an interesting difference regarding effects of *Vrs1* on enzymatic activity. In Harrington × Morex there is a large QTL for diastatic power (with the Harrington allele increasing diastatic power) and no QTL peak for α-amylase activity, whereas in Calicuchima-sib × Bowman, there is a large QTL for α-amylase activity (with Calicuchima-sib increasing activity), and no QTL peak for diastatic power. Thus, there may be more than one locus affecting enzymatic activity linked to *Vrs1* (one locus affecting α-amylase activity and another one affecting the activity of some other starch degrading enzyme). Depending on the alleles carried by the two parents of a mapping population, one or both of these loci would be detectable as QTL. In any case, the results for Harrington × Morex and Calicuchima-sib × Bowman confirm that the two-rowed allele contributes larger kernel weight, increased grain protein, and decreased starch-degrading capacity.

Langridge et al. (1996) found a kernel weight QTL at the same area (BIN 10), in the population derived from the cross between Chebec and Harrington, both of which are two-rowed. Thus, the effect of this region on kernel weight goes beyond the determination of spike types.

A multiple effect of *Vrs1* on quantitative traits, mostly agronomic, has also been described in other studies involving two-rowed by six-rowed crosses (Powell et al. 1990; Kjaer et al. 1996). As noted by Allard (1988) 'the two-row six-row locus affects developmental processes in ways that leave few quantitative characters untouched.' To visualize the importance of the loci that determine inflorescence architecture on grain and malt quality traits, we superimposed them on the graph from the principal components analysis for Harrington × Morex. In Figure 1e, the loci *Vrs1* and *int-c* are represented by vectors proportional to the regression coefficients of these loci on the principal components. The first principal component in Harrington × Morex was mostly related to *Vrs1* and *int-c*; together, they explain 59% of the variance for this component.

In Harrington × TR306, only one QTL was detected on chromosome 2 (2H). At that QTL, near the end of the map for the 'minus' arm, the Harrington allele had a small positive effect on kernel weight. This is in contrast to the other QTL for kernel weight in this cross, at which the TR306 alleles increased kernel weight. Blenheim × E224/3 and

Derkado × B83-12/21/5 also have kernel weight QTL on this chromosome, but it is not possible to be precise about their location.

In the analysis of the six-rowed subpopulation of Harrington × Morex, a significant peak for diastatic power appeared in the interval *CDO64-B15* (BIN 7), which encompasses the β-amylase locus *Bmy2*.

Chromosome 3 (3H)

Few QTL were detected on chromosome 3 (3H) in the three North American crosses. In Steptoe × Morex, there are three contiguous peaks for kernel plumpness; these could indicate the presence of one or several QTL. They are close to markers *ABG399*, *PSR156*, and *MWG571A*, in BINs 6, 7, and 9, respectively. Steptoe alleles were associated with greater plumpness. Han et al. (1995) also found a QTL for malt β-glucan (Steptoe providing the 'high' allele) around *Dor4* in BINs 5-6. In Harrington × TR306, the region around *Ugp2* (BIN 4) has a QTL for extract β-glucan and wort viscosity, with alleles from the malting parent (Harrington) giving larger, unfavorable values.

There was much more QTL activity on this chromosome for the European crosses, as these involve polymorphism for the *denso* locus, also known by its accepted symbol *sdw1*. This dwarfing allele gene was incorporated into many European cultivars from Triumph. Blenheim and Derkado both carry the dwarfing allele. This locus and/or other factors closely linked to it, affect the development of the plant beyond mere height. Thomas et al. (1991) reported that this locus affected many quantitative characters, including heading date, kernel weight and plant yield.

The *denso* region affected grain protein content in the mapping populations, with effects being more evident for Derkado × B83-12/21/5 than for Blenheim × E224/3. This difference may simply be due to lower statistical power in the Blenheim × E224/3 mapping experiment, which included only 59 progeny lines, compared to 156 for the Derkado × B83-12/21/5 experiment. In both cases, the *denso* allele (and/or alleles linked to it) reduced grain protein content considerably. This is a favorable circumstance, since the agronomic advantages of the dwarfing gene come with an associated malting bonus of lower grain protein content. Other favorable effects of the Derkado allele were a reduction of grain β-glucan and an increase of the soluble/total protein ratio. On the other hand, Derkado alleles at this position decreased kernel weight. This was an unfavorable effect from the *denso* parent. In our analyses, the *denso* region had no significant effects on traits other than grain protein content but in the original mapping report for the Blenheim × E224/3 population, Powell et al. (1997) mentioned QTL for diastatic power and milling energy loss (both with negative effects from the Blenheim allele). If the QTL for diastatic power is real, then the correlation between grain protein content and diastatic power observed in Blenheim × E224/3 (Figure 1f) could be partly due to QTL linked in coupling at or near *denso* (Thomas et al. 1996).

Studies with other populations have also found malting QTL around the *denso* region. Oziel et al. (1996) found a QTL for soluble/total protein ratio in Dicktoo × Morex at the *ABG004-ABC174* interval. The *denso* locus is included in this interval, though the dwarfing allele was probably not present in this cross. In Galleon × Haruna Nijo, Langridge et al. (1996) detected QTL for β-amylase activity on this chromosome, not far from *denso*. These researchers also reported QTL for various enzymes in Clipper × Sahara, but in a more central region of the chromosome. Bezant et al. (1997) found a QTL for malt extract at the *denso* locus in Blenheim × Kym, with the favorable allele coming from Kym. They described this finding as surprising, as *denso* is present in all current UK spring malting barley varieties, and speculated about the possibility of adding positive malt extract alleles like this to a *denso* background.

It is fortunate that *denso*, being such an important locus for plant architecture, comes associated with mostly favorable malting characters. However, it can reduce kernel weight in some cases, as demonstrated in Derkado × B83-12/21/5. It would be interesting to elucidate if this association is due to linkage or pleiotropy. If due to linkage, it could be broken by repeated crossing.

Chromosome 4 (4H)

The analysis of Steptoe × Morex showed an abundance of malting quality QTL on chromosome 4 (4H). Ullrich et al. (1997) reported two malt extract QTL, at intervals *WG622-BCD402B* (BIN 1-3) and *ABG484-ABG472* (BIN 6-8). The first one, with the largest effect on malt extract, was in the same region as QTL for α-amylase activity, diastatic power, dormancy and malt β-glucan. In the same region, near *WG622*, there is a QTL for milling energy loss in Blenheim × E224/3. In Harrington × Morex, there are some QTL for malt extract, α-amylase activity, and soluble/total protein ratio in a neighboring region, with favorable alleles from Morex. In Dicktoo × Morex, Oziel et al. (1996) detected a QTL for grain protein content near *WG1026b* (BIN 5) but it is not clear whether it is in the same region. These similar results from several crosses point to the presence of one or more important loci, with effects on modification features. A second malt extract QTL in Steptoe × Morex coincides with a QTL for β-glucanase activity in malt (Han et al. 1995). Our analyses of Steptoe × Morex and Blenheim × E224/3 gave the same results.

There is a QTL for grain protein content in the same region (intervals spread over BINs 8 to 10, centered around BIN 9, markers *ABG319A* and *MWG655C*) in Steptoe × Morex and Harrington × TR306, but not in Harrington × Morex. This combination of results could mean that the TR306 allele confers the highest protein content, with the Harrington and Morex alleles being intermediate and equivalent, and the Steptoe allele giving the lowest grain protein content. Results from Bleinheim × Kym reinforce the existence of a QTL in this region (*psb37-MWG042*, BIN 9 or 10) affecting grain protein content, which in this case is also linked to a malt extract QTL. In Derkado × B83-12/21/5, there is also a QTL-rich region, between *mlo* (very close to *MWG042* in BIN 10), and *HVM67* (very close to Bmy1 in BIN 13), with effects on grain protein content, soluble/total protein ratio, viscosity, and the largest effect on malt extract. There are not enough markers to assess whether this area is the same region as in Blenheim × Kym.

In Harrington × Morex, there is another QTL affecting grain protein content, in the vicinity of the int-c locus. This locus is the second determinant of inflorescence type, after *Vrs1*, affecting the size, morphology and fertility of lateral florets. Interestingly, it also had the second largest effect largest effect on grain protein content (with the QTL at *Vrs1* having the largest effect). Thus, endosperm constitution is apparently rather dependent on the loci that control inflorescence type. Moreover, there was an epistatic interaction between *Vrs1* and *int-c*, (detectable using the epistasis feature of MQTL). At both loci, the Harrington allele was responsible for higher protein content, but the increase at the int-c region was only evident in the six-rowed subpopulation (i.e., with the Morex allele at *Vrs1*). At this same position, there are QTL for malt extract, soluble/total protein ratio and α-amylase activity, all with positive effects from the Morex allele; and for kernel weight only in the two-rowed sub-population, with a positive effect from the Harrington allele.

In Steptoe × Morex, there is a QTL for diastatic power near the β-amylase locus *Bmy1* on the 'minus' arm of chromosome 4(4H). The same QTL was detected in Dicktoo × Morex (Oziel et al. 1996). In both cases, the feed parent (Steptoe or Dicktoo) contributed the larger-value allele. In Calicuchima-sib × Bowman, Hayes et al. (1996) detected QTL for α-amylase activity and diastatic power coincident with *Bmy1*.

Chromosome 5 (1H)

There are apparently three QTL-rich regions on Chromosome 5 (1H). It is especially rich in malt extract QTL for Steptoe × Morex, Harrington × TR306, and Harrington × Morex. For malt extract, Hayes et al. (1993) reported five poorly resolved peaks for malt extract in Steptoe × Morex; their support intervals spanned the entire length of the chromosome. In the present analysis, four of these peaks exceed the significance threshold. Comparing the Steptoe × Morex scans to those for Harrington × Morex and Harrington × TR306, it appears that there are two regions with QTL for malt extract. One is on the 'plus' arm (around marker Ica1, BIN 6), with higher malt extract conferred by the Harrington allele in Harrington × Morex, and by the Morex allele in Steptoe × Morex. In Steptoe × Morex, a grain β-glucan QTL is also evident at this position. This last QTL was reported by Han et al. (1995), who also detected a large QTL for β-glucanase activity on BINs 9 to 10, near *Glb1*, a β-glucanase isoenzyme locus. Powell et al. (1997) reported QTL on chromosome 5 (1H) for milling energy, malt extract, and diastatic power, but only those for diastatic power are significant in our analyses. It is not possible to assign precise locations for these QTL.

The 'minus' arm of chromosome 5 (1H) also has a region with QTL effects on malt extract. There are large peaks for Harrington × Morex (*MWG943-MWG706*, BINs 8 to 12, with the highest peak in BIN 12) and Harrington × TR306 (*Dor3-MWG733A*, very close to *MWG706*, BINs 11 to 12, with peak in BIN 12). In Steptoe × Morex, this area has a small but significant peak (*His3B*, BIN 11). It is not known whether these peaks represent the same QTL in all three populations. If they do, it is interesting that the Harrington allele was more favorable than the Morex allele and the TR306 allele (coming from a feed-type parent) was even more favorable than the Harrington allele. Two QTL detected in Blenheim × Kym for grain protein and malt extract could be also ascribed to this region. They were 12 and 22 cM proximal to *MWG733*, which is in BIN 12.

The third interesting QTL region on this chromosome is near two genes (*Hor1* and *Hor2*, BIN 3 and 2, respectively) coding for hordein proteins. Hordeins are the main storage proteins of barley. There are no significant QTL in this region in Steptoe × Morex, Harrington × TR306 or Harrington × Morex, even though there was molecular polymorphism at the hordein loci in Steptoe × Morex and Harrington × TR306 (Hayes et al. 1993; Tinker et al. 1996). There were grain protein QTL in this region for Dicktoo × Morex (interval *Hor2-Hor1*) and nearby for Blenheim × Kym. Langridge et al. (1996) reported QTL at *Hor1*, one for α-amylase in Clipper × Sahara, and another for kernel weight in Chebec × Harrington.

Chromosome 6 (6H)

Chromosome 6 (6H) has few QTL for grain or malt quality, except in Harrington × TR306, which has a cluster of QTL around *Amy1*, on BIN 9. This region affected malt extract, grain protein content, diastatic power, viscosity and, above all, α-amylase activity. The lack of effect for α-amylase activity in other populations suggests lack of polymorphism at the *Amy1* locus in the other cross combinations. In Calicuchima-sib × Bowman and Dicktoo × Morex, QTL were detected for malt extract (*Ldh1-ABG458* interval) and grain protein content (*ABG387-WG223* interval), respectively, both in BINs 5 to 6. These two locations may be very close; the two intervals largely overlap. Bezant et al. (1997) found QTL for grain protein content (actually nitrogen content) scattered along BINs 3 to 7 and malt extract (BINs 3 to 4) in Blenheim × Kym. In Blenheim × E224/3, Powell et al. (1997), reported a QTL for milling energy loss.

Chromosome 7 (5H)

In some populations, chromosome 7 (5H) carries QTL with important effects on grain and malt quality. For Harrington × TR306, this chromosome has two 'hot-spots,' one at each end. One of these, (near marker *Act8B*, BIN 1) affected kernel weight and plumpness, malt extract, grain protein content, and malt β-glucan, with TR306 contributing toward higher kernel weight and plumpness; and Harrington contributing toward lower grain protein content, lower malt β-glucan and higher extract. Thus, selection for plump or heavy kernels could result in correlated responses toward lower extract and higher grain protein content. In contrast, this region has no significant QTL for grain and malt quality in Steptoe × Morex. In Harrington × Morex, QTL could not be mapped in this region due to a lack of polymorphic markers. For Blenheim ×

Kym, Bezant et al. (1997) reported a grain nitrogen QTL on the same end of the chromosome, but there are no common markers to verify its position relative to that of the QTL detected in Harrington × TR306.

In Steptoe × Morex, there are QTL peaks for grain protein content, soluble/total protein ratio and α-amylase activity near Ltp1 (BIN 5), but these are poorly resolved, and extend into BIN 9. Langridge et al. (1996) reported a QTL for α-amylase activity close to markers in BIN 9. In Dicktoo × Morex, this region exhibited QTL for most grain and malt quality traits, with peaks near *Lpt1* (BINs 4 to 6), and *BCD298-ABC302* (BIN 7) (Oziel et al. 1996).

In Blenheim × E224/3, Powell et al. (1997) and Thomas et al. (1996) detected numerous QTL on this chromosome. The most important was a milling energy QTL on the linkage group they named 7a, corresponding to the 'plus' arm. In Derkado × B83-12/21/5, there are QTL for kernel weight, milling energy, and malt extract. These are on the same chromosome arm as similar QTL in Harrington × TR306, but it is not possible to determine whether they are in the same region, due to lack of common markers.

Perhaps the most interesting region is at the end of the 'minus' arm of chromosome 7 (5H), near *MWG851B* (BIN 15). Here, the Harrington allele had very large effects on nearly all malting quality traits in Harrington × TR306. Some of these were also evident in Harrington × Morex, in which the Harrington allele substantially increased α-amylase activity and soluble/total protein ratio, and had a small effect on malt extract. It could be speculated that, among the Harrington × Morex progeny, the modification capacity is high enough that it is no longer a limiting factor for malt extract, and that malt extract is principally affected by other imbalances. Mather et al. (1997) speculated that this QTL region might affect a key step in the germination process (possibly water uptake or gibberellic acid synthesis) resulting in a cascade of effects through the malting process. Indeed, Oberthur et al. (1995) had previously reported a QTL for dormancy at a similar position (near *MWG851B*) in Steptoe × Morex. That QTL was hypostatic to another major dormancy QTL on the central part of the chromosome (*Ale-ABC309*). Furthermore, Langridge et al. (1996) reported a dormancy QTL in the same region in Chebec × Harrington. If seed dormancy were involved, differences in germination might be responsible for the malt quality QTL observed in Harrington × Morex and Harrington × TR306. However, since dormancy is transient and greatly affected by the environment, it is not possible to determine what role it may have played in the original Harrington × TR306 experiment. Data from more recent experiments show no indication of effects in this chromosome region for either germination or water sensitivity (reduction in germination in the presence of excess water). These experiments include one in which a QTL were mapped for germination speed and water sensitivity using new samples of the original Harrington × TR306 doubled haploid lines (Iwasa et al., 1999), and a validation study involving new Harrington × TR306 doubled haploid lines (Spaner et al. 1999; Igartua et al. 2000). In the latter, data on germination and water sensitivity (unpublished) indicate no differences between lines carrying the Harrington allele and lines carrying the TR306 allele.

DISCUSSION

The development of a new malting barley cultivar is a very elaborate technological feat. The new cultivar must conform to many requirements of growers, maltsters, and brewers. It is not surprising that there are many QTL affecting the traits reviewed in this report or that there are many complex interactions among traits. With many loci involved, it seems unlikely that the 'ideal' cultivar has been developed for any geographic region or end use. The potential for improvement is reinforced by repeated findings that parents with inferior grain and malt quality can be the source of at least some favorable QTL alleles for quality traits.

With the many fragments of information generated by mapping experiments that have been conducted to date, it might be possible to assemble a genotype carrying the most favorable combination of alleles from all sources, using marker-assisted breeding and/or genetic engineering techniques. In practice, it would be challenging to construct such a cultivar, and it would be impossible to accurately predict its performance, given our incomplete understanding of epistatic interactions, and our lack of knowledge about the genetic basis (linkage and/or pleiotropy) for specific interrelationships among traits.

Fortunately, as breeders well know, it is not necessary to develop an 'ideal' cultivar, only something that is better than current cultivars. That is more an achievable objective, and could be facilitated by using the kind of information reviewed here. Some mapping populations are of direct interest as breeding populations, while for others QTL information must be extrapolated and used with new, unmapped parents. Despite the low repeatability of results across populations that is evident in the comparative analysis presented here, it is possible to identify regions of the genome that show consistent effects over traits, and could be targeted through marker-assisted selection. Marker-assisted improvement of a few QTL regions, or even just of one QTL region, could be worthwhile.

Finally, several of the practical challenges faced by breeders when developing improved malting cultivars can be reviewed under the light of the information presented in this article. These include:

Maintaining or increasing kernel weight and plumpness while improving malt quality: This objective seemed very difficult for some crosses, but achievable for others. Tight associations between increased kernel weight and unfavorable traits were detected in Harrington × TR306, Harrington × Morex, and Derkado × B83-12/21/5. In Harrington × TR306, TR306 contributed most of the high kernel weight alleles. The one with largest effect, on chromosome 7 (5H), and a smaller one on chromosome 1 (7H), coincided in position (and allelic phase) with grain protein QTL, providing an explanation for the observed positive correlation between kernel weight and grain protein content. In Harrington × Morex, Vrs1 on chromosome 2 (2H) seemed to be the main factor underlying the correlation between kernel weight and grain protein content. In Derkado × B83-12/21/5, the B83-12/21/5 allele that conferred larger weight in the denso region on chromosome 3(3H) also conferred higher grain protein. Even if the associations in these three populations were due to linkage, it would be very costly to break them. In contrast, kernel weight and grain protein content seemed to have a more independent inheritance in both Steptoe × Morex and Blenheim × E224/3.

Keeping grain protein content low in two-rowed by six-rowed crosses: The results of the mapping experiments conducted with two-rowed by six-rowed crosses illustrate the need for caution when selections are to be made from wide crosses. Further work is needed to understand the epistatic interactions and negative associations among traits. Both parents of Harrington × Morex had very good grain and malt quality characteristics, and one might have expected the progeny to exhibit transgressive segregation, perhaps resulting in some progeny with overall superiority to either parent. Positive transgressive segregation was found for some traits, but the negative transgressive segregation for grain protein content appeared to dominate, resulting in low malt extract. Furthermore, there was an epistatic interaction between *Vrs1* and *int-c* that gave very high grain protein when Harrington alleles were present at each both loci. Selection of a Harrington × Morex line with an acceptable quality profile would require very careful control of grain protein levels. This might be achieved with the aid of marker-assisted selection. For selection of a two-rowed line, it would be necessary to recover Harrington alleles in QTL regions on chromosomes 1(7H), 2(2H), and 7(5H), to counteract the increase in grain protein content due to the Harrington allele(s) in the Vrs1 region. For a six-rowed line, the Morex allele at *int-c* should be incorporated, because the combination of the Harrington allele at *int-c* with the Morex allele at *Vrs1* produces a large increase in grain protein. Clearly, these approaches would not be practical, compared to crossing and selecting within germplasm groups, unless there were other advantages to be gained (e.g., agronomic advantages and/or disease resistance) from by combining genes from different germplasm groups.

To decrease grain protein content without losing diastatic power: The positive correlation between diastatic power and grain protein content has been established in a number of works (summarized by Peterson and Foster 1973). The nature of this correlation is partially environmental, but some works have found a genetic correlation between these two traits (Rasmusson and Glass 1965; Smith 1990; Goblirsch et al. 1996). This seems to be the case for the progenies of the widely used low protein cultivar Karl. This six-rowed cultivar has been repeatedly used as a parent in the US midwest, but failed to produce new releases, as the progenies with low grain protein had unacceptable dark kernel color, low diastatic power, or both (Goblirsch et al. 1996). In two-rowed by six-rowed crosses, the association between high grain protein and high diastatic power has been observed in several crosses attempted by Swanston (1983) and was evident in Harrington × Morex. In Harrington × Morex, the correlation observed was a consequence of the prevalent effect of the *Vrs1* region. The Harrington (two-rowed) allele(s) in that region increased both grain protein content and diastatic power. In a study with isolines involving variation only at the *Vrs1*, *int-c*, and the lax-dense spike locus, McGuire and Hockett (1983) reported effects that were very similar to those found in Harrington × Morex. The two-rowed allele increased grain protein content and diastatic power, and there was an interaction between *Vrs1* and *int-c* for the two traits. In crosses such as Harrington × Morex, it would be difficult to successfully select for both grain protein content and diastatic power. However, the information reviewed here suggests that the correlation between these two traits is population-dependent. No common QTL for the two traits were found in Harrington × TR306, Steptoe × Morex, and few were found in Blenheim × E224/3.

To decrease wort β-glucan without an excessive increase in soluble/total protein: In both Harrington × TR306 and Harrington × Morex, there were two main QTL for malt β-glucan. In each population, one of these was coincident with a soluble/total protein QTL, but the other one was independent. When they were coincident, the allele that decreased β-glucan increased soluble/total protein. Thus there should be at least some room to manipulate both traits in the desired direction. For Steptoe × Morex, the two traits seem even more independent; Han et al. (1995) found six QTL for malt β-glucan and only one of them was close to a QTL for soluble/total protein ratio.

REFERENCES

Allison, M.J., I.A. Cowe, and R. McHale. 1976. A rapid test for the prediction of malting quality of barley. *J. Heredity* 79:225:238.

Bamforth, C.W. and A.H.P. Barclay. 1993. Malting technology and the uses of malt. In A.W. MacGregor and R.S. Bhatty (eds.) *Barley: Chemistry and Technology.* Am. Assoc. Cereal Chemists, Inc., St. Paul, Minnesota, pp. 297-354.

Bezant J.H., D.A. Laurie, N. Pratchett, J. Chojecki, and M.J. Kearsey. 1997. Mapping of QTL controlling NIR predicted hot water extract and grain nitrogen content in a spring barley cross using marker-regression. *Plant Breeding* 116:141-145.

Burger, W.C. and D.E. LaBerge. 1985. Malting and brewing quality in barley. *In* D.C. Rasmusson (ed.) *Barley. Agron. Monograph #26,* ASA, CSSA, and SSSA, Madison, WI, pp. 367-401.

Chalmers, K.J., U.M. Barua, C.A. Hackett, W.T.B. Thomas, and R. Waugh. 1993. Identification of RAPD markers linked to genetic factors controlling the milling energy requirement of barley. *Theor. Appl. Genet.* 87:314-320.

Day, A.D., E.E. Down, and K.J. Frey. 1955. Association between diastatic power and certain visible characteristics and heritability of diastatic power in barley. *Agron. J.* 47:163-165.

Dickson, A.D. 1979. Barley for malting and food. In *Barley. Agriculture Handbook 338,* US Department of Agriculture, Washington, pp. 136-246.

Eagles, H.A., A.G. Bedgood, J.F. Panozzo, and P.J. Martin. 1995. Cultivar and environmental effects on malting quality of barley. *Aus. J. Agric. Res.* 46:831-844.

Edney, M.J. 1999. Canadian malting barley varieties for the future. *Proceedings of the Canadian Barley Symposium,* February 23-25, Winnipeg, Manitoba, Canada, pp. 113-119. <http://www.cgc.ca/Pubs/Barleysymp99/proceedings.pdf>.

Fincher, G.B. and B.A. Stone. 1993. Physiology and biochemistry of germination in barley. In A.W. MacGregor and R.S. Bhatty (eds.) *Barley: Chemistry and Technology.* Am. Assoc. Cereal Chemists, Inc., St. Paul, Minnesota, pp. 247-295.

Goblirsch, C.A., R.D. Horsley, and P.B. Schwarz. 1996. A strategy to breed low-protein barley with acceptable kernel color and diastatic power. *Crop Sci.* 36:41-44.

Han, F., S.E. Ullrich, S. Chirar, S. Menteur, L. Jestin, A. Sarrafi, P.M. Hayes, B.L. Jones, T.K. Blake, D.M. Wesemberg, A. Kleinhofs, and A. Kilian. 1995. Mapping of β-glucan content and β-glucanase activity loci in barley grain and malt. *Theor. Appl. Genet.* 91:921-927.

Hayes, P.M., B.H. Liu, S.J. Knapp, F. Chen, B. Jones, T. Blake, J. Franckowiak, D. Rasmusson, M. Sorrells, S.E. Ullrich, D. Wesenberg, and A. Kleinhofs. 1993. Quantitative trait locus effects and environmental interaction in a sample of North American barley germplasm. *Theor. Appl. Genet.* 87:392-401.

Hayes, P.M., D. Prehn, H. Vivar, T. Blake, A. Comeau, I. Henry, M. Johnston, B. Jones, B. Steffenson, C.A. St. Pierre, and F. Chen. 1996. Multiple disease resistance loci and their relationship to agronomic and quality loci in a spring barley population. *J. Quantitative Trait Loci*, 2(2) <http://www.ncgr.org/ag/jag>.

Igartua, E., M. Edney, B. Rossnagel, D. Spaner, W.G. Legge, G.J. Scoles G.J., P.E. Eckstein, G.A. Penner, N.A. Tinker, K.G. Briggs, D.E. Falk, and D.E. Mather. 2000. Marker-based selection of QTL affecting grain and malt quality in two-row barley. *Crop Sci* 40: (in press).

Iwasa, T., H. Takahashi, and K. Takeda. 1999. QTL mapping for water sensitivity in barley seeds. *Bull Res. Inst. Bioresour. Okayama Univ.* 6:21-28.

Kjaer, B. and J. Jensen. 1996. Quantitative trait loci for grain yield and yield components in a cross between a six-rowed and a two-rowed barley. *Euphytica* 90: 39-48.

Kleinhofs, A. 1999. Description and chromosome BIN location of barley genes and markers. <http://barleygenomics.wsu.edu/>.

Langridge, P., A. Karakousis, J. Krestchmer, S. Manning, R. Boyd, C.D. Li, S. Logue, R. Lance, and SARDI. 1996. <http://greengenes.cit.cornell.edu/WaiteQTL>.

Li, C.D., X.Q. Zhang, P. Eckstein, B.G. Rossnagel, and G. Scoles. 1999. A polymorphic microsatellite in the limit dextranase gene of barley (*Hordeum vulgare* L.). *Molec. Breeding* 5:569-577.

MacGregor, A.W., L.J. Macri, S.W. Schroeder, and S.L. Bazin. 1994. Purification and characterisation of limit dextranase inhibitors from barley. *J. Cereal Sci.* 20:33-41.

MacGregor, A.W. 1999. Improving malting quality through biotechnology. *Proceedings of the Canadian Barley Symposium*, February 23-25, Winnipeg, Manitoba, Canada, pp. 41-46. <http://www.cgc.ca/Pubs/Barleysymp99/proceedings.pdf>.

Márquez-Cedillo, L.A., P.M. Hayes, B.L. Jones, A. Kleinhofs, W.G. Legge, B.G. Rossnagel, K. Sato, S.E. Ullrich, D.M. Wesenberg, and the NABGMP. 2000. QTL analysis of malting quality in barley based on the doubled haploid progeny of two elite North American varieties representing different germplasm groups. *Theor. Appl. Genet.* (in press).

Mather D.E., N.A. Tinker, D.E. LaBerge, M. Edney, B.L. Jones, B.G. Rossnagel, W.G. Legge, K.G. Briggs, R.B. Irvine, D.E. Falk, and K.J. Kasha. 1997. Regions of the genome that affect grain and malt quality in a North American two-row barley cross. *Crop Sci.* 37:554-554.

McGuire, C.F. and E.A. Hockett. 1983. Relationship of V, I, and L alleles with malting quality of Bonneville barley. *Field Crops Res.* 7:51-60.

Mitsui, T. and K. Itoh. 1997. The alpha-amylase multigene family. *Trends in Plant Science*, 2:255-261.

Oberthur, L., T.K. Blake, W.E. Dyer, and S.E. Ullrich. 1995. Genetic analysis of seed dormancy in barley (*Hordeum vulgare* L.). *J. Quantitative Trait Loci*, <http://www. ncgr.org/ag/jag>.

Oziel, A., P.M. Hayes, F.Q. Chen, and B. Jones. 1996. Application of quantitative trait locus mapping to the development of winter habit malting barley. *Plant Breeding*. 115:43-51.

Peterson, G.A. and A.E. Foster. 1973. Malting barley in the United States. *Advances in Agronomy*, 25: 327-378.

Powell, W., R.P. Ellis, and W.T.B. Thomas. 1990. The effects of major genes on quantitatively varying characters in barley. III. The two row/six row locus (V-v). *Heredity*, 65:259-264.

Powell, W., W.T.B. Thomas, D.M. Thomson, J.S. Swanston, and R. Waugh. 1992. Association between sDNA alleles and quantitative traits in doubled haploid populations of barley. *Genetics* 130:187-194.

Powell, W., W.T.B. Thomas, E. Baird, P. Lawrence, A. Booth, B. Harrower, J.W. McNicol, and R. Waugh. 1997. Analysis of quantitative traits in barley by the use of amplified fragment length polymorphisms. *Heredity* 79:48-59.

Rasmusson, D.C. and R.L. Glass. 1965. Effectiveness of early generation selection for four quality characters in barley. *Crop Sci.* 5:389-391.

Smith, D.B. 1990. Barley seed protein and its effects on malting and brewing quality. *Plant Varieties and Seeds* 3:63-80.

Spaner, D., B.G. Rossnagel, W.G. Legge, G.J. Scoles, P.E. Eckstein, G.A. Penner, N.A. Tinker, K.G. Briggs, D.E. Falk, J.C. Afele, P.M. Hayes, and D.E. Mather. 1999. Verification of a quantitative trait locus affecting agronomic traits in two-row barley. *Crop Sci.* 39:248-252.

Swanston, J.S. 1983. An alternative approach to breeding for high diastatic power in barley. *Euphytica* 32:919-924.

Swanston, J.S. 1990. The use of milling energy to predict increases in hot water extract in response to the addition of gibberellic acid during steeping. *J. Inst. Brew.* 96:209-212.

Takahashi, R., J. Hayashi, and I. Moriya. 1976. Basic studies on breeding barley by the use of two-rowed and six-rowed varietal crosses. *Proc. Third Int. Barley Genetics Symp.*, Munich, pp. 662-667.

Thomas, W.T.B., W. Powell, and J.S. Swanston. 1991. The effects of major genes on quantitatively varying characters in barley. IV. The GPert and denso loci and quality characters. *Heredity* 66:381-389.

Thomas, W.T.B., W. Powell, J.S. Swanston, R.P. Ellis, K.J. Chalmers, U.M. Barua, P. Jack, V. Lea, B.P. Forster, R. Waugh, and D.B. Smith. 1996. Quantitative trait loci for germination and malting quality characters in a spring barley cross. *Crop. Sci.* 36: 265-273.

Tinker, N.A., and D.E. Mather. 1995. Methods for QTL analysis in progeny replicated in multiple environments. *J. Quantitative Trait Loci*, <http://www.ncgr.org/ag/jag>.

Tinker, N.A., D.E. Mather, T.K. Blake, K.G. Briggs, T.M. Choo, L. Dahleen, S.M. Dofing, D.E. Falk, T. Ferguson, J.D. Franckowiak, R. Graf, P.M. Hayes, D. Hoffman, R.B. Irvine, A. Kleinhofs, W. Legge, B.G. Rossnagel, M.A. Saghai

Maroof, G.J. Scoles, L.P. Shugar, B. Steffenson, S. Ullrich, and K.J. Kasha. 1996. Loci that affect agronomic performance in two-row barley. *Crop Sci.* 36:1053-1062.

Ullrich, S.E., F. Han, and B.L. Jones. 1997. Genetic complexity of the malt extract trait in barley suggested by QTL analysis. *J. Am. Soc. Brew. Chem.* 55:1-4.

Wainwright, T. and G.K. Buckee. 1977. Barley and malt analysis-a review. *J. Inst. Brew.* 83:325-347.

Zale, J.M., J.A. Clancy, S.E. Ullrich, B.L. Jones, and P.M. Hayes. 2000. *Barley Genetics Newsletter*, 30. <http://wheat.pw.usda.gov/ggpages/bgn/30>.

Zwickert-Menteur, S., L. Jestin, and G. Branland. 1996. Amy2 polymorphism as a possible marker of β-glucanase activity in barley (*Hordeum vulgare* L.). *J. Cereal. Sci.* 24:55-63.

Quality Improvement in Oat

Douglas C. Doehlert

SUMMARY. Oats (*Avena sativa* L.) are a nutritious, high protein grain crop with important food, feed and value-added applications. For commercial purposes, oat quality is frequently expressed as a grade, based on test weight, foreign matter, and the physical appearance of the grain. More detailed quality analyses may include evaluations for percent groat, kernel size and uniformity, and groat composition. Compositional components of economic importance include protein, fat, β-glucan and antioxidant concentrations. Exact quality specification for quality requirements can vary widely among applications. Feed applications favor higher protein and fat concentrations, and lower fiber. Food applications favor lower fat concentrations, and higher β-glucan. Production of high quality oat grain is dependent on the planting of high quality seed with current local adaptation, sound cultural practices, and favorable weather conditions. It is particularly important that seed planted produce plants resistant to current races of pathogens. Disease infestation is a common source of quality loss. Whereas, unfavorable environmental conditions can frequently result in quality loss, breeders continue to strive for environmental stability of quality traits through genetic improvement of oats. *[Article copies available for a fee from The Haworth Document Delivery Service: 1-800-HAWORTH. E-mail address: <getinfo@haworthpressinc.com> Website: <http://www.HaworthPress. com>]*

KEYWORDS. Antioxidants, *Avena*, β-glucan, grain protein, grain fat, groat composition, groat percentage, kernel size uniformity, test weight

Douglas C. Doehlert is affiliated with the USDA-ARS Wheat Quality Laboratory, Harris Hall, North Dakota State University, Fargo, ND 58105 USA (E-mail: doehlert@ plains.nodak.edu).

[Haworth co-indexing entry note]: "Quality Improvement in Oat." Doehlert, Douglas C. Co-published simultaneously in *Journal of Crop Production* (Food Products Press, an imprint of The Haworth Press, Inc.) Vol. 5, No. 1/2 (#9/10), 2002, pp. 165-189; and: *Quality Improvement in Field Crops* (ed: A. S. Basra, and L. S. Randhawa) Food Products Press, an imprint of The Haworth Press, Inc., 2002, pp. 165-189. Single or multiple copies of this article are available for a fee from The Haworth Document Delivery Service [1-800-HAWORTH, 9:00 a.m. - 5:00 p.m. (EST). E-mail address: getinfo@haworthpressinc.com].

Improved oat (*Avena sativa* L.) grain quality benefits the producer as well as the processor, by improving the value of a crop being produced and improving the value of products generated from the grain. As a Scottish proverb puts it "What's good to get is good to give." But a discussion of quality improvement in oat requires first a definition of quality. As a rule, requirements for quality vary with the end use of the crop. Because oat is used both as forage and as a grain, quality requirements can vary widely. This review will deal only with grain quality. Diverse applications of the grain to different end uses can sometimes dictate divergent definitions of quality (Lea, 1955). Thus, a brief discussion of end-uses of oat grain will precede the improvement of oat grain quality characteristics.

OAT GRAIN APPLICATIONS

Animal Feed: Over 70% of the oat grain harvested in the world is fed to animals. Oats are a good source of protein, fiber and minerals. It contains more protein per kilogram than maize, but has fewer calories. However, oat groats, or hull-less oats contain more metabolizable energy per kilogram than corn. Oats are more commonly fed to animals in colder climates where corn and soybeans cannot be grown as well. The high quality protein and fat content make oats desirable for feed, especially for horses, mules, dairy cattle, and turkeys, and to lesser extents to hogs, beef cattle, and sheep. The presence of the hull represents a major negative feature contributing to lower energy of oats as feed (Hoffman and Livezey 1987; Cuddeford 1995; Hoffman 1995). Therefore, oats are less desirable as a feed for finishing and fattening animals.

Oats are commonly fed on the same farms where they are grown, so grain quality frequently does not become a market issue. Oats that cannot be sold for milling are likely to be used as animal feed. Nevertheless, high test weight and high protein are both important factors in determining oat feed quality. High fat and lower beta-glucan concentrations are usually more desirable for animal feed, because of improved energy content. Similarly, high groat percentage is desirable for feed because of the low energy content of the largely indigestible hull, although hulls are also used in some feed formulations (Eastham and Baker 1922; Lathrop and Bohstedt 1938). Unlike many other applications, hull color is frequently important for oats bound for horse feed. Hobby and racehorse owners are frequently willing to pay high premiums for white oats with large kernels. This bias may be largely aesthetic, or may be related to an improved ability to detect fungal infestations of white oats, which may be detrimental to highly prized animals (Hoffmann and Livezey 1987; Forsberg and Reeves 1992).

Food Uses: Human consumption remains an important use for oats produced in the world. The high nutritional value of oat protein and the health-benefiting properties of its soluble fiber have been two important factors contributing to the continuing popularity of oats as a food. In Western countries, oats are consumed primarily as oatmeal or ready-to-eat cereals, but are also consumed as oat flour, meat product extenders, granola, baby food, oat bran, and snack foods (Hoffmann and Livezey 1987; Ranhotra and Gelroth 1995). Oats are widely consumed in China as noodles or in Chinese dumplings, which consist of oat flour dough, stuffed with vegetables and then steamed (Zhang 1997).

Millers processing oats for human consumption generally require a high test weight, and prefer grain with higher groat percentages, larger groats, and uniform size. Some food applications have minimum protein and/or β-glucan requirements, and maximal fat contents, generally to meet labeling requirements.

Value Added Applications: Many current and developing applications for oats and oat products involve further processing for non-traditional products (Webster 1986). Oat starch and protein are isolated for use in cosmetics, β-glucan is isolated as a food supplement, antioxidants are purified as food ingredients, and polar lipids are purified for pharmaceutical and food applications. Oats bound for these applications may have additional quality requirements above those of oats bound for traditional food and feed applications. For example, a processor may want only oat cultivars containing higher than normal concentrations of the target compounds. Special cultivars may be developed especially for specific value-added applications.

QUALITY CHARACTERISTICS

Grading: Quality of commercial oats is frequently expressed as a grade, such as those described by the United States Grains Standard Act (USDA 1978). Grades are determined by minimum limits on test weight and sound oats, and maximum limits on heat damaged kernels, foreign material, and wild oats. Special grades are assigned to bleached, bright, smutty, ergoty, garlicky, thin, heavy, and extra-heavy oats.

In general, oats bound for commercial milling should not contain discolored, weather stained or unsound kernels. Oats should not contain more than 15% moisture before processing and should have a thousand kernel weight of more than 27 g. They should have test weights more than 53 kg/hl, groat percentages of more than 74%, and no more than 10% of the grain should pass through a 2-mm slotted sieve. They should contain no more than 0.8% double oats, no more than 1% foreign matter, and no more than 3% wild oats (Ganssmann and Vorwerck 1995). Different processors may have differing requirements.

Test Weight: Test weight is the mass of a given volume of grain and represents the density of the packed grain. It is the most universally used measure for oat quality. Higher test weight translates to higher quality. The popularity of this test is largely due to the ease of the measurement, the effectiveness of the test in predicting groat percentage and milling yield, and in detecting grain damaged by adverse environmental conditions, disease problems or by poor cultural practices. Test weight also provides a definitive value by which the volume required to store or ship a given mass of oats can be calculated. Some weaknesses of test weight as a quality measure also have been pointed out. Greig and Findlay (1907) suggested little relation between test weight and milling yield. Zavitz (1927) suggested that test weight was more influenced by kernel shape than by meaningful quality characteristics. Sword (1948-49) as well as Peek and Poehlman (1949) indicated that certain cultivars with poorer test weights proved to have better quality by milling yield tests. They acknowledged that oats with high test weights were very unlikely to be poor in quality, but they added an appropriate caveat that test weight is only a guide, not an accurate index. Nevertheless, test weight is frequently the primary, if not the sole criterion for determining the quality of oat grain (Stoa 1922; Sword 1948-49; Hoffmann and Livezey 1987). Even though breeders have consistently selected for high test weight oats, production of high test weight oats is frequently elusive. This discussion will evaluate components comprising test weight and review reports indicating factors affecting test weigh in crop production.

Because test weight is the mass of a given volume of grain, it is affected by both the *kernel density* and by a *packing factor*. Kernel density appears to be controlled largely by groat percentage, groat density and the tightness of the hull wrapping around the groat. Essentially the groat is more dense than the hull, so the greater proportion of the oat mass and volume which is groat, the greater the kernel density. Numerous studies have indicated positive correlations between test weight and groat percentage (Stoa et al. 1936; Atkins 1943; Bartley and Weiss 1951; Pomeranz et al. 1979; Doehlert et al. 1999). A regression analysis presented by Pomeranz et al. (1979) indicated that less that half of the variation in test weight could be accounted for by variation in groat percentage. Results presented by Doehlert and McMullen (2000) indicated that groat percentage, groat starch concentration and groat weight accounted for nearly 70% of the variation in test weight in oats from six environments. Although a suggested relationship between groat composition and test weight might indicate variation in groat density to influence test weight, analyses by Murphy and Frey (1962) measuring groat density by acetone displacement suggested that groat density was fairly constant (1.36 g mL^{-1}) among 27 genotypes tested. Root (1977) found greater variation in groat density, reporting values from 1.08 to 1.56 g mL^{-1}. Root concluded that groat density was not entirely constant and could influence test weight.

The kernel shape and size determine the packing factor. A number of investigators have reported relationships between oat kernel shape and test weight (Love 1914; Zavitz 1927; Barbee 1935; Brηckner 1956; Root 1979; Forsberg and Reeves 1985; Doehlert et al. 1999). These reports largely concluded that longer kernels generated lower test weights, whereas shorter plumper kernels generated higher test weights, largely because of the way they pack. Longer kernels allowed for more air spaces between kernels and resulted in lower test weights. Cutler (1940) illustrated this concept by showing that test weight of a given oat sample could be increased 20 to 45% by clipping off the tips of oat kernels, by mechanically rubbing and polishing the oat kernels. Also, plumper kernels represent grains that have filled out better, producing a denser kernel (Love 1914; Forsberg and Reeves 1985). Correlation analyses (Table 1) derived from data presented by Doehlert et al. (1999) also indicated a negative correlation between test weight and kernel length, but failed to show a significant correlation between test weight and kernel width. Test weight was highly correlated with groat percentage, whole oat area uniformity, and several components derived from digital image analysis. Doehlert et al. (1999) also attempted to derive an estimation of kernel density independent of any packing factor from digital image data. The mass of the sample of oats used for the image was divided by the number of kernels in the sample to obtain a mean oat mass, equivalent to a thousand-kernel weight. This mean kernel mass was divided by the mean area of the whole oat images derived from image analysis. Although this value was correlated with both test weight and

TABLE 1. Physical oat kernel characteristics correlated with test weight. Oat and groat dimensional information determined from digital image analysis. Recalculated from results obtained in Doehlert et al. (1999).

Factor	Correlation with test weight
	r
Groat Percentage	0.742**
1000-Kernel weight	0.363
Whole oat image area	−0.302
Whole oat image area CV[1]	−0.626**
Oat mass:area ratio	0.809**
Whole oat length	−0.408*
Whole oat width	−0.049
Groat image area	0.161
Groat image area CV[1]	−0.238
Groat length	0.373*
Groat width	−0.108
Groat mass:area ratio	0.442*
Groat:oat area ratio	0.735**

[1]CV = coefficient of variation, an estimation of size uniformity.

groat percentage, test weight was more highly correlated with groat percentage (Doehlert et al. 1999).

Environmental factors are well known to affect oat grain test weight (Zavitz 1927; Stoa et al. 1936; Atkins 1943; Sappenfield and Poehlman 1952; Frey and Wiggans 1956; Mackey 1959; Ohm 1976). Cultural factors, such as early planting, higher seeding rates and lower levels of nitrogen fertilization have been documented to improve test weight. Environmental stress originating from high temperatures, drought, or lodging are also known to reduce test weight of harvested grain. Murphy et al. (1940) and Atkins (1943) have documented reductions in test weight with crown rust infection, indicating the influence of biotic stress. Doehlert, McMullen and Hammond (unpublished manuscript) also indicated significant test weight decreases in environments heavily infested with crown rust. Regression analyses suggested that warm spring temperatures to facilitate early planting, good rains during the vegetative growth stages, and reduced rain during the grain filling period to reduce the risk of crown rust infection, were the most important environmental factors contributing to high test weight of oat grain. Gullord and Aastveit (1987) reported significant genotype by environment interactions affecting test weight, indicating differences among genotypes in their environmental responses affecting test weight. Such interactions offer challenges to breeders striving to generate cultivars that will produce high test weight grain in many environments. However, test weight is a highly heritable trait (Pawlisch and Shands 1962; Wesenberg and Shands 1973) and remains a top priority with most oat breeding programs, along with grain yield, straw strength, and disease resistance.

Groat Percentage: The oat groat or caryopsis is typically covered with a lemma and palea, or hull, after harvest. This hull must be removed before the groat can be processed for food purposes. The proportion of the groat (the oat caryopsis without the hull) to the whole oat is known as the groat percentage. The groat percentage is important because it represents the economic yield an oat miller can derive from a given lot of oat grain. It also provides an estimate of the digestible portion of the grain, if being fed to animals.

Groat percentage, also referred to as groat proportion, caryopsis percentage, and kernel percentage or reported as hull percentage, has long been recognized as an important indicator of oat quality (Love et al. 1925; Zavitz 1927; Stoa et al. 1936; Atkins 1943; Bartley and Weiss 1951). In contrast, Peek and Poehlman (1949) considered test weight to be a more valuable oat quality evaluation tool because hand dehulling of oats to measure groat percentage was considered too tedious. Zavitz (1927) considered groat percentage more valuable than test weight for evaluating oat quality because it better reflects feeding value. Stoa et al. (1936) suggested that early maturing oats were superior in groat percentage, and rust susceptible lines were generally high in percent hull. These conclusions were also supported by the findings of Bunch and Forsberg (1989). Sword (1948-49) considered groat percentage to be of primary importance in evaluating oat quality because of the low value of oat hulls and because a high groat percentage is the first essential in milling yield. Ganssmann and Vorwerck (1995) also provided evidence indicating the primary importance of groat percentage to milling yield. Groat percentage is a quantitatively inherited trait with a broad sense heritability of 36 to 92% (Wesenberg and Shands 1973; Ronald et al. 1999). Stuthman and Granger (1977) found a narrow sense heritability of 34 to 72%.

The studies of Bartley and Weiss (1951) indicated strong environmental effects on groat percentage and demonstrated correlations between groat percentage and yield, test weight, and kernel weight. A study by Hutchinson et al. (1952) concluded that environment had a much stronger effect on groat percentage than did genotype, because of the large variation observed with single cultivars. Another environmen-

tal study (Doehlert, McMullen and Hammond, unpublished) indicated
that improved groat percentages were associated with oat crops sown
earlier in the spring, and with cooler temperatures during the grain filling period. Uniform abundant rainfalls, warmer spring temperatures
and higher levels of solar radiation were also associated with improved
groat percentages.

Numerous studies have demonstrated that the primary kernel of the
multi-floret oat spikelet (Figure 1) is about 4 to 6% lower in groat percentage than the smaller secondary or tertiary kernels (Zade 1915;
Villers 1935; Atkins 1943; Hutchinson et al. 1952; Bartley and Weiss
1951; Youngs and Shands 1974). However, Palagyi (1983) found that
genotypes with higher frequencies of tertiary kernels had lower groat
percentage, and suggested that tertiary kernels compete with primary
and secondary kernels for assimilate, preventing them from filling

FIGURE 1. Drawing of oat spikelet showing the multifloret habit of the spikelet
and relative size differences between the primary, secondary, and tertiary florets.

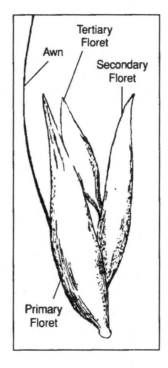

properly. Souza and Sorrells (1988) attempted to select for higher groat percentage by means of mass selecting by kernel density, using a gravity table. They concluded that selection against the presence of tertiary kernels would be most effective for improving milling quality.

A number of different methods are commonly used to evaluate groat percentage on an experimental basis. The most basic of these is hand dehulling, where oat grains are stripped of their hulls by hand. Mechanical methods of oat dehulling include the impact dehuller (Cleve 1948; Stuke 1955; Ganssmann and Vorwerck 1995), the compressed air dehuller (Fraser 1944; Kittlitz and Vetterer 1972) and the wringer type dehuller (Love and Craig 1944). The impact dehuller feeds grain into a spinning rotor. The impact of the grain with wall as it is expelled from the rotor knocks the hull from the groat. The hulls are then removed by aspiration. The compressed air dehuller uses a stream of pressurized air to apply mechanical shock to a sample of oat grain, which knocks the hull off the groat. Hulls are removed by aspiration. Compressed air dehullers have only recently been marketed commercially and are just starting to become widely used among oat research laboratories. In wringer-type dehullers, oat kernels are fed by hand into a roller, which exudes the hull. Hand dehulling is thought to provide a ceiling value for groat percentage, because no groats are lost in mechanical treatments. However, mechanical dehullers, especially impact and compressed air type dehullers, allow for the dehulling of larger samples, which is impractical by hand dehulling. Impact dehullers are commonly used by commercial millers to dehull oat, although an additional method, known as stone hulling, is used by some commercial milling operations (Ganssmann and Vorwerck 1995).

A recent study from Doehlert et al. (1999) compared groat percentage values as obtained by hand dehulling, impact dehulling and compressed air dehulling. The results indicated that dehulling conditions could have a profound effect on groat percentage data gathered. In mechanical dehulling, insufficient mechanical stress resulted in incomplete dehulling, whereas excessive mechanical stress resulted in groat breakage. Excessive aspiration, required to remove hulls from the groats, can remove groats as well as hulls. Insufficient aspiration left excess hulls with groats, resulting in an overestimation of groat percentage. Groat percentages obtained by the various methods correlated well, although absolute values differed. Additional processing to remove excess hulls from the groat preparation improved the correlation of mechanical dehulling with hand dehulling.

Milling Yield: The milling yield is an index used by oat millers to indicate the mass of whole oats required to produce 100 kg (or pounds) of finished product. It is frequently expressed in mass units, or can be expressed as a percentage. Groat percentage is the most important factor affecting milling yield (Greig and Findlay 1907; Salisbury and Wicher 1971; Ganssmann and Vorwerck 1995). Milling yield is also strongly affected by the presence of foreign materials including mixed grains in with the oats and by the presence of excessive amounts of unmillable oats. Unmillable oats include thin oats (less than 2 mm wide), light oats (removed by aspiration), and double, or bosom oats (Salisbury and Wichser 1971; Ganssmann and Vorwerck 1995). These are removed prior to milling largely through size sorting. An additional factor affecting mill yield is the largely mill specific factor controlling loss of fines or middlings (Salisbury and Wichser 1971). Although the mill can sell most by-products from oatmeal milling for animal feed, by-products do not command a premium price as does the oat flakes.

Kernel Size: Oat kernel size can be evaluated by mass (thousand kernel weight), or by length, width, depth and/or volume. Villers (1955) presented excellent definitions of oat length, width and depth. Although Brηckner et al. (1956) indicated that the oat kernel width to depth ratio was fairly constant at 1.25, it should be noted that in studies separating kernel sizes by slotted sieves, oat kernels are separated by their depth. In contrast, studies analyzing oat kernel size by digital image analysis use width, not depth, as the differentiating feature. Such specifics are particularly important if comparing values obtained by different methods.

As described earlier, oat kernel size and shape are known to affect test weight. Longer kernels tend to reduce test weight, whereas plumper kernels tend to increase test weight, because of their influence on the packing factor. As might be expected, oat mass is highly correlated with oat length and less strongly with oat width (Doehlert et al. 1999).

Some food applications require specific sizing of oat groats. Frequently, larger groats are preferred. Larger groats can be processed into larger oat flakes, such as those required for old-fashioned oatmeal. Horse feed application also demand larger oats for their premium market, although there is little nutritional difference among different sized oats. Size uniformity is probably more important to the general oat milling industry.

Oat kernels have inherent non-uniformity in their size because of the multi-floret habit of the oat spikelet (Figure 1). Oat kernels derived from primary florets are larger in size than secondary kernels, which are larger than tertiary kernels, if present. As a result, kernel size in oats usually represents a multimodal population (Symons and Fulcher 1988).

Oat dehulling is usually most efficient with a uniform sized kernel sample (Brηckner 1953; Ganssmann and Vorwerck 1995). With an impact dehuller, larger kernels require a faster rotor speed for most efficient dehulling. The application of that faster speed to smaller kernels results in excessive groat breakage and reduces milling yield. Size uniformity is even more important to the stone dehuller, where oats are dehulled by rubbing them between two disks (Ganssmann and Vorwerck 1995). In order to optimize mill yield and dehulling efficiency, oat millers routinely sort oats according to size (Hachmann 1947; Peek and Poehlman 1949; Deane and Commers 1986; Salisbury and Wichser 1971; Ganssmann and Vorwerck 1995). Oversized double or bosom oats, as well as undersized, or pin oats, are removed and sold as feed. The remaining milling oats are then further sorted into two or three additional size classes according to width and/or length (see Deane and Commers 1986 for size specifications). These size classes are then dehulled separately. Ideally, millers would want oats delivered that would sort into size classes of about equal proportions. This goal is particularly difficult when mills process oats of many different varieties, grown in many different geographic regions.

One of the earliest attempts to quantify oat kernel size uniformity was made by Milatz (1933), who evaluated uniformity from the difference in mass between primary and secondary kernels from oat spikeletes. By expressing this difference as a ratio of the thousand-kernel mass, he normalized the size difference to the mean kernel mass of the sample. Ratios of 1.38 to 1.96 were considered to represent more uniformly sized samples, whereas ratios of 1.99 to 2.98 were considered to represent less uniformly sized samples. Hηbner (1951a) separated oat samples into different size fractions with slotted sieves and used the mass distributions to evaluate uniformity. He provided an early report indicating that smaller kernels have higher groat percentages than the larger kernels from the same sample and observed that all size fractions maintained a fairly constant length: width ratio of 3.63 to 4.29. Sword (1948-49) provided an earlier report on the use of slotted sieves for oat size uniformity analysis. In addition to the proportion of grain recovered from each sieve (3.0, 2.5, and 2.0 mm), he evaluated the thousand kernel weight of each fraction. He expressed the opinion that oat samples having large differences in thousand-kernel mass between grain recovered from the 2.5 and 2.0 mm sieves were considered to have poor uniformity. Small differences in mean kernel mass between these two samples was considered to indicate greater uniformity. Brŋckner et al. (1956) further developed the oat size sieving method. They offered the "kernel size value," which was an index of quality based on the kernel size distributions. The kernel size value was defined as the percentage of kernels held back by the 2.5 mm sieve multiplied by 1.2, plus the percentage of kernels that were held back by the 2.2 mm sieve multiplied by 0.8, plus the percentage of kernels that were held back by the 2.0 mm sieve multiplied by 0.4. This calculation places greater value on kernels over 2.5 mm, because of their greater value in the milling industry. This index has been applied in industry, particularly in Germany (Ganssmann 1964; Meyer and Zwingelberg 1981).

Symons and Fulcher (1988) introduced the use of digital imaging for oat quality analysis. This method involves the analysis of a digital image taken of a sample of oat kernels against a light background, by a computer. Once a calibration factor indicating distance per pixel is entered into an image analysis program, the computer can derive a variety of size measurements, including length, width, image area, perimeter distance and aspect (length/width). An example of an analyzed digital image is shown in Figure 2. The findings of Symons and Fulcher (1988) emphasized the multi-modal nature of oat kernel size, derived from the multi-floret habit of the oat spikelet. The presence of tertiary florets is generally considered to be undesirable, because they compete with the primary and secondary florets for photosynthate, resulting in poorer fill by the primary and secondary florets, and lower overall test weight (Palagyi, 1983). Doehlert et al. (1999) used whole oat and groat size as determined by digital image analysis for the analysis of physical characteristics affecting groat percentage. In that study, oat size uniformity was evaluated from the coefficient of variation in the image area of whole oat kernels. By this measure, greater amounts of variation (higher coefficient of variation) were associated with a lower groat percentage. A graphical example of oat size distributions is provided in Figure 3.

Groat weight is highly correlated with whole oat weight (Doehlert and McMullen 2000). Strong environmental influences on groat mass have been documented (Doehlert and McMullen 2000). Crown rust infection in particular have been linked to reduced groat weights (Murphy et al. 1940; Doehlert and McMullen 2000). Environmental factors associated with higher groat weights include cooler summers, less precipitation during grain fill (to prevent crown rust infection), and higher levels of solar radiation during the growing season (Doehlert, McMullen

FIGURE 2. Digitized image of dark oat kernels against a light background after computer analysis for kernel length, kernel width, and image area. A transparent ruler is used for length calibration. The computer calculates pixels per unit length from the calibration. The computer then calculates the length and width of each kernel from the number of pixels in each axis as drawn on the image. Image area is calculated from the total number of dark pixels per kernel. Each kernel is numbered. This type of analysis allows for the rapid analysis of kernel dimensions and statistical analysis of kernel size uniformity (Doehlert, unpublished results).

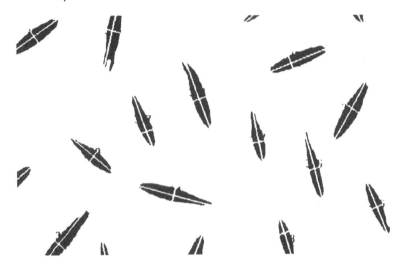

and Hammond, unpublished manuscript). The study of Gullord and Aastveit (1987) indicated significant genotype × environment interactions affecting groat weight.

GROAT COMPOSITION

Oat groat composition is of importance in determining the nutritional quality of oats (Krarup 1904; Berry 1920; Honcamp et al. 1928a; 1928b; Hill 1933; Hηbner 1951b; Asp et al. 1992) and can be considered to comprise the "inner quality" of oats (Ganssmann 1964). Peterson (1992) and Welch (1995) have provided more detailed reviews of oat composition. Oat groat compositional components of nutritional or industrial significance include starch, protein, fat, soluble fiber (β-glucan), and antioxidants. Examples of oat groat compositions from ten cultivars

178   *QUALITY IMPROVEMENT IN FIELD CROPS*

FIGURE 3. Frequency distributions of oat kernel length and cross-sectional area for three oat cultivars. Values are derived from digital image analysis of 900 kernels of each (Doehlert, unpublished results).

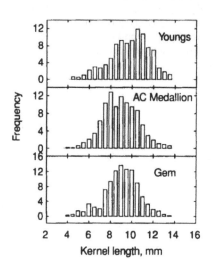
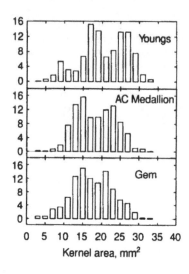

commonly grown in the north central United States and south central Canada are provided in Table 2.

Protein: Among the compositional components of oat, protein concentration often is ranked highly in importance because of its nutritional significance. Oat groats may contain from 124 to 244 g kg^{-1} protein (Peterson 1992). Oat protein contains an excellent nutritional balance for protein and is limited only by lysine (Zarakadas et al. 1982; Ma 1983). Also, the nutritional balance of amino acids in oat groats is independent of the groat protein concentration (Peterson 1976). The high concentration of high quality protein makes oat groats a nutritionally important cereal for human consumption and for animal feed. Efforts to genetically improve oat protein concentration have indicated relatively high heritability of this trait (Frey et al. 1954; Frey et al. 1955; Campbell and Frey 1972; Ohm and Patterson 1973a; 1973b; Frey 1975; Sraon et al. 1975). Unfortunately, increases in groat protein concentration are usually associated with decreases in grain yield (Jenkins 1969; Forsberg et al. 1974; Lyrene and Shands 1974; Sraon et al. 1975; Takeda and Frey 1979). Studies have shown genotypic and environmental effects on oat protein concentration (Jenkins 1969; Forsberg et al. 1974). In particular, nitrogen supply strongly affects oat protein concentration

TABLE 2. Groat composition (dry basis) of ten oat cultivars grown at three locations in North Dakota over three years (Doehlert and McMullen, unpublished results).

Cultivar	Starch	Protein	Oil	β-Glucan	Ash
			$g\ kg^{-1}$		
AC Marie	613	147	78.2	51.7	19.6
Bay	607	195	46.2	54.1	21.1
Hazel	587	187	70.0	56.0	22.0
Hytest	598	196	53.1	53.6	21.7
Jerry	633	171	49.6	45.0	20.3
Marion	595	164	65.5	59.9	21.4
Paul	589	183	70.9	52.6	20.0
Riel	611	174	58.9	48.4	19.4
Robert	636	155	59.4	45.4	20.1
Whitestone	632	158	62.7	49.6	19.8
LSD (0.05)	19	4	1.4	3.8	0.5

(Portch et al. 1968; Ohm, 1976; Welch and Yong 1980; Brinkman and Rho 1984; Humphreys et al. 1994; Welch and Leggett 1997). Increased nitrogen supply increases protein concentration in the grain, and to a lesser extent, increases grain yield. Increased nitrogen also decreases test weight slightly, and greatly increases lodging.

Lipids: Hammond (1983) and Youngs (1986) have previously reviewed oat lipids and lipid contents. Oats contains much higher oil concentrations than do other small grains (Youngs 1986). Higher oil content is an advantage for animal feeding because of its higher caloric content. However, in food applications, higher oil concentrations are deleterious because of their potential for rancidity and production of off-flavors. Oat grain destined for human consumption is routinely steamed before milling to stabilize the lipids by inactivating lipase activity in the oat (Deane and Comers 1985; Vorwerck 1988). Oat fat concentration is highly heritable and recurrent selection programs (Schipper and Frey 1991; Branson and Frey 1989) have generated oat lines with over 18% oil (Peterson and Wood 1997; Frey and Holland 1999). Environmental effects on groat oil concentration have also been documented (Brown et al. 1966; Saastamoinen et al. 1989; Welch 1975; Humphreys et al. 1994). Cooler growth environments have been reported to stimulate oil accumulation in groats (Beringer 1971; Saastamoinen et al. 1989). There also are several reports of negative correlations between protein concentration and oil concentration among different oat genotypes

(Brown et al. 1966; Forsberg et al. 1974). This relationship has been disputed (Youngs and Forsberg 1979), and cultivars with both high protein and oil concentrations have been developed. Kianian et al. (1999) demonstrated the association of a groat oil concentration quantitative trait locus (QTL) with an acetyl-CoA carboxylase gene, suggesting a possible role of this enzyme, which catalyzes the first step in *de novo* fatty acid synthesis, in the control of lipid accumulation in oat.

The composition of oat oil has become a matter of increasing interest for value-added applications. Oats are a rich source of polar lipids (Price 1975; Sahasrabudhe 1979). Glycolipids, such as digalactosyl-diacylglycerol, can be purified with yields of six grams per 1000 g of oat bran (Andersson et al. 1997). This polar lipid has potential for pharmaceutical applications in the formation of liposomes for drug delivery systems (Blom et al. 1996).

b-*Glucan:* The soluble fiber component of oat groats is primarily composed of (1→3), (1→4)-β-D-glucan, commonly referred to as β-glucan. The β-glucan component of oat has garnered increasing interest in recent years because of studies that indicated that β-glucans associated with oat bran in diets appear to be responsible for the lowering of blood cholesterol in both animals and humans (see reviews by Shinnick and Marlett 1993; Anderson and Bridges 1993). In the United States, the Food and Drug Administration has approved health claims on oat food products that contain adequate amounts of β-glucan, along with low fat levels, indicating that a low fat diet including the oat product can reduce the risk of heart disease (FDA 1995; 1996).

Numerous studies have emerged reporting genetic variation in β-glucan concentration in oats (Xman 1987; Welch and Lloyd 1989; Welch et al. 1991; Lim et al. 1992; Saastamoinen et al. 1992; Cho and White 1993; Miller et al. 1993; Peterson et al. 1995). These reports largely indicate a range of 3.2 to 7.0% dry basis for β-glucan concentrations in groats. A number of studies have also investigated environmental variation in oat β-glucan concentrations (Welch and Lloyd 1989; Peterson 1991; Welch et al. 1991; Brunner and Freed 1994; Humphreys et al. 1994; Jackson et al. 1994; Peterson et al. 1995). Although strong genotypic differences in β-glucans can be demonstrated consistently, environmental effects have been more difficult to document (Peterson 1991; Peterson et al. 1995). For example, three studies examined the effects of nitrogen fertilization on β-glucan accumulation. One study reported significant nitrogen × location and nitrogen × year interactions affecting β-glucan concentration in groats, whereas, the other two studies found no significant main effect or interaction effects of nitrogen on β-glucan concentration (Brunner and Freed 1994; Humphreys et al. 1994; Jackson et al. 1994). Several studies have suggested that drought conditions may influence β-glucan accumulation in oat (Peterson 1991; Welch et al. 1991; Brunner and Freed 1994; Peterson et al. 1995), but no study has conclusively demonstrated this.

Starch: Starch is the major storage component in oat. However, because most of the value of oat lies in the non-starch components, its concentration is usually not considered in quality analyses. Paton (1986) reviewed properties of oat starch.

Antioxidants: Peters and Musher (1937) first proposed the use of oat flour as an antioxidant. They suggested the powdering of wax paper wrappers of butter and meat with oat flour to slow the rancidity of fats. Duve and White (1991) demonstrated the high antioxidant activity of oat oil. This antioxidant activity may represent a value-added product from oat that may have varied application in stabilizing other lipid-based food products (Tian and White 1994a; 1994b; Forssell et al. 1990). Higher antioxidant activities are localized in the outermost layers of the oat groat, and significant amounts of activity are lost during the steam treatment that is routine to oat processing for human consumption (Handelman et al. 1999; Emmons et al. 1999).

Antioxidant activity that is extracted with the oil is composed of various tocols (Peterson 1995) as well as a variety of phenolic compounds (Daniels and Martin 1961; Daniels et al. 1963; Daniels and Martin 1964; Daniels and Martin 1965; Daniels and Martin 1968; Collins 1989; Duve and White 1991; Emmons and Peterson 1999). As the value of antioxidants in oats becomes better established, new cultivars may be developed with especially high antioxidant activities to supply natural alternatives for the food ingredient industry.

OAT QUALITY IMPROVEMENT

One of the most important genetic factors affecting oat grain quality is disease resistance. Cultivars with the highest grain quality potential will produce poor quality grain if plantings become heavily diseased. In particular, disease will adversely affect yield, test weight, thousand kernel weight, groat percentage, and to lesser extents, groat composition. Producers generally realize advantage if they plant recently released cultivars adapted to their region in order to take advantage of current disease resistance genes. Producers are also well advised to plant oats as early in the growing season as possible and to maintain sound cultural practices in order to obtain the best quality grain possible.

Oat breeding efforts continue to strive to provide improved quality oats for evolving markets. Divergent quality requirements for food and feed applications may eventually require that specialty cultivars be developed with specific target markets. High test weights and high groat percentage will continue to be priorities for all cultivars. However, cultivars with higher β-glucan and lower fat may be developed for the human food applications, whereas cultivars with higher fat and protein, and less fiber may be developed for animal feed applications. Industrial applications may demand the development of even more specialized oat cultivars with elevated antioxidant concentrations, or with elevated levels of some specific lipid, or other compound of exceptional value.

The environment will continue to add a significant level of uncertainty to the production of quality oat grain. Even though many oat breeding programs make great efforts to assure environmental stability of quality characteristics, the production of the highest quality oat grain will always be at the mercy of the elements, as are most agricultural efforts.

REFERENCES

Anderson, J. W., and S. R. Bridges. 1993. Hypocholestrolemic effects of oat bran in humans. In *Oat Bran*, ed. P.J. Wood, 139-157. St. Paul: American Association of Cereal Chemists.

Andersson, M. B. O., M. Demirbuker, and L. G. Blomberg. 1997. Semi-continuous extraction/purification of lipids by means of supercritical fluids. *J. Chromat.* 785: 337-343.

Asp, N. G., B. Mattsson, and G. Onning. 1992. Variation in dietary fibre, β-glucan, starch, protein, fat and hull content of oats grown in Sweden in 1987-1989. *Eur. J. Clin. Nutr.* 46:31-37.

Atkins, R. E. 1943. Factors affecting milling quality in oats. *J. Am. Soc. Agron.* 35:532-539.

Barbee, O. E. 1935. Markton and other varieties of oats. *Bull. Wash. St. Coll. Exp. Stn.* 314.

Bartley, B. G., and M. G. Weiss. 1951. Evaluation of physical factors affecting quality of oat varieties from Bond parentage. *Agron. J.* 43:22-25.

Beringer, H. 1971. Influence of temperature and seed ripening on the *in vivo* incorporation of $^{14}CO_2$ into lipids of oat grains (*Avena sativa* L.). *Plant Physiol.* 48:433-436.

Berry, R. A. 1920. Composition and properties of oat grain and straw. *J. Agric. Sci.* 10:360-414.

Blom, M., L. Andersson, A. Carlsson, B. Herslof, L. Zhou, and A. Nilsson. 1996. Pharmacokinetics, tissue distribution and metabolism of intravenously administered digaltosyldiaclycerol and monogalactosyldiacylglycerol in the rat. *J. Liposome Res.* 6:737-753.

Branson, C. V., and K. J. Frey. 1989. Recurrent selection for groat oil contents in oat. *Crop Sci.* 29:1382-1387.

Brinkman, M. A., and Y. D. Rho. 1984. Response of three oat cultivars to N fertilizer. *Crop Sci.* 24:973-977.

Brown, C. M., D. E. Alexander, and S. G. Carmer. 1966. Variation in oil content and its relation to other characters in oat (*Avena sativa* L.). *Crop Sci.* 6:190-191.

Brηckner, G. 1953. Der Einfluss der Korneigenschaften auf die Sch≅lung des Hafers. *Die Mηhle* 90:434-436.

Brηckner, G., C. Nernst, M. Rohrlich, and E. Timm. 1956. Technologische und chemische Eigenschaften von Hafersorten. *Jahrb. Versuch. Getreid. Berlin* 1954-1956: 19-38.

Brunner, B. R., and R. D. Freed. 1994. Oat grain β-glucan content as affected by nitrogen level, location and year. *Crop Sci.* 34:473-476.

Bunch, R. A., and R. A. Forsberg. 1989. Relationship between groat percentage and productivity in an oat head-row series. *Crop Sci.* 29:1409-1411.

Campbell, A. R., and K. J. Frey. 1972. Inheritance of groat-protein in interspecific oat crosses. *Can. J. Plant Sci.* 52:735-742.

Cho, K. C., and P. J. White. 1993. Enzymic analysis of β-glucan content in different oat genotypes. *Cereal Chem.* 70:539-542.

Cleve, H. 1948. Das Hamringverfahren bei der Hafersch≅lung. *Getreide Mehl Brot* 2:32-36.

Collins, F. W. 1989. Oat phenolics: Avenanthramides, novel substituted N-cinnamoyl-anthranilate alkaloids from oat groats and hulls. *J. Agric. Food Chem.* 37:60-66.

Cuddeford, D. 1995. Oats for animal feed. In *The Oat Crop: Production and Utilization*, ed. R. W. Welch, pp. 321-368, London: Chapman & Hall.

Cutler, G. H. 1940. Effect of "clipping" or rubbing of the oat grain on the weight and viability of the seed. *J. Am. Soc. Agron.* 32:167-175.

Daniels, D. G. H., and H. F. Martin. 1961. Isolation of a new antioxidant from oats. *Nature* 191:1302.

Daniels, D. G. H., and H. F. Martin. 1964. Structures of two antioxidants isolated from oats. *Chem. Indust.* 1964:2058.

Daniels, D. G. H., and H. F. Martin. 1965. Antioxidants in oats: Diferulates of long-chain diols. *Chem. Indust.* 1965:1763.

Daniels, D. G. H., and H. F. Martin. 1968. Antioxidants in oats: Glyceryl esters of caffeic and ferulic acids. *J. Sci. Food Agric.* 19:710-712.

Daniels, D. G. H., H. G. C. King, and H. F. Martin. 1963. Antioxidants in oats: Esters of phenolic acids. *J. Sci. Food Agric.* 14:385-390.

Deane, D., and E. Commers. 1986. Oat cleaning and processing. In *Oats: Chemistry and Technology*, ed. F. H. Webster, pp. 371-412. St. Paul: American Association of Cereal Chemists.

Doehlert, D. C., and M. S. McMullen. 2000. Environmental and genotypic effects on oat milling characteristics and groat hardness. *Cereal Chem.* 77:148-154.

Doehlert, D. C., M. S. McMullen, and R. R. Baumann. 1999. Factors affecting groat percentage in oat. *Crop Sci.* 39:1858-1865.

Duve, K. J., and P. J. White. 1991. Extraction and identification of antioxidants in oats. *J. Am. Oil Chem. Soc.* 98:365-370.

Eastham, A., and L. V. Baker. 1922. Oat hulls and their use in feeding stuffs. *Circ. Can. Dep. Agric. Seed Branch* 11.

Emmons, C. L., and D. M. Peterson. 1999. Antioxidant activity and phenolic contents of oat groats and hulls. *Cereal Chem.* 76:902-906.

Emmons, C. L., D. M. Peterson, and G. L. Paul. 1999. Antioxidant capacity of oat (*Avena sativa* L.) extracts. 2. *In vitro* antioxidant activity and contents of phenolic and tocol antioxidants. *J. Agric. Food Chem.* 47:4894-4898.

Food and Drug Administration. 1996. Food labeling: Health claims; oats and coronary heart disease. *Fed. Reg.* 61:296-328.

Food and Drug Administration. 1997. Food labeling; health claims; oats and coronary heart disease; final rule. *Fed. Reg.* 62:3583-3601.

Forsberg, R. A., and D. L. Reeves. 1992. Breeding oat cultivars for improved grain quality. In *Oat Science and Technology*, eds. H. G. Marshall and M. E. Sorrells, 751-775, Madison: American Society of Agronomy.

Forsberg, R. A., V. L. Youngs, and H. L. Shands. 1974. Correlations among chemical and agronomic characteristics in certain oat cultivars and selections. *Crop Sci.* 14:221-224.

Forssell, P., M. Cetin, G. Wirtanen, and Y. Malkki. 1990. Antioxidative effects of oat oil and its fractions. *Fat Sci. Technol.* 92:319-321.

Fraser, C. R. 1944. The air jet method of dehulling seeds. *Can. J. Res. F* 22:157-162.

Frey, K. J. 1975. Heritability of groat-protein percentage in hexaploid oats. *Crop Sci.* 15:277-279.

Frey, K. J., and J. B. Holland. 1999. Nine cycles of recurrent selection for increased groat-oil content in oat. Crop Sci. 39:1636-1641.

Frey, K. J., and S. C. Wiggans. 1956. Cultural practices and test weight in oats. *Proc. Iowa Acad. Sci.* 63:259-265.

Frey, K. J., M. C. Shekleton, H. H. Hall, and E. J. Benne. 1954. Inheritance of niacin, riboflavin, and protein in two oat crosses. *Agron. J.* 46:137-139.

Frey, K. J., H. H. Hall, and M. C. Shekleton. 1955. Inheritance of protein, niacin, and riboflavin in oats. *J. Agric. Food Chem.* 3:946-948.

Ganssmann, W. 1964. Vergleichende Untersuchungen der Qualität von Industriehafer. *Die Mηhle* 101:776-779.

Ganssmann, W., and K. Vorwerck. 1995. Oat milling, processing and storage. In *The Oat Crop: Production and Utilization*, ed. R. W. Welch, pp. 369-408, London: Chapman & Hall.

Greig, R. B., and W. M. Findlay. 1907. The milling properties of oats. *J. Bd. Agric. London* 14:257-268.

Gullord, M., and A. H. Aastveit. 1987. Developmental stability in oats (*Avena sativa* L.) II. Quality characteristics. *Hereditas* 107:65-74.

Hachmann, W. 1947. Hafermullerei: Gr ssensortierung vor dem Sch≅len. *Getreide Mehl Brot* 1:7-9.

Handelman, G. J., G. Cao, M. F. Walter, Z. D. Nightingale, G. L. Paul, R. L. Prior, and J. B. Blumberg. 1999. Antioxidant capacity of oat (*Avena sativa* L.) extracts. 1. Inhibition of low-density lipoprotein oxidation and oxygen radical absorbance capacity. *J. Agric. Food Chem.* 47: 4888-4893.

Hammond, E. G. 1983. Oat lipids. In *Lipids in Cereal Technology*, ed. P. J. Barnes, pp. 331-352, New York: Academic Press.

Hill, D. D. 1933. The chemical composition and grades of barley and oat varieties. *J. Amer. Soc. Agron.* 25:301-311.

Hoffman, L. A. 1995. World production and use of oats. In *The Oat Crop: Production and Utilization*, ed. R. W. Welch, pp. 34-61, London: Chapman & Hall.

Hoffman, L. A., and J. Livezey. 1987. *The U.S. Oats Industry*. Agricultural Economic Report, Number 573, U.S. Department of Agriculture, Economic Research Service.

Honcamp, F., W. Schramm, and H. Stotz. 1928a. Weitere Untersuchungen uber die chemische Zusammensetzung und den N≅hrwert von Gelbhafer und Weisshafer. *Z. Tierzucht. Zηchtungbiol.* 11:433-444.

Honcamp, F., W. Schramm, and H. Wiessmann. 1928b. Untersuchungen uber die chemische Zusammensetzung und den N≅hrwert von Gelbhafer sowie von den einzelnen Sortierungen derselben bie der Schule'schen Reinigungsanlage. *J. Landwirtsch.* 76:113-126.

Hηbner, R. 1951a. Vierj≅hrige Untersuchungen ηber Kornqualit≅t und Leistungeschaften des Hafers. I Teil: Morphologisch-physikalische Untersuchngen. *Z. Acker. Pflanzenbau.* 93:44-78.

Hηbner, R. 1951b. Vierj≅hrige Untersuchungen ηber Kornqualit≅t und Leistungseigenschaften des Hafers. II. Tiel: Chemische Untersuchungen, Ertrasleistungen und Schlussbetrachtung. *Z. Acker. Pflanzenbau.* 93:169-197.

Humphreys, D. G., D. L. Smith, and D. E. Mather. 1994. Nitrogen fertilizer and seedling date induced changes in protein, oil and β-glucan contents of four oat cultivars. *J. Cereal Sci.* 20:283-290.

Hutchinson, J. B., N. L. Kent, and H. F. Martin. 1952. The kernel content of oats. Variation in percentage kernel content and 1,000 kernel weight within the variety. *J. Nat. Inst. Agric. Bot.* 6:149-160.

Jackson, G. D., R. K. Berg, G. D. Kushnak, T. K. Blake, and G. I. Yarrow. 1994. Nitrogen effects on yield, β-glucan content, and other quality factors of oat and waxy hulless barley. *Commun. Soil Sci. Plant Anal.* 25:3047-3055.

Jenkins, G. 1969. Grain quality in hybrids of *Avena sativa* L. and *A. byzatina* C. Koch. *J. Agric. Sci.* 72:311-317.

Kianian, S. F., M. A. Egli, R. L. Phillips, H. W. Rines, D. A. Sommers, B. G. Gegenbach, F. H. Webster, S. M. Livingston, D. M. Wesenberg, D. D. Stuthman, and R. G. Fulcher. 1999. Association of a major groat oil content QTL and an acetly-CoA carboxylase gene in oat. *Theor. Appl. Genet.* 98:884-894.

Kittlitz, E. V., and H. Vetterer. 1972. Untersuchungen zur maschinellen Entspelzung von Hafer mit einem Druckluftspelzer. *Dtsch. Mηllerzeitung* 70:245-246.

Krarup, A. V. 1904. Untersuchung ηber die Erblichkeit und Variabilit≅t biem Hafer, mit besonderer Rηcksicht auf die Isolierung fettreicher Typen fur die Hafergrηtzefabrikation. *Biedermanns Zentralbl. Agric. Chem.* 33:94-106.

Lathrop, A. W., and G. Bohstedt. 1938. Oat mill feed: Its usefulness and value in livestock rations. *Bull. Univ. Wisc. Exp. Stn.* 135.

Lea, H. 1955. Quality in oats and oat products from the oatmeal and provender miller's point of view. *J. Nat. Inst. Agric. Bot.* 7:404-410.

Lim, H. S., P. J. White, and K. J. Frey. 1992. Genotype effects on β-glucan of oat lines grown in two consecutive years. *Cereal Chem.* 69:262-265.

Love, H. H. 1914. Oats for New York. *Bull. Cornell Univ. Agric. Exp. Stn.* 343.

Love, H. H., and W. T. Craig. 1944. A machine for dehulling oats. *J. Am. Soc. Agron.* 36:264-266.

Love, H. H., T. R. Stanton, and W. T. Craig. 1925. Improved oat varieties for New York and adjacent states. *Dep. Circ. U.S. Dep. Agric.* 353.

Lyrene, P. M., and H. L. Shands. 1974. Groat protein percentage in *Avena sativa* L. factuoids and in a factuoid × *A. sterilis* L. cross. *Crop Sci.* 14:765-767.

Ma, C. Y. 1983. Chemical characterization and functionality assessment of protein concentrates from oats. *Cereal Chem.* 60:36-42.

MacKey, J. 1959. Hafer (*Avena sativa* L.) III. Genetics of some agronomic characteristics of oats. In *Handbuch der Pflanzenzuchtung. Vol. II: Zuchtung der Getreidearten* eds. H. Kappert and W. Rudorf, pp. 494-531, Berlin: Paul Parey.

Meyer, D., and H. Zwingelberg. 1981. Untersuchungen zur Verwendung von inl≅ndischem Hafer in der Sch≅lmηllerei. *Getreide Mehl Brot* 35:230-234.

Milatz, R. 1933. Neue Hafersortenmerkmale. *Angew. Bot.* 15:481-518.

Miller, S. S., D. J. Vincent, J. Weisz, and R. G. Fulcher. 1993. Oat β-glucans: An evaluation of eastern Canadian cultivars and unregistered lines. *Can. J. Plant Sci.* 73:429-436.

Murphy, C. F., and K. J. Frey. 1962. Inheritance and heritability of seed weight and its components in oats. *Crop Sci.* 2:509-512.

Murphy, H. C., L. C. Burnett, C. H. Kingsolver, T. R. Stanton, and F. A. Coffman. 1940. Relation of crown-rust infection to yield, test weight and lodging of oats. *Phytopathology* 30:808-819.

Ohm, H. W. 1976. Response of 21 oat cultivars to nitrogen fertilization. *Agron. J.* 68:773-775.

Ohm, H. W., and F. L. Patterson. 1973a. Estimation of combining ability, hybrid vigor and gene action in *Avena* spp. L. *Crop Sci.* 13:55-58.

Ohm, H. W., and F. L. Patterson. 1973b. A six-parent dialle cross analysis for protein in *Avena sterilis* L. *Crop Sci.* 13:27-30.

Palagyi, A. 1983. Tertiary seed proportions in the grain yield of several oat varieties. *Cereal Res. Comm.* 11:269-274.

Pawlisch, P. E., and H. L. Shands. 1962. Breeding behavior for bushel weight and agronomic characteristics in early generations of two oat crosses. *Crop Sci.* 2:231-237.

Paton, D. 1986. Oat starch: Physical, chemical, and structural properties. In *Oats: Chemistry and Technology*, ed. F. H. Webster, pp. 93-120, St. Paul: American Association of Cereal Chemists.

Peek, J. M., and L. M. Poehlman. 1949. Grain size and hull percentage as factors in the milling quality of oats. *Agron. J.* 41:462-466.

Peters, F. N., and S. Musher. 1937. Oat flour as an antioxidant. *Ind. Eng. Chem.* 29:146-151.

Peterson, D. M. 1976. Protein concentration, concentration of protein fractions, and amino acid balance in oats. *Crop Sci.* 16:663-666.

Peterson, D. M. 1991. Genotype and environmental effects on oat β-glucan concentration. *Crop Sci.* 31:1517-1520.

Peterson, D. M. 1992. Composition and nutritional characteristics of oat grain products. In *Oat Science and Technology*, eds. H. G. Marshall and M. E. Sorrells, pp. 265-292, Madison: American Society of Agronomy.

Peterson, D. M. 1995. Oat tocols: Concentration and stability in oat products and distribution within the kernel. *Cereal Chem.* 72:21-24.

Peterson, D. M., and D. F. Wood. 1997. Composition and structure of high-oil oat. *J. Cereal Sci.* 26:121-128.

Peterson, D. M., D. M. Wesenberg, and D. E. Burrup. 1995. β-Glucan content and its relationship to agronomic characteristics in elite oat germplasm. *Crop Sci.* 35: 965-970.

Pomeranz, Y., G. D. Davis, J. L. Stoops, and F. S. Lai. 1979. Test weight and groat-to-hill ratio in oats. *Cereal Foods World* 24:600-602.

Portch, S., A. F. MacKenzie, and H. A. Steppler. 1968. Effect of fertilizers, soil drainage class and year upon protein yield and content of oat. *Agron. J.* 60:672-674.

Price, P. B., and J. G. Parsons. 1976. Lipids of seven cereal grains. *J. Am. Oil Chem. Soc.* 52:490-493.

Ranhotra, G. S., and J. A. Gelroth. 1995. Food uses of oats. In *The Oat Crop: Production and Utilization* ed. R. W. Welch, pp. 409-432, London: Chapman & Hall.

Ronald, P. S., P. D. Brown, G. A. Penner, A. Brule-Babel, and S. Kibite. 1999. Heritability of hull percentage in oat. *Crop Sci.* 39:52-57.

Root, W. R. 1977. A study of oat (*Avena sativa* L.) caryopsis morphology and its effect on certain oat quality characteristics. M.S. thesis, University of Wisconsin, Madison.

Root, W. R. 1979. The influence of oat (*Avena sativa* L.) kernel and caryopsis morphological traits on grain quality characteristics. Ph.D. diss., University of Wisconsin, Madison.

Saastamoinen, M., J. Kumpulainen, and S. Nummela. 1989. Genetic and environmental variation in oil content and fatty acid composition of oats. *Cereal Chem.* 66:296-300.

Saastamoinen, M., S. Plaami, and J. Kumpulainen. 1992. Genetic and environmental variation in β-glucan contents of oats cultivated or tested in Finland. *J. Cereal Sci.* 16:279-290.

Sahasrabudhe, M. R. 1979. Lipid composition of oats (*Avena sativa* L.). *J. Amer. Oil Chem. Soc.* 56:80-84.

Salisbury, D. K., and W. R. Wichser. 1971. Oat milling: Systems and products. *Bull. Assoc. Oper. Mill.* 1971:3242-3247.

Sappenfield, W. P., and J. M. Poehlman. 1952. Effect of date of seeding on the yield and test weight of oat varieties. *Bull. Univ. Missouri Exp. Stn.* 499.

Schipper, H., and K. J. Frey. 1991. Observed gains from three recurrent selection regimes for increased groat-oil content of oat. *Crop Sci.* 31:1505-1510.

Shinnick, F. L., and J. A. Marlet. 1993. Physiological responses to dietary oats in animal models. In *Oat Bran* ed. P. J. Wood, pp. 113-137, St. Paul: American Association of Cereal Chemists.

Souza, E. J., and M. E. Sorrells. 1988. Mechanical mass selection methods for improvement of oat groat percentage. *Crop Sci.* 28:618-623.

Sraon, H. S., D. L. Reeves, and M. D. Rumbaugh. 1975. Quantitative gene action for protein content in oat. *Crop Sci.* 15:668-670.

Stoa, T. E. 1922. Varietal trials with oats in North Dakota. *Bull. N. Dak. Agric. Exp. Stn.* 164.

Stoa, T. E., R. W. Smith, and C. M. Swallers. 1936. Oats in North Dakota. *Bull. N. Dak. Agric. Exp. Stn.* 287.

Stuke, H. 1955. Eine Schnellmethock zur Bestinimung des Spelzengehaltes beim Hafer. *Der Znchter* 25:90-92.

Stuthman, D. D., and R. M. Granger. 1977. Selection for caryopsis percentage in oats. *Crop Sci.* 17:411-414.

Sword, J. 1948-1949. Milling values of oat varieties. I. 1946 results. *Scot. Agric* 28:137-148.

Symons, S. J., and R. G. Fulcher. 1988. Determination of variation in oat kernel morphology by digital image analysis. *J. Cereal Sci.* 7:219-228.

Takeda, K., and K. J. Frey. 1979. Protein yield and its relationship to other traits in backcross populations from an *Avena sativa* × *A. sterilis* cross. *Crop Sci.* 19: 623-628.

Tian, L. L., and P. J. White. 1994a. Antioxidant activity of oat extract in soybean and cottonseed oils. *J. Amer. Oil Chem. Soc.* 71:1079-1086.

Tian, L. L., and P. J. White. 1994b. Antipolymerization activity of oat extract in soybean and cottonseed oils under frying conditions. *J. Amer. Oil Chem. Soc.* 71: 1087-1094.

United States Department of Agriculture. 1978. *The Official United States Standards for Grain*, Federal Grain Inspection Service, Washington: U.S. Government Printing Office.

Villers, P. J. R., Jr. 1935. A genetic study of the inheritance of various characters in certain *Avena* hybrids. *Bull. Dept. Agric. Stellenbosch S. Afr. Sci.* 140.

Vorwerck, K. 1988. Hydrothermische Behandlung von Hafer. *Getriede Mehl Brot* 42:199-202.

Webster, F. H. 1986. Oat utilization: Past, present, and future. In *Oats: Chemistry and Technology*, ed. F. H. Webster, pp. 413-426, St. Paul: American Association of Cereal Chemists.

Welch, R. W. 1975. Fatty acid composition of grain from winter and spring sown oats, barley and wheat. *J. Sci. Food Agric.* 26:429-435.

Welch, R. W. 1995. The chemical composition of oats. In *The Oat Crop: Production and Utilization* ed. R. W. Welch, pp. 279-320, London: Chapman & Hall.

Welch, R. W., and L. M. Leggett. 1997. Nitrogen content, oil content, and oil composition of oat cultivars (*A. sativa*) and wild *Avena* species in relation to nitrogen fertility, yield and partitioning of assimilates. *J. Cereal Sci.* 26:105-120.

Welch, R. W., and J. D. Lloyd. 1989. Kernel (1→3) (1→4)-β-D-glucan content of oat genotypes. *J. Cereal Sci.* 9:35-40.

Welch, R. W., and Y. Y. Yong. 1980. The effect of variety and nitrogen fertilizer on protein production in oats. *J. Sci. Food Agric.* 31:541-548.

Welch, R. W., J. M. Leggett, and J. D. Lloyd. 1991. Variation in the kernel (1→3), (1→4)-β-D-glucan content of oat cultivars and wild *Avena* species and its relationship to other characteristics. *J. Cereal Sci.* 13:173-178.

Wesenberg, D. M., and H. L. Shands. 1973. Heritability of oat caryopsis percentage and other grain quality components. *Crop Sci.* 13:481-484.

Xman, P. 1987. The variation in chemical composition of Swedish oats. *Acta Agric. Scand.* 37:347-352.

Youngs, V. L. 1986. Oat lipids and lipid-related enzymes. In *Oats: Chemistry and Technology,* ed. F. H. Webster, pp. 205-226, St. Paul: American Association of Cereal Chemists.

Youngs, V. L., and R. A. Forsberg. 1979. Protein-oil relationships in oats. *Crop Sci.* 19:798-802.

Youngs, V. L., and H. L. Shands. 1974. Variation in oat kernel characteristics within the panicle. *Crop Sci.* 14:578-580.

Zade, A. 1915. Methoden zur Bestinmung des Spelzenanteils beim Hafer. *Frηhlings. Landw. Z.* 64:295-311.

Zarkadas, C. G., H. W. Hulan, and F. G. Proudfoot. 1982. A comparison of the amino acid composition of two commercial oat groats. *Cereal Chem.* 59:323-327.

Zavitz, C. A. 1927. Forty years' experiments with grain crops. *Bull. Ont. Dep. Agric.* 332.

Zhang, D. 1997. Characterization of oat flour rheological properties and the application of oat flour in bread baking. Ph.D. diss. North Dakota State University.

Improving the Nutritional Value of Cool Season Food Legumes

K. E. McPhee

F. J. Muehlbauer

SUMMARY. Cool season food legumes such as pea, lentil, chickpea, faba bean, grasspea and lupin have been consumed in animal and human diets since domestication and have been cultivated for over 9000 years. Due to their nutritional value they continue to make up a substantial portion of diets in developing countries worldwide. Seeds are composed of protein, starch, fiber, lipids, vitamins and minerals as well as several antinutritional compounds. Protein and starch combined account for 50-75% of seed mass. Similar amounts of protein and starch are present in each species and range from 15-40% and 35-53%, respectively. Fiber accounts for 1-22% of the seed mass. Chickpea is quite variable for fiber content, but faba bean, grasspea and lupin all contain approximately 15% fiber while pea and lentil have substantially lower levels, 4-8%. Protein composition is deficient in the sulfur amino acids, methionine and cysteine, but contains adequate levels of lysine making the legumes an excellent dietary complement to cereals. As a result these crops have been produced and consumed together for centuries. Despite the high nutritional value of these crops, they contain several antinutritional factors (ANF) such as phytic acid, protease inhibitors, heamagglutinins, tannins, alkaloids, raffinose oligosaccharides and antigenicity factors at relatively low levels. These compounds serve important roles in pest resis-

K. E. McPhee (E-mail: kmcphee@mail.wsu.edu) and F. J. Muehlbauer (E-mail: muehlbau@wsu.edu) are affiliated with USDA-ARS Grain Legume Genetics and Physiology Research Unit, and the Department of Crop and Soil Sciences, 303 Johnson Hall, Washington State University, Pullman, WA 99164-6434 USA.

[Haworth co-indexing entry note]: "Improving the Nutritional Value of Cool Season Food Legumes." McPhee, K. E., and F. J. Muehlbauer. Co-published simultaneously in *Journal of Crop Production* (Food Products Press, an imprint of The Haworth Press, Inc.) Vol. 5, No. 1/2 (#9/10), 2002, pp. 191-211; and: *Quality Improvement in Field Crops* (ed: A. S. Basra, and L. S. Randhawa) Food Products Press, an imprint of The Haworth Press, Inc., 2002, pp. 191-211. Single or multiple copies of this article are available for a fee from The Haworth Document Delivery Service [1-800-HAWORTH, 9:00 a.m. - 5:00 p.m. (EST). E-mail address: getinfo@haworthpressinc.com].

tance or plant survival, but reduce digestibility and palatability when consumed. Genetic analysis of the antinutritional factors has shown that many of the ANFs are controlled by single genes and has allowed geneticists to reduce accumulation of the ANFs, thereby improving nutritional value. Legumes will continue to serve as the primary source of protein in areas of the world where meat is not readily available. Continued research toward a better understanding of the genetic control of legume quality will give geneticists the opportunity to improve the dietary value of these crops. *[Article copies available for a fee from The Haworth Document Delivery Service: 1-800-HAWORTH. E-mail address: <getinfo@haworthpressinc. com> Website: <http://www.HaworthPress.com>]*

KEYWORDS. Food legumes, seed quality, antinutritional factors, protein, starch, trypsin inhibitor

Cool season food legumes, including pea (*Pisum sativum* L.), lentil (*Lens culinaris* Medik.), chickpea (*Cicer arietinum* L.), faba bean (*Vicia faba* L.), grasspea (*Lathyrus sativus* L.), and lupin (*Lupinus albus* L. and *L. angustifolius* L., *L. luteus* L., and *L. mutabilis* Sweet [tarwi]), have high nutritional value and have been consumed in human diets or used for animal feed since domestication. Often referred to as the poor man's food, food legumes have been cultivated for over 9000 years. As these crops were domesticated, man selected seeds with more desirable color, size, and shape. In addition, certain quality traits such as reduced antinutritional factors and improved flavor were selected. It was not until the mid-twentieth century that substantial scientific effort was placed on seed quality and the potential for improvement. Increased emphasis on quality was brought about by greater awareness of these crops as suitable feed for livestock and an increased reliance on these crops as protein sources for human consumption in developing countries. The potential use of grain legumes in industrial products such as starch, high protein concentrates has brought about greater attention to quality.

The food legumes originated in the Mediterranean basin and have spread throughout the world as man traveled to new regions. Carbonized remains in that region indicate that pea, lentil, and chickpea were cultivated with cereals as early as the seventh millennium B.C. (Ladizinsky and Adler, 1976; Smartt, 1990; Williams et al., 1974; Zohary, 1972, 1973). The eastern Mediterranean region to Asia Minor into central

Asia are considered important areas where these crops were likely domesticated; however, the eastern Mediterranean seems most important for the evolution and domestication of the cool season food legumes (Khvostova, 1975; Makasheva, 1973; Smartt and Hymowitz, 1985).

Food legumes have two to three times greater protein content than cereals. The biological cost for producing one unit of protein is relatively high compared to starch and may be the reason that yields of the grain legumes have not kept pace with the increases in wheat and rice.

Food legumes are consumed as a protein source in nearly all countries worldwide and are widely traded in world markets. The Food and Agriculture Organization reports that over 59 million metric tons of pulses were produced in 1999, a 43% increase since 1969 (FAO, 2000). Faba bean is the only cool season legume to experience a net loss in production since 1969. This is primarily due to reduced area of production (Table 1). Dry pea also experienced a net loss in production area; however, total world production increased by 2 million MT. The increased production is largely due to increased areas sown in Canada and Australia. Total production and area sown has increased for chickpea, lentil and lupin.

Most food legumes are harvested as dry seed and prepared for consumption by reconstitution consumed in a variety of preparations. Specialty types in which the pods and immature seed are consumed have also been developed, i.e., snap peas and beans. Other specialty uses include a Middle Eastern tradition of consuming chickpea seed after roasting immature pods over a hot fire. Consumption of mature seed is traditionally in soups or similar dishes after the seed has been soaked

TABLE 1. World production and area harvested of six cool season food legumes in 1999.

Crop	Scientific Name	Production (000's MT)	Net Change Since 1969	Area Harvested (000's ha)	Net Change Since 1969
Chickpea	*Cicer arietinum* L.	9,244	+3,364	12,034	+2,454
Faba bean	*Vicia faba* L.	3,561	−701	2,224	−2,502
Grasspea*	*Lathyrus sativus* L.	--	--	--	--
Lentil	*Lens culinaris* Medik.	3,112	+2,024	3,292	+1,579
Lupin	*Lupinus albus* L.	1,534	+857	1,328	+385
Pea	*Pisum sativum* L.	11,699	+2,398	5,946	−2,043

*Data unavailable

and softened. Other preparations common in the Indian subcontinent include grinding the seed to prepare flour used in making Dhal or mixing it with cereal flour to make bread. Feeding grain legumes to livestock, particularly hogs and poultry, is a common practice due to low cost and high protein content of the seed. Individual seed components isolated through fractionation are also used in various industrial applications. For example, starch is used in the textile and printing industry as well as in some adhesives.

The broad range of uses for food legumes including food, feed and industrial uses, presents a diversity of quality considerations on each crop (Table 2). Though the food legumes can be used as fodder for animal feeding with a separate set of quality considerations, this chapter will only discuss the quality of dry seed. The quality factors for food and feed include both nutritional and antinutritional factors. The food legumes serve as excellent sources of protein, fiber, carbohydrate, and some minerals and vitamins. Though food legumes are highly nutritious, they also contain several antinutritional factors which limit their use. These include trypsin and chymotrypsin inhibitors, alkaloids, raffinose oligosaccharides, antigenicity factors, tannins, phytic acid, and lectins (phytohemagglutinins). The food legumes containing each of these antinutritional factors is presented in Table 3.

TABLE 2. Quality factors for food legumes consumed for human food and animal feed.

Agronomic	Nutritive Factors	Antinutritional Factors	Industrial Uses
Seed size	Starch	Alkaloids	Oil content
Size distribution	Protein	Antigenicity Factors	Starch
Seed color	– methionine	Trypsin/Chymotrypsin	
Seed shape	Cooking quality	Vicine/Convicine	
Hard seededness	– Time	Lectins	
Decortication loss	– Taste	Oligosaccharides	
Water absorption	– Texture	Tannins	
Seed damage	– Smell	Phytates	
Seed coat integrity	– Appearance		
	– After-taste		
	Oil		
	Vitamins		
	Minerals		

TABLE 3. Distribution and physiological effects of some antinutritional factors (ANF) in legume seeds (modified from Liener, 1989).

ANF	Distribution	Physiological Effect
Proteins		
Protease inhibitors	Most legumes	Depressed growth; pancreatic hypertrophy/hyperplasia; acinar nodules
Lectins	Most legumes	Depressed growth; death
Amylase inhibitors	Most legumes	Interference with starch digestion
Amino acid analogues		
β-N-oxalyl-α,β-propionic acid	*Lathyrus sativus*	Lathyrism
Glycosides		
Vicine/convicine	*Vicia faba*	Hemolytic anemia
Oligosaccharides	Most legumes	Flatulence
Saponins	Most legumes	Affects intestinal permeability
Miscellaneous		
Phytate	Most legumes	Interference with mineral availability
Tannins	Most legumes	Interference with protein digestibility
Alkaloids	Lupins	Depressed growth

Seed quality is a function of exterior appearance, internal composition, and functionality for both agronomic and dietary uses. Exterior appearance or color of the seed is often the basis by which consumers make their purchase decisions. The whole seed or split cotyledons must have a bright color and lack any form of blemish or irregularity such as chips, cracks or dirt. Chips and cracks in the seed coat or testa of seeds used to establish the crop the following year are not desirable due to increased leakage of cellular components into the soil surrounding the seed attracting pathogenic fungi. Attack from the pathogenic fungi reduce stand establishment and seedling vigor. Seed shape and size are also important for consumer acceptance.

Other factors regarding the seed include the thickness and/or permeability of the seed coat. Hard-seededness is a primative trait that legume species required for survival in periods of drought. It is undesirable when cooking legumes because of the long time required for soaking. In addition, if the hard seed are not removed prior to preparation, pain and damage to the tooth surface are possible. Hard-seededness is partially controlled genetically, but is highly influenced by the growing environment. Selection of seed types with lower tendency toward hardseededness over time has greatly reduced this negative aspect but not completely eliminated it. Rapid terminal drought conditions during the seed filling period will reduce seed size and increase the number of hard seed. The later pods on many indeterminate plant types, for example in pea, will have reduced seed size and contribute to hard-seededness.

CANNED FOOD LEGUMES

Pea, lentil, chickpea, and faba bean are regularly processed and canned product prior to consumption. Several quality factors are required for canning. These include high quality seed with a minimum of foreign material and intact seed coats. When legume seeds are canned it is necessary for the seeds to maintain integrity. This provides for a clear brine solution and an attractive product. Increased salt concentration in the brine can improve seed coat integrity during canning. Increased hydration capacity of legume seed is also important to the canning industry for obvious monetary considerations. Cooking time, taste, texture, smell, appearance, and after-taste are also important considerations and can differ among genotypes.

Seed color is particularly important when canning green cotyledon dry pea. Many cultivars are susceptible to bleaching which is caused by the initiation of germination during maturation and prior to harvest. Theses processes breakdown the chlorophyll pigment in the cotyledons giving them a white appearance which is undesirable in canned peas. Seed color is also important when canning chickpea. The production environments can have a profound influence on the canning characteristics of chickpea including the final color as well as the clarity and viscosity of the brine solution (personal observation of the authors).

DRY SEED USES

Food legumes provide a rich source of nutrition in animal rations and for many groups of people around the world. They are prepared in a wide variety of dishes and procedures including whole and split dry seeds, milled and as fresh immature seeds. Decorticating and splitting the grain legumes into the component cotyledons is common when preparing dishes for human consumption. Ease of decortication is an important quality factor to the processing industry. Removal of the entire seed coat is desired while small fragments or "whitecaps" are undesirable. In addition, chipped edges of the cotyledons during splitting also detracts from visual appeal and overall quality. Some processors will polish the cotyledons after splitting to enhance their color and eye appeal.

Physical quality considerations for seed used in the animal feeding industry are not nearly as strict as those for human consumption. Seed composition is the primary concern when feeding livestock.

SEED COMPOSITION AND CONCENTRATION

Food legume seeds are composed of starch, protein, fiber, and several other constituents which are minor in amount, but are important nutritionally. Seed composition is influenced by genetic differences among cultivars and variation in production environment (Hulse 1994). The range of values for the major seed components are presented in Table 4.

Starch is present in the legume seed in greater amounts than any other component. It is stored in amyloplasts within the cotyledons and serves as a carbohydrate source for the germinating seedling. Starch is broken down into simple sugars by heat which is applied in many of the preparatory procedures used for legumes. Starch content ranges from 35 to 53 percent (Table 4).

TABLE 4. Seed composition of six cool season food legumes.

Seed Fraction	Chickpea	Faba Bean	Grasspea	Lentil	Lupine	Pea
Starch (%)	37-50[3]	41-53	34.8	35-53		37-49
Protein (%)	15-30[1]	23-39	28.2	20-31	35-38	21-33
		28.2[2]	25-32			
Lipids (%)	4-10	1.6,0.8[2]	0.6	1.2	9	2.4
		1.5-2.5				
Fiber (%)	1.2-13.5[1]	5-8.5	15	3.8-4.6	12	4.6-7
		15-22				
Ash (%)	2-4.8	3	3	2.1	2.6	2.6
Vitamin A (µg)	0-225	130	70 IU			

[1] Hulse, 1994
[2] Li-juan, 1993
[3] Reddy et al., 1984
[4] Duke, 1981

Protein, the second most concentrated seed component, ranges from 15 to 38 percent (Table 4). There are ten essential amino acids which human and non-ruminant animals can not produce internally. Methionine and cysteine, two of the essential amino acids, are generally low in the food legumes. The food legumes do, however, have sufficient lysine making them an excellent complement to cereal grains. They are also generally deficient in the other essential amino acids (Table 5).

Fiber in the legume seed is primarily in the seed coat or testa and comprises approximately 1.0 to 13.5 percent of the seed weight. This depends largely on the size of the seed. Smaller seeds will have a larger proportion of fiber compared to large seeds. The food legumes are generally low in lipid content. Chickpea contains the greatest lipid content of the food legumes and can be up to 10 percent (Table 4). The other legumes range from < 1.0 to 5 percent. Lupine has also been reported to have relatively high levels of lipid (Table 4). Ash content ranges from 2-5 percent.

Chickpea protein (and possibly other legumes) have 60-80% albumin and 15-25% globulin. Albumins are important due to their enzymatic and metabolic function. They also have a high content of lysine and sulfur amino acids. Resistance to digestive proteases is attributed to structure and antinutritional factors. Disruption of disulfide bonds was the major factor affecting chickpea protein digestibility (Clemente et al., 2000). Faba bean protein digestibility is limited by the structural properties and not by the binding of polyphenols (Carbonaro et al., 2000).

TABLE 5. Essential amino acid composition of six food legumes.

Amino Acid	Chickpea	Faba Bean	Grasspea	Lentil	Pea
Arginine	8.8	9.6	7.8	6.9	9.6
Cysteine	1.2	1.4	1.3	1.9	1.7
Isoleucine	4.4	4.1	6.7	3.8	4.1
Leucine	7.6	7.6	6.6	7.1	7.0
Lysine	7.2	6.4	7.4	7.3	7.2
Methionine	1.4	0.7	0.6	1.6	1.0
Phenylalanine	6.6	4.2	4.2	4.6	4.6
Threonine	3.5	3.6	2.3	3.5	3.8
Tryptophan	0.8	0.9	0.4	0.9	0.8
Valine	4.6	4.6	4.7	4.2	4.6

Adapted from Deshpande and Demodoran, 1990 (amino acid/gm seed protein)

ANTINUTRITIONAL FACTORS

Several compounds contained in legume seeds cause negative physiological effects when consumed by animals or in human diets (Table 6). These effects range from producing an offensive taste to very severe and life threatening physiological disorders. Despite their antinutritional effects many of them serve important roles in nature. Lectins, alkaloids, and tannins contribute to disease and insect resistance. While it is a goal to reduce the content of these factors in seed it must be kept in mind that alternative methods of pest resistance must be available either through enhanced genetic resistance or chemicals.

TABLE 6. Content of antinutritional factors in six food legumes.

Antinutritional Factor	Chickpea	Faba Bean	Grasspea	Lentil	Lupine	Pea	Heat Labile
Alkaloids	n/a	n/a	n/a	n/a	9.6%[3]	n/a	
Trypsin/Chymotrypsin (TIU/mg DM)	10.43[9]	3.3-6.2	15.05[7]	2-5	<1	1.7-11.4[1] 2.0-15.0[2] 1.8-14.6[4]	Y
Lectins (HU)	180	640	-	2-8	2.4-7.5 g/kg[6]	100-400	Y
α-Galactosides (%)	4.84[9]	3-4.3	-	2.7	0.1-10[11]	2.3-6.3[5] 4.7-6.0[6]	N
Tannins	0.31[10]	0.72[10]	0.18%[7]	0.70[10]	0.20[10]	0.62[10]	N
Phytic acid (mg/g)	-	1-1.2%	3.05[7]	5.9[12]	-	7.0-15.6 μmol/g[5] 6.3-9.9[6]	N
Vicine/Convicine (μg/g)[10]	52/51	4136/ 2219	-	114.7/ 94.7	34.7/ 97.0	41.5/ 105.7	Y
b-ODAP (μg/g)	n/a	n/a	89.5[7] 0.270%[8]	n/a	n/a	n/a	

[1] Leterme et al., 1992
[2] Cousin et al., 1993
[3] Stobiecki et al., 1992
[4] Muel et al., 1998
[5] Frias et al., 1998
[6] Gelencser et al., 1998
[7] Grela and Winiarska, 1998
[8] Malek et al., 2000
[9] Frias et al., 2000
[10] Saini, 1993
[11] Kuo et al., 1988
[12] Gorospe et al., 1992

Lectins (Phytohemagglutinins)

Lectins are sugar-binding proteins which agglutinate cells and precipitate glycoconjugates (Etzler, 1985). Lectins adhere to the glycoproteins present on the inner lining of the small intestine and inhibit absorption of food particles. Legume lectins are divided into two groups, one- and two-chain lectins. The two chains are the product of proteolytic cleavage of a precursor protein (van Driessche, 1988). Lectin proteins are stored in protein bodies within the cell (Grant and van Driessche, 1993). Lectins may have some involvement with host defense mechanisms based on the homology between arcelin and phytohemagglutinin (Osborn et al., 1988).

Tannins

Tannins are water-soluble phenolic compounds commonly found in the testa of food legume seed and produce a bitter flavor when eaten. Tannins are classified as hydrolysable or condensed based on their chemical properties. Condensed tannins are flavonoid polymers and have individual monomers joined through carbon bonds and resist hydrolysis. Hydrolysable tannins are gallic or hexahydroxydiphenic acid esters of glucose (Butler and Bos, 1993). The tannins are capable of binding proteins and reduce digestibility in the gut. Despite their antinational characteristics the tannins also function to deter birds and insects and provide resistance to some environmental stresses. Reduced tannin in lentil has resulted in loss of flavor. Tannins also reduce cellulase activity in ruminant animals limiting their ability to digest the seeds. Tannins do not affect ruminants to the same degree.

Trypsin and Chymotrypsin Inhibitors

Trypsin and chymotrypsin inhibitors are part of a larger group of protease inhibitors found in food legumes. They are found in all food legumes and inhibit the activity of trypsin and chymotrypsin in the digestive system reducing protein digestibility. Inhibition of enzyme activity has been linked to increased production of pancreatic enzymes and subsequent pancreatic enlargement in rats.

Trypsin inhibitor activity (TIA) is reported in trypsin inhibitor units (TIU). One trypsin unit is defined as an increase in 0.01 absorbance units at 410 nm per 10 ml of reaction mixture (Gueguen et al., 1993). Trypsin inhibitor activity in food legumes ranges from < 2-15 TIU

(Table 6). The level of TIA is controlled both genetically and variation in growing environment (Leterme et al., 1992).

Alkaloids

Alkaloid content in the food legumes is most often associated with lupins (*Lupinus albus* and *L. angustifolius* L.). Lupin seed can contain up to 9.6% (dry weight basis) of alkaloid (Stobeicki et al., 1992). Alkaloids produce a bitter taste and are diverse in chemical composition. Papadoyannis and von Baer (1993) summarized alkaloids as: (1) having nitrogen in their basic components, (b) comprised of complex components derived from various amino acids, and (c) providing a characteristic color and fluorescence reaction with many reagents and produce pharmacological effects on various organs of humans and animals. Eight different categories of alkaloids have been established based on chemical structure: (1) pyridine, piperidine, and quinolizidine, (2) tropane, (3) quinoline, (4) isoquinoline, (5) bisbenzylisoquinoline, (6) indole, (7) steroidal, and (8) purine alkaloids (Papadoyannis and Baer, 1993).

RFOs

The raffinose oligosaccharides (RFOs) have successive galactose molecules attached via an alpha 1-4 linkages to sucrose. Four RFOs have been identified in food legumes, raffinose, stachyose, verbascose, and ajugose; each has one additional galactose molecule attached beginning with one for raffinose and four for ajugose. The human and monogastric gut lacks the alpha 1-4 galactosidase enzyme necessary to metabolize these sugars. As a result, they are passed into the intestine where enteric bacteria metabolize them and produce methane, carbon dioxide, and hydrogen gas as byproducts. Accumulation of these gases and subsequent release causes gastro-intestinal and social discomfort. These sugars are important biologically to maintain cell integrity during cold and desiccating conditions and are used as a carbon source during germination.

Soaking reduced alpha-galactosides 16-27%, soaking plus cooking reduced RFOs 45-58% and dry heating reduced RFOs 46% in chickpea (Frias et al., 2000).

Phytic Acids

Phytic acid is the storage form of phosphorus in the legume seed and is present as a mixed salt with various cations (Gorospe et al., 1992).

Phytates in the food legumes reduce mineral absorption through forming complexes with calcium and magnesium. Rickets, a bone disorder, can also result in severe cases from the removal of these minerals from the diet. Phytates also reduce protein and lipid bioavailability. Phytases are activated during germination and are responsible for the hydrolysis of phytates. Germination of seeds prior to food preparation has been used as one way to reduce the antinutritional effect of phytates.

Vicine/Convicine/DOPA (Dihydroxyphenylalanine)

Favism or haemolytic anemia is caused by a lack of glucose-6-phosphate dehydrogenase and is particularly serious in infants. Vicine and convicine are hydrolysed to produce divicine and isouramil which produce free radicals in the blood stream and reduce glutathione in erythrocytes (Jansman and Longstaff, 1993).

Antigenicity Factors

Antigenic or allergenic factors in legume seeds are often undigested or partially digested proteins which incite allergic or hypersensitive reactions when they are absorbed through the intestinal wall and passed into the lymph system where they stimulate the immune system (Gueguen et al., 1993). Allergic and hypersensitive reactions in young livestock reduce intake and digestibility and in severe cases may cause diarrhea and death.

GRAIN LEGUME CROPS

Pea

Pea is consumed in many forms including dry and immature seed as well as immature pods. The multiple uses of peas require numerous quality considerations for marketed products. This chapter will consider only those considerations related to dry seed used for human and animal consumption (Table 2). Dried pea seed contains between 21 and 33% protein, 37 and 49% starch, and 4-7% fiber which is primarily in the testa. Other minor components include vitamin A, niacin, riboflavin, thiamine, and minerals. The antinutritional factors which peas contain are listed in Table 3 and include trypsin/chymotrypsin inhibitors, raffinose series oligosaccharides, lectins and in the Austrian winter and maple pea types, tannins are found in the testa.

Approximately 70% of world production enters the animal feed market while the remaining 30% is used for human consumption. The major antinutritional factors present in pea are destroyed by heat or germination processes during preparation. The effect of RFOs can be minimized if the soak water is thrown out and the seed rinsed with tap water prior to cooking. Numerous mutants affecting the starch biosynthetic pathway have been characterized. The most important is the *r* locus which affects the starch branching enzyme. Cultivars homozygous for the dominant allele have round seeds while homozygous recessive cultivars have wrinkled seeds.

Trypsin inhibitor (TI) is higher in winter compared to spring pea cultivars (Valdebouze et al., 1980). In addition, 'Progreta' and 'Maro', two marrowfat peas contain approximately three times as much TI as other cultivars.

Lentil

Numerous types of lentil are produced and consumed worldwide including those with large seeds and yellow cotyledons and those with small seeds with red or yellow cotyledons and with the seedcoats variously pigmented. Nearly all the lentil produced in the world is for human food in various preparations. However, the fodder from lentil harvest is valued as an animal feed and often commands prices equal or sometimes exceeding that of the grain. In South Asia where the majority of the world's lentil crop is produced, it is primarily used to make "dhal," a pulse that has had the seedcoats removed and the cotyledons split. Dhal is used to prepare a form of soup that is consumed mostly with rice. In the Middle East and North Africa lentil is most often used as a whole pulse in soups and in mixtures with rice often called "mujadurra." Decorticated and split lentil is also used in soup making. There are numerous other preparations made from whole lentil including stews with lamb and mixtures with other vegetables. In developed countries, whole lentils with seedcoats intact are used in a variety of dishes including soups, stews, salads, and various mixtures that include rice. Removing the seedcoats of lentil improves appearance, color appeal of dishes made from lentil and also the taste, although this is a matter of personal opinion. Lentils are sometimes sprouted and used in soups and stews, but this use is not common.

Chickpea

Chickpeas are conveniently divided into two major types: "desi" and "kabuli." Desi types are generally small seeded with an angular shape, rough seed surfaces and are variously pigmented. Desi types are the principal type produced in South Asia and East Africa and can be tan, green or brown with various markings on the seedcoats, or the seeds can be solid black. Desi types are produced in greater quantities in the world when compared to kabuli types and most are produced in South Asia. Nearly all the chickpeas produced in the world are used for food rather than as feed. Desi chickpeas are most often used to make dhal, a process of decorticating the seeds and splitting the cotyledons. Dhal is made to shorten cooking time, reduce fiber and improve appearance of the preparation. Chickpea can also be made into flour that can blended with wheat flour in bread making. Other uses of ground chickpea include falafel, a preparation that uses a ground pulse, as the main ingredient, mixed with green vegetables such as onion, parsley and spiced with coriander, garlic, peppers, and cumin and deep fried. Falafel is popular in the Middle East and South Asia. Other uses for chickpea flour are as a medium to deliver spices and other flavors to food preparations.

Kabuli types are generally large seeded (greater than 35 grams per 100 seeds) but can range from very small (less than 20 grams per 100 seeds) to very large (greater than 70 grams per 100 seeds). Kabuli types are usually light tan or cream colored with thin seedcoats and their shape is sometimes described as "ramshead." Kabuli types are common in the Middle East, the Mediterranean region, North Africa and North America and South America. Kabuli chickpeas vary in size, shape and seedcoat coloration. "Café" or the light cream colored types which are large seeded (greater than 45 grams per 100 seeds) are the most popular. "Spanish White" type varieties have exceptionally large seeds (greater than 56 grams per 100 seeds) with white seedcoats which gives them a unique and attractive appearance. This Spanish White type is popular in Mexico, Spain, and other countries of the Mediterranean region. Other kabuli types include those that are small seeded and round, almost pea shaped, and some that have rough seed surfaces. Uses for kabuli chickpeas other than in main dishes include roasting to make snack items that are popular in the Mediterranean region, the Middle East, and South Asia. For the roasting process, the seeds are conditioned by adding water, roasting for a short time and adding flavorings during or immediately after roasting. Some kabuli type varieties have orange cotyledons and "salmon" colored seedcoats.

Most chickpeas are consumed as whole chickpeas that have been soaked, cooked and added to soups and stews. In the U.S., canned chickpeas are a popular component in restaurant salad bars. To be suitable for canning, chickpeas must soak up their weight in water, retain their color and integrity during processing and the liquid in the can must remain clear.

Faba Bean

Faba bean, sometimes called broad bean, is consumed either as a fresh cooked vegetable or as a dry pulse in various preparations. Faba bean is a staple pulse in Egypt, Morocco, Ethiopia and other Middle East countries where it is used to prepared a number of popular dishes including but not restricted to "Foul akhdar," "Foul matbookh," "Foul medamis," and "Foul nabet" (Gabrial, 1982). The former two preparations are made from the immature seeds and pods, respectively; while the latter two preparations are made from the dried seeds. Foul medamis is consumed as a breakfast food and is prepared by slowly boiling the seeds for 10 to 12 hours followed by adding salt, cotton seed oil, and lemon juice. Foul nabet is prepared by first germinating the seeds followed by boiling for about one hour, followed by adding some fried garlic. Falafel is also a popular preparation made from faba bean and is made similarly to that made from other pulses. When used in animal feed mixtures, faba bean is an important protein supplement; however, the proportion of the ration is usually restricted to 15 to 20%.

Favism is a severe and sometimes fatal haemolytic anemia in humans that is associated with consuming faba beans. It is generally thought that concentrations of vicine/convicine and the aglycones, divicine, isouramil, and dihydroxyphenylalanine (L-DOPA) in faba bean seed is the cause of the problem. These compounds are troublesome because they are not completely inactivated by cooking. Other compounds in faba bean, particularly haemagglutinins, appear to cause blood disorders; fortunately, they are mostly destroyed by cooking (Williams et al., 1994). Genetic variation for concentrations of vicine/convicine and L-DOPA exists and is reportedly highly heritable (Bjerg et al., 1985). However, as far as is known there are no breeding programs underway that are designed to reduce concentrations of these compounds in new faba bean varieties.

Grasspea

Grasspea is primarily an animal feed but is often used as food in times of famine. Grasspea is know to have harmful concentrations of the neurotoxin beta-oxalyl amino-alanine (BOAA) which causes irreversible paralysis of the lower limbs. For the safe use of grasspea as a food, these neurotoxins must be removed during food preparation. BOAA, an amino acid analogue that is the primary cause of lathyrism, has been the subject of intensive plant breeding to reduce the concentrations in grasspea seed to non-harmful amounts. Considerable progress has been made in the national plant breeding programs of Bangladesh, India, Canada, and Ethiopia. ICARDA has also been involved in the development of low BOAA germplasm of grasspea.

The value of the crop is its ability to survive and produce under conditions of extreme drought as well as water logging. To develop the crop as a valuable food legume to be grown in regions of drought, poor soils and wet conditions would be a considerable advance for food security in those marginal areas. Grasspea has many of the advantages of a food legume including high protein concentration, a favorable balance of essential amino acids and adaptation to marginal areas.

Lupin

Four *Lupinus* species, *L. albus*, *L. agustifolius*, *L. luteus*, and *L. mutabilis*, have been domesticated and are currently viable crops throughout the world; while *L. cosentinii* appears to have potential for domestication (Pate et al., 1985). By far the most important of these species is *L. agustifolius* which is grown on nearly two million acres in Australia. *Lupinus albus* is important in northern Europe and *L. mutabilis* is important in South America.

GENETIC CONTROL OF ACCUMULATION

Genetic control of the compositional quality in the grain legumes has not been well studied. Only a few studies can be found in the literature. As quality issues in the grain legumes becomes more important greater understanding of the genetic and environmental influence on seed quality will develop. Research is currently underway in several of the grain legumes to reduce or eliminate the antinutritional factors, i.e., alkaloids that cause lathyrism in grasspea.

Duc and Cubero (1998) reported that two recessive genes (*zt1* and *zt2*) eliminate tannin production in *Vicia faba*. These genes are comparable to the genes in pea and lentil that result in the absence of tannin in the testa. Kabuli chickpea seed generally lack tannins in the seedcoat.

Vicine and convicine in faba bean is controlled by a single recessive gene (*vc*) which reduces seed content 20-fold (Duc and Cubero, 1998). *vc* increases susceptibility to pathogens and parasites, therefore, extra care must be taken to select for pest resistance in this type. Sixdenier and Duc (1989) reported that the genetic effect was greater than the environmental effect suggesting that improvement could be achieved through breeding.

It is likely that very little of the genetic variation available in germplasm collections for nutritional parameters has been explored or used. (This statement may be generally true, but the tannin genes have been of general use in chickpea and pea. Also, the r genes in pea are used extensively in processing peas.)

IMPROVEMENT THROUGH BREEDING

Sufficient variation for the antinutritional factors (i.e., trypsin inhibitors, tannins, lectins, etc.) is present in grain legume germplasm to allow significantly reduced accumulation. Low heritability for several of the ANFs gives them relatively low priority compared to other more profitable traits and those with greater potential for improvement. Several examples of near elimination have been reported in the literature, i.e., tannin in faba bean (Bond and Smith, 1989). It must be kept in mind that many antinutritional compounds serve as natural insecticides and fungicides in plant defense systems. Reduction in these compounds must be accompanied by careful evaluation and determinations of the effects on susceptibility to pathogens and insects and ultimately the effects on yield. An example of successful reduction of an antinutritional factor through breeding is alkaloid content in lupin. Some 'sweet' lupin contains less than 0.25 percent alkaloid.

Though molecular approaches using transgenes is an option, their functional benefits will depend on substantial testing to identify any pleiotropic effects. The allergenic potential of transgenes to enhance protein quality/quantity will need to be elucidated prior to final acceptance and large scale production. Traditional plant breeding will continue to be the primary mode of crop improvement even beyond genetic engineering.

The advent of molecular DNA technology has produced the term, molecular breeding. Molecular breeding can take two meanings: (1) selection of desirable genotypes through the use of tightly linked DNA markers, and (2) incorporation of novel gene combinations through genetic engineering and transformation. These technologies have provided plant breeders with the ability to insert useful genes into genotypes lacking those genes naturally. Genetic engineering of plant species focused initially on disease resistances governed by a single gene. These traits were of high agronomic importance and were relatively easy to work with. More complex traits such as yield and end-use quality have lagged behind due the multigenic control of many of these characters. Recent understanding of the biochemical pathways and synthesis have allowed scientists to identify rate limiting steps and target specific sites within the biosynthetic pathway.

Improved protein quality has been accomplished through incorporation of genes encoding specific storage proteins which contain enhanced levels of individual limiting amino acids. These genes can also be over expressed to increase the overall seed protein content.

CONCLUSION

Grain legumes are an important source of nutrition in human and animal diets worldwide and are grown and consumed in nearly every country. Selection during and subsequent to domestication has improved the nutritional value of the seed and reduced the presence of ANFs. Despite this improvement, very little is known of the genetic control governing many seed components. The need for increasing amounts of high quality nutritious food sources to feed the ever increasing world population has put greater emphasis on understanding the genetic factors governing seed quality. Rapid improvements in seed quality will be realized as knowledge of seed quality increases and more crop improvement programs begin selecting directly for improved nutritional status of the grain legumes.

REFERENCES

Bjerg, B., J.C.N. Knudsen, O. Olsen, M.H. Poulsen, and H. Sorensen. 1985. Quantitative analysis and inheritance of vicine and convicine content in seeds of *Vicia faba* L. Journal of Plant Breeding. 94:135-148.

Bond, D.A. and D.B. Smith. 1989. Possibilities for the reduction of antinutritional factors in grain legumes by breeding. J. Huisman, T.F.B. van der Poel, and I.E. Liener (Eds.). In: Proceedings of the First International Workshop on 'Antinutritional Factors (ANF) in Legume Seeds,' November 23-25, 1988, Wageningen, The Netherlands. pp. 285-296.

Butler, L.G. and K.D. Bos. 1993. Analysis and characterization of tannins in faba beans, cereals and other seeds. A literature review. In: Recent Advances of Research in Antinutritional Factors in Legume Seed. A.F.B. van der Poel, J. Huisman and H.S. Saini (Eds.) Wageningen Pers, Wageningen, The Netherlands. pp. 81-90.

Carbonaro, M., G. Grant, M. Cappelloni, and A. Pusztai. 2000. Perspectives into factors limiting *in vivo* digestion of legume proteins: antinutritional compounds or storage proteins. J. Agric. Food Chem., 48:742-749.

Clemente, A., J. Vioque, R. Sanchez-Vioque, J. Pedroche, J. Bautista, and F. Millan. 2000. Factors affecting the *in vitro* protein digestibility of chickpea albumins. J. Sci. Food Agric., 80:79-84.

Cousin, R., D. Tome, and T. Gaborit. 1993. The genetic variation in trypsin inhibitor activity among varieties of pea (*Pisum sativum* L.). In: Recent Advances of Research in Antinutritional Factors in Legume Seed. A.F.B. van der Poel, J. Huisman and H.S. Saini (Eds.) Wageningen Pers, Wageningen, The Netherlands. pp. 173-177.

Deshpande, S.S. and S. Damodaran. 1990. Advances in Cereal Science and Technology, 10: 147-241.

Duke, J.A. 1981. Handbook of Legumes of World Economic Importance. Plenum Press, New York, New York. 343 p.

Etzler, M.E. 1985. Plant lectins: molecular and biological aspects. Ann. Rev. Pl. Physiol. 36:209-234.

FAO. 2000. <http://apps1.fao.org/cgi-bin/nph-db.pl?subset=agriculture>.

Frias, J., C. Vidal-Valverde, C. Sotomayor, C. Diaz-Pollan, and G. Urbano. 2000. Influence of processing on available carbohydrate content and antinutritional factors of chickpeas. Eur. Food Res. Technol., 210:340-345.

Frias, J., G. Vicente, M. Pacual-Montaner, P. Morales and C. Vidal-Valverde. 1998. An assessment of variation of antinutritional factors in pea seeds. pp. 166-167. In: Proceedings of the 3rd European Conference on Grain Legumes. November 14-19, 1998. Valladolid, Spain.

Gabrial, G.N. 1982. The role of faba beans in the Egyptian diet. In: G. Hawtin and C. Webb. (Eds.) World Crops: Production, Utilization, Description. Volume 6. Faba Bean Improvement. Martinus Nijhoff Publishers. The Hague, The Netherlands, pp. 309-315.

Gelencser, E., G. Hajos, Z. Zdunczyk, and L. Jedrychowski. 1998. Content of lectins in respect to the levels of protein and other antinutritional factors in Polish pea cultivars. pp. 170-171. In: Proceedings of the 3rd European Conference on Grain Legumes. November 14-19, 1998. Valladolid, Spain.

Gorospe, M.J., C. Vidal-Valverde, and J. Frias. 1992. The effect of processing on phytic acid content of lentils. In: Proceedings of the 1st European Conference on Grain Legumes. June 1-3, 1992. Angers France. pp. 425-426.

Grant, G. and E. van Driessche. 1993. Legume lectins: physicochemical and nutritional properties. In: Recent Advances of Research in Antinutritional Factors in Legume Seed. A.F.B. van der Poel, J. Huisman and H.S. Saini (Eds.) Wageningen Pers, Wageningen, The Netherlands. pp. 219-233.

Grela, E.R., and A. Winiarska. 1998. Influence of different conditions of extrusion on the antinutritional factors content in grass pea (*Lathyrus sativus* L.) seeds. pp. 182-183.

In: Proceedings of the 3rd European Conference on Grain Legumes. November 14-19, 1998. Valladolid, Spain.

Gueguen, J., M.G. van Oort, L. Quillien, and M. Hessing. 1993. The composition, biochemical characteristics and analysis of proteinaceous antinutritional factors in legume seeds. A review. In: Recent Advances of Research in Antinutritional Factors in Legume Seed. A.F.B. van der Poel, J. Huisman and H.S. Saini (Eds.) Wageningen Pers, Wageningen, The Netherlands. pp. 9-30.

Hulse, J.H. 1994. Nature, composition and use of food legumes. In: Expanding the Production and Use of Cool Season Food Legumes, F.J. Muehlbauer and W.J. Kaiser (Eds.). Kluwer Academic Publishers. pp.77-97.

Jansman, A.J.M. and M. Longstaff. 1993. Nutritional effects of tannins and vicine/convicine in legume seeds. In: A.F.B. van der Poel, J. Huisman and H.S. Saini (Eds.), Recent Advances of Research in Antinutritional Factors in Legume Seed. Wageningen Pers, Wageningen, The Netherlands.

Khvostova, V.V. 1975. Genetics and Breeding of Peas. Nauka Publ., Novosibirsk. (Translated from Russian by B.R. Sharma, Oxonian Press, New Delhi, 1983.

Kuo, T.M., J.F. Van Middleswarth, and W.J. Wolf. 1988. Contents of raffinose oligosaccharides and sucrose in variouse plant seeds. Journal of Agriculture and Food Chemistry, 36:32-36.

Ladizinsky, G. and A. Adler. 1976. The origin of chickpea (*Cicer arietinum* L.). Euphytica, 25:211-217.

Leterme, P., T. Monmart, and A. Thewis. 1992. Varietal distribution of trypsin inhibitors in peas. In: Proceedings of the 1st European Conference on Grain Legumes. June 1-3, 1992. Angers France. pp. 417-418.

Liener, I.E. 1989. Antinutritional factors in legume seeds: state of the art. In: Proceedings of the First International Workshop on 'Antinutritional Factors in Legume Seeds,' November 23-25, 1988. Wageningen, The Netherlands. J. Huisman, T.F.B. van der Poel and I.E. Liener (eds.). Pudoc, Wageningen, The Netherlands.

Li-juan, L., Y. Zhao-hai, Z. Zhow-jie, X. Ming-shi, and Y. Han-qing. 1993. Faba bean in China: State-of-the-art review. International Center for Agricultural Research in the Dry Areas, Aleppo, Syria. pp. 144.

Makasheva, R. Kh. 1973. The pea Kolos Publ., Leningrad. (Translated from Russian by B.R. Sharma, 1983. Oxonian Press, New Delhi.)

Malek, M.A., A. Afzal, M.M. Rahman, and A. Salahuddin. 2000. *Lathyrus sativus*: A crop for harsh environments. In: R. Knight (Ed.). Linking Research and Marketing Opportunities for Pulses in the 21st Century. Kluwer Academic Publishers, The Netherlands, pp. 369-373.

Muel, F.B. Carrouee and F. Grosjean. 1998. Trypsin inhibitors activity of pea cultivars: New data and a proposal strategy for breeding programme. Proceedings of the 3rd European Conference on Grain Legumes. Valladolid, Spain, 14-19 November.

Osborn, T.C., D.C. Alexander, S.S. Sun, C. Cardona, and F.A. Bliss. 1988. Insecticidal activity and lectin homology of arcelin seed protein. Science, 240:207-210.

Papadoyannis, I.N. and D. von Baer. 1993. Analytical techniques used for alkaloid analysis in legume seeds. In: A.F.B. van der Poel, J. Huisman and H.S. Saini (Eds.), Recent advances of research in antinutritional factors in legume seed. Wageningen Pers, Wageningen, The Netherlands, pp. 131-145.

Pate, J.S., W. Williams, and P. Farrington. 1985. Lupin (*Lupinus* spp.). In: R.J. Summerfield and E.H. Roberts (Eds.). Grain Legume Crops. Collins, London, pp. 699-746.

Reddy, N.R, M.D. Pierson, S.K. Sathe, and D.K. Salunkhe. 1984. Chemical, nutritional and physiological aspects of dry bean carbohydrates: A review. Food Chemistry, 13:25-68.

Saini, H.S. 1993. Distribution of tannins, vicine and convicine activity in legume seeds. In: Recent Advances of Research in Antinutritional Factors in Legume Seed. A.F.B. van der Poel, J. Huisman and H.S. Saini (Eds.) Wageningen Pers, Wageningen, The Netherlands. pp. 95-100.

Sixdenier, G. and G. Duc. 1989. Evaluation of genotype and location effects on vicine and convicine contents in *Vicia faba* L. FABIS Newsletter, 24:6-7.

Smartt, J. 1990. Pulses of a classical world. In: Grain legumes. Cambridge University Press, Cambridge, UK, pp. 176-244.

Smartt, J. and T. Hymowitz. 1985. Domestication and evolution of grain legumes. In: R.J. Summerfield and E.H. Roberts (eds.) Grain Legume Crops. Collins Professional and Teck. Books, London, pp. 37-72.

Stobiecki, M., M. Markiewiz, Z. Michalski, and K. Gulewicz. 1992. New concept of the bitter lupin seeds utilization. In: Proceedings of the 1st European Conference on Grain Legumes. June 1-3, 1992. Angers France. pp. 423-424.

Valdebouze, P., E. Bergeron, T. Gaborit, and J. Delort-Laval. 1980. Content and distribution of trypsin inhibitors and hemagglutinins in some legume seeds. Canadian Journal of Plant Science, 60: 695-701.

van Driessche, E. 1988. Structure and function of leguminous lectins. In: H. Franz (ed.), Advances in Lectin Research 1, VEB Verlag Volk and Gesundheit, Berlin. pp. 73-134.

Williams, J.T., A.M.C. Sanchez, and M.T. Jackson. 1974. Studies on lentils and their variation. I. The taxonomy of the species. SABRAO Journal, 6:133-145.

Williams, P.C., R.S. Bhatty, S.S. Deshpande, L.A. Hussein, and G.P. Savage. 1994. Improving nutritional quality of cool season food legumes. In: F.J. Muehlbauer and W.J. Kaiser (eds.), Expanding the Production and Use of Cool Season Food Legumes. Kluwer Academic Publishers. Dordrecht, The Netherlands, pp. 113-129.

Zohary, D. 1972. The wild progenitor and place of origin of the cultivated lentil, *Lens culinaris*. Economic Botany, 26:236-332.

Zohary, D. 1973. The origin of cultivated cereals and pulses in the near East. Chromosomes Today, 4:307-320.

Improved Quality of Lentil
Using Molecular Marker Technology

P. W. J. Taylor
R. Ford

SUMMARY. A number of molecular marker techniques have been applied to improve the quality of cultivated lentil, aiding in the breeding and release of better adapted varieties and in quality assurance of export product. Continued increase in lentil production is dependent on the identification of present and potential limiting factors to the progression of lentil breeding. In particular, this chapter is concerned with limitations related to the threat from disease, limitations on the availability of genetic resources and quality assurance issues. The application of molecular markers in the development of strategies to help combat such threats is discussed. *[Article copies available for a fee from The Haworth Document Delivery Service: 1-800-HAWORTH. E-mail address: <getinfo@haworthpressinc. com> Website: <http://www.HaworthPress.com> © 2002 by The Haworth Press, Inc. All rights reserved.]*

KEYWORDS. Lentil, plant breeding, molecular markers, mapping, ascochyta blight, diversity, vetch, quality assurance

P. W. J. Taylor is Senior Research Fellow, and R. Ford is Research Fellow, Molecular Plant Genetics and Germplasm Development Group, Joint Centre for Crop Improvement, Department of Crop Production, The University of Melbourne, Victoria, 3010, Australia.

Address correspondence to: P. W. J. Taylor, Molecular Plant Genetics and Germplasm Development Group, Joint Centre for Crop Improvement, Department of Crop Production, The University of Melbourne, Victoria, 3010, Australia (E-mail: paulwjt@ unimelb.edu.au).

[Haworth co-indexing entry note]: "Improved Quality of Lentil Using Molecular Marker Technology." Taylor, P. W. J., and R. Ford. Co-published simultaneously in *Journal of Crop Production* (Food Products Press, an imprint of The Haworth Press, Inc.) Vol. 5, No. 1/2 (#9/10), 2002, pp. 213-226; and: *Quality Improvement in Field Crops* (ed: A. S. Basra, and L. S. Randhawa) Food Products Press, an imprint of The Haworth Press, Inc., 2002, pp. 213-226. Single or multiple copies of this article are available for a fee from The Haworth Document Delivery Service [1-800-HAWORTH, 9:00 a.m. - 5:00 p.m. (EST). E-mail address: getinfo@haworthpressinc.com].

INTRODUCTION

Lentil is an important pulse crop worldwide and is mostly grown for human consumption as a rich supply of carbohydrate and protein. Lentil cultivars are adapted to soils unsuited for lupins and to areas too dry for chickpea and field pea thus play an important role in rotation with cereals and canola as a disease and weed break and as a nitrogen-fixer for maintaining soil fertility. Lentil seed, either as whole seed or split seed, is also an important export crop for Australia, Turkey, Canada, and India.

Fungal diseases are major biotic factors limiting production and quality of lentils. Disease such as ascochyta blight (*Ascochyta lentis*) (Figure 1), botrytis grey mould (*Botrytis cinerea*), fusarium wilt (*Fusarium oxysporum*), and anthracnose (*Colletotrichum truncatum*) reduce the on-farm productivity, stability of supply, and the marketability of grain. Currently in Australia there is an average annual loss of yield of 10% due to ascochyta blight and blemished lentil grain caused by ascochyta blight infection of the seed attracts a lower price on the international market. This will continue unless better control measures are developed such as the release of new resistant genotypes to the industry. Implementation of resistant lentil cultivars will provide higher yields and lead to premium quality export seed. A greater understanding of the mechanisms underlying resistance will improve the ability of lentil breeders to devise efficient breeding systems.

New breeding technologies that utilize molecular markers, such as DNA fingerprinting and marker assisted selection, will assist breeders to develop new resistant genotypes. Molecular markers are short sequences of DNA used to tag genes or unique DNA sequences in genomes.

Molecular markers closely linked to genes governing resistance to diseases such as ascochyta blight and anthracnose, can be used to identify resistant germplasm during glasshouse backcrossing programs and during field testing. The development of rapid, routine molecular tests will enable lentil breeders to quickly screen seedlings from crosses and identify those that contain the resistance genes. These tests will be reliable and efficient, as screening trials will not depend on infection by the pathogens either by natural or artificial sources, and will not be subject to variable environmental influence or plant growth stage. Growers will readily replace susceptible cultivars with adapted resistant cultivars in regions where disease losses continue to be significant. Implementation of resistant genotypes will also enable exporters to reliably supply pre-

FIGURE 1. Disease caused by *Ascochyta lentis* infection under field conditions on: a. pod and b. leaf.

a.

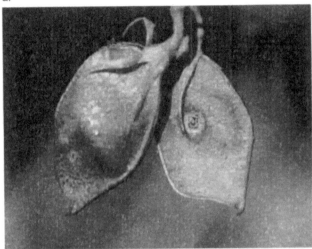

b.

mium quality pulse types, such as light colored green lentil, where the need for immaculate grain appearance without blemishes from disease is paramount.

DNA fingerprinting provides a practical and efficient method for determining the level of genetic diversity within lentil breeding germplasm and for assessing batch quality assurance in export lentil seed.

Genetic variation is necessary to achieve genetic progress in cultivated crops. This will result in improved quality of seed and disease resistance. Assessing batch quality assurance in export lentil seed, such as split, red lentils, is achieved by certifying the level of variety purity. Split, red lentil, an important export pulse, is morphologically identical and hence, difficult to differentiate from split vetch (*Vicia sativa*), a low value stock feed product that contains a human neurotoxin. Recent incidences of wrongly labeled split, red lentil being exported as split vetch has affected the international trade in Australian lentils. Also, in Australia, these incidences of mislabeling have caused public concern as to the identification and quality of split, red lentil being sold in local retail outlets.

Through molecular marker technology, easy-to-use tests can be developed for crop export shipment certification and for applications such as seed classification. Molecular markers are a quicker and far more precise method for identifying species than biochemical and visual methods. Molecular marker technology also has a wider scope for use in genotype fingerprinting and in the future, may be used to identify between genetically modified grain legumes and non-genetically modified lines.

IMPROVED DISEASE RESISTANCE IN LENTIL

Ascochyta blight of lentil (*Ascochyta lentis*) is the major biotic factor limiting lentil production in Australia. The pathogen is present in all lentil-growing regions, with the potential to cause over 50% loss of income through yield decrease and seed discoloration. The search for resistance genes and their introduction into high yielding lentil cultivars, in combination with chemical and cultural methods, is currently the most economically viable and sustainable source of disease control. Therefore, there is an ongoing search for new sources of resistance to this pathogen from within the currently available germplasm. Furthermore, in order to avoid hybridization barriers and unwanted genetic drag it would initially be more advantageous to utilize resistance from within the cultivated species than to explore inter-specific introgression.

Several techniques for screening cool season food legumes for resistance to foliar fungal diseases have previously been reported (Porta-Puglia et al., 1994; Nasir and Bretag, 1997). Although field screening provides an economical method for testing for a reaction to a fungal pathogen and a relatively large number of plants may be tested at once, the result may be somewhat unreliable due to environmental influence and the load of the inoculums. In contrast, glasshouse based screening techniques provide a more reliable method for detailed testing of disease response due to controlled environmental factors and optimal inoculum concentrations (Portu-Puglia et al., 1994).

Disease scoring systems are also an important element of the disease assessment, open to both subjectivity and objectivity from the scorer. Several scoring systems have been employed for assessing *A. lentis* infection of lentil (Ahmed and Morrall, 1996; Nasir and Bretag, 1997a; Ford, 1999).

Knowledge of the variability of virulence of the pathogen and the level of genetic diversity within the pathogen population is an important prerequisite to identifying resistance genes, especially where one or two genes control resistance. The pathogenic variability among isolates of *A. lentis* has been studied by several researchers (Kemal and Morrall, 1995; Ahmed et al., 1996; Nasir and Bretag, 1997b). All of these studies showed that there were differences in the reaction of lentil differentials to *A. lentis* isolates. Therefore, in order to identify any potentially resistant genotypes, a range of *A. lentis* isolates that are diverse in pathogenicity should be used in the screening procedure.

Isolates from different *A. lentis* populations have been classified into virulence forms or pathotype groups. Ahmed et al. (1996) tested 100 isolates collected from Western Canada and 13 other countries and grouped them into 3 pathotypes including weak, intermediate and highly virulent. In Australia, Nasir and Bretag (1997b) examined variation for virulence among 39 Australian isolates and divided them into six distinct pathotypes with different virulence patterns on six lentil lines. Pathotype six was most virulent, causing disease on all differentials while pathotype one was avirulent on all differentials. In another study, Nasir (1998) divided 219 international isolates into 14 pathotypes.

Fungal populations that possess high levels of genetic variation are likely to adapt to a new environment more rapidly than those that do not (McDonald and McDermott, 1993). Therefore, it is important to assess the amount and distribution of genetic variation that is maintained within a pathogen species (within and between populations) in order to predict the likelihood that a control measure will remain effective. Together with an understanding of the host/pathogen interaction, this information is essential for designing disease risk reduction strategies.

The genetic variation of *A. lentis* in Australia was studied by Ford et al. (2000a) who found that variation among *A. lentis* isolates collected from Australia was of similar magnitude to variation among isolates originating from outside Australia. There was a significant increase in the level of genetic variation in the isolates of the pathogen collected from lentil growing regions in Western Australia than in other geographical locations within southern and eastern parts of Australia. This variation was thought to have occurred through multiple introductions into Western Australia of the pathogen from overseas through the import of lentil seed used in breeding programs.

Sexual recombination is known to occur within the reproductive cycle of *Ascochyta lentis*. This is a major cause of genetic variability for this pathogen, thus greatly enhancing its potential for adaptation. *A. lentis* is heterothallic and compatible mating types are required for the production of fertile pseudothecia. Two *A. lentis* mating types were initially identified in the USA (Kaiser and Hellier, 1993). Subsequently, these were both identified in Canada and other major lentil growing regions of the world (Ahmed et al., 1996; Kaiser et al., 1997). Both mating types have also recently been observed within Australia (Nasir, 1998). The introduction of isolates of compatible mating types into lentil growing areas is likely to result in an increased potential for the evolution of the teleomorph (*Didymella*) and thus increased pathogenic variability (Porta-Puglia, 1996).

Indeed, there is clear evidence that *A. lentis* populations are different between geographical areas and have changed over time with the creation of more virulent isolates. Ahmed et al. (1996) studied the pathogenic variability of Canadian *A. lentis* isolates and reported that the cultivar Laird was more susceptible to isolates from Saskatchewan collected in the 1992 growing season than from those collected in 1978. The isolates collected in 1978 induced mainly leaf lesions on Laird seedlings, whilst the isolates collected in 1992 induced severe stem lesions and dieback on Laird. Furthermore, although the cultivar ILL5588 was shown to be resistant to all 84 *A. lentis* isolates collected within western Canada (Ahmed et al., 1996), a more recent study showed that 21 out of 39 Australian isolates were able to overcome this resistance (Nasir, 1998). This suggests the need for additional sources of resistance if Australian breeding lines are to remain resistant to potentially more virulent pathotypes of *A. lentis*.

For effective exploitation of resistance germplasm, it is important to understand the mechanisms of inheritance of the gene(s) controlling resistance. Several studies have been conducted to understand the genetics of resistance to both seed and foliar infection of lentil by *A. lentis*. Sakr (1994) reported that foliar resistance in Laird was governed by a single recessive gene and in ILL5684 resistance was governed by two genes; one dominant and one recessive. Resistance to *A. lentis* in the accession PI 339283 was proposed to be controlled by two dominant genes (Vakulabharanam et al., 1997).

One source of resistance recently exploited in Australian lentil breeding programs was ILL5588 (Michael Materne, pers. comm.). Several studies have been conducted to determine the genetics of the resistance in this accession. Andrahennadi (1997), Vakulabharanam et al. (1997), and Ford et al. (1999) reported that a single dominant gene controlled resistance to foliar infection. Another cultivar reported to contain resistance is Indianhead and according to Andrahennadi (1994; 1997), foliar resistance was conditioned by one recessive gene, whilst seed resistance was conditioned by two recessive genes. Nguyen et al. (2000) found that foliar resistance in Indianhead was conditioned by either one recessive gene or two recessive genes depending on the isolate of *A. lentis* used to inoculate the plants.

In the recent study by Nguyen et al. (2000), the accession ILL7537 was found to be the most resistant among the three putatively resistant accessions (ILL5588, Indianhead and ILL7537) tested under optimal conditions for pathogenic infection in Australia. ILL7537 was highly resistant to all of the Australian *A. lentis* isolates tested with an overall mean disease score of 1.7 (1-9 scale). Resistance was determined to be conditioned by two dominant complimentary genes. This was consistent with Nasir and Bretag (1997a) who tested the reaction of 22 accessions from the 1996 Lentil International Ascochyta Blight Nursery (LIABN-96) to 39 Australian *A. lentis* isolates. The authors reported that ILL7537 maintained a mean disease score of less than 3 against all isolates. In another study, ILL7537 was screened against 219 isolates, collected worldwide and shown to be resistant to all isolates (Nasir, 1998). Therefore, ILL7537 may present a good source for resistance to ascochyta blight within the lentil breeding programs in Australia and worldwide. Furthermore, this source of resistance may be durable since it is polygenic and was maintained against a wide range of pathogenic isolates.

Using bulked segregant analysis, several random amplified polymorphic DNA (RAPD) markers were identified flanking the major dominant resistance gene (AbR_1) in ILL5588 (Ford et al., 1999) (Figure 2). One set of sequence characterised amplified region (SCAR) primers, designed to the internal sequence of one of the RAPD markers, produced a single polymorphic marker ($SCW19_{700}$), which was 11.5 cM

FIGURE 2. Linkage map of RAPD markers flanking the AbR$_1$ resistance locus in cultivar ILL5588. Distance is measured in centiMorgans.

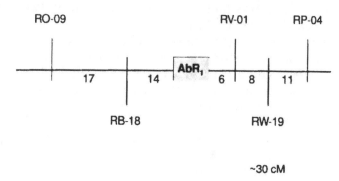

away from the *AbR*$_1$ locus. This marker was subsequently found in resistant cultivars derived from ILL5588 but absent in the resistant cultivars ILL7537 and Indianhead (Ford et al., 2000b). This result indicated that genes controlling resistance in ILL7537 and Indianhead were different to those in ILL5588.

F$_3$ populations from a cross containing ILL7537 have recently been established that segregated for each of the single dominant resistant genes (unpublished). Preliminary results using bulked segregant analysis identified one RAPD marker linked in coupling phase and one in repulsion phase to one of the two dominant resistance genes in ILL7537. The use of molecular markers linked to these resistance genes and the others in ILL5588 and Indianhead will facilitate marker-assisted selection of resistant germplasm and gene pyramiding which will lead to durable field resistance against *A. lentis*.

ENHANCING GERMPLASM DIVERSITY IN BREEDING PROGRAMS

The genetic variation that exists within breeding germplasm is an essential prerequisite for the genetic improvement of any crop. For lentil, as for many other crops, accessions within existing core collections aim to include the genetic diversity of the species in a condensed yet fully representational manner (Erskine et al., 1989). However, programs developed for breeding superior cultivars are often focused more on obtaining the maximum genetic gain for the incorporation of a few agronomic and/or economically important traits, rather than the importance of maintaining a long-term broad genetic base.

Traditional lentil cultivation relied mainly on local landraces of lentil (*L. culinaris* ssp. *culinaris*) whereas, modern lentil cultivars are derived from only a small number of improved landraces that are adapted for specific environmental conditions. Consequently, the current practice of not including more genetically diverse landraces, available from germplasm collections, in breeding programs may have led to a decrease of genetic diversity in those breeding programs (Ladizinsky, 1993). A comprehensive understanding of the genetic variation within any lentil breeding program is important for the efficient selection of parents, for introgression of genetic material into superior cultivated lines and for the implementation of an effective genetic conservation program for the cultivated species.

Genetic diversity in cultivated lentil has been studied using a variety of biochemical methods. These included seed protein electrophoresis (Ladizinsky, 1979b), isozyme analysis (Pinkas et al., 1985; Hoffman et al., 1986; Ferguson and Robertson, 1996) and sodium dodecyl-sulfate polyacrylamide gel electrophoresis (SDS-PAGE; Ahmad et al., 1997a). However, there has been a move away from analysis at the protein level in favor of analysis at the genome level due to the number of loci that may be compared and moreover, the limited number of polymorphic loci revealed with biochemical techniques. The use of DNA-based molecular markers allows for a direct genetic comparison between related individuals without any concern for environmental factors or effects on gene expression levels. Also, DNA techniques allow for the development of a theoretically unlimited number of polymorphic marker loci (Jones et al., 1997).

The genetic diversity of *L. c.* ssp. *culinaris* has been studied with many DNA-based molecular marker systems including: restriction fragment length polymorphism (RFLP) analysis of nuclear DNA (Havey and Muehlbauer, 1989) and chloroplast DNA (Muench et al., 1991; Mayer and Soltis, 1994), amplified fragment length polymorphism (AFLP) analysis (Sharma et al., 1996) and RAPD analysis (Abo-elwafa et al., 1995; Sharma et al., 1995; Ford et al., 1997). RFLP analysis of the chloroplast DNA of two accessions within *L. c.* ssp. *culinaris* showed less than 3% diversity, defined as the fraction of conserved fragments over a total of 90 fragments assessed (Muench et al., 1991). Also, the assessment of 399 restriction sites within the chloroplast DNA of 114 cultivated accessions revealed a single polymorphism (Mayer and Soltis,

1994), which indicated a high level of similarity between these accessions using this technique.

Molecular marker technologies involving the polymerase chain reaction (PCR) and based on specific DNA sequences, such as the coding and non-coding regions of the ribosomal genes, or non-specific sequences such as RAPD analysis have been used for the identification of many plant species and cultivars (Graham et al., 1994; Ko and Henry, 1994; Taylor et al., 1995). These techniques are now commonly used to estimate the genetic diversity within and between plant species and cultivars (Newbury and Ford-Lloyd, 1993; Morell et al., 1995).

In *Lens*, several studies using RAPD analysis have been used to estimate the inter- and intraspecific variation in selected cultivars and accessions included in ICARDA (International Center for Agricultural Research in the Dry Areas) in Syria, USDA (United States Department of Agriculture), New Zealand and Australian breeding programs (Abo-elwafa et al., 1995; Sharma et al., 1995, 1996; Ahmad et al., 1996; Ford et al., 1997). International germplasm breeding programs have been shown to contain relatively low levels of intraspecific diversity among the accessions tested, with maximum levels of 32%-ICARDA, 10%-USDA; 44%-New Zealand; and 46%-Australia.

The mean level of intraspecific diversity measured from the VIDA collection was 23%, which is in contrast to the level of genetic variation that exists among collections of wild relative species such as *L. odemensis* (mean = 45%) and *L. nigricans* (mean = 35%; Abo-elwafa et al., 1995). Indeed, one way to overcome this problem may be to reintroduce variation into the breeding germplasm through introgression of genetic material from wild relative species. Abo-elwafa et al. (1995) proposed that the cultivated species had passed through a "genetic bottleneck" during domestication, whereby intensive selection for individuals containing desirable traits such as larger grain size had led to the exclusion of much genetic variation.

IMPROVED QUALITY ASSURANCE OF EXPORT SEED

The export of Australian lentil into several Indian and Middle Eastern markets has recently been compromised due to uncertainties in the purity of the product. In one incident a shipment of premium split, red lentil was rejected by Saudi Arabian custom officials after arriving at port on the basis that they may be split, red vetch (*Vicia sativa*). The spherical cotyledons of the Blanche Fleur cultivar of *Vicia sativa* L.

(common vetch) are morphologically similar to that of split red lentil, particularly when coated with vegetable oil. Vetch is grown and exported as a high protein stock feed but contains a neurotoxin that affects humans such as causing blindness.

Other cases have occurred where containers of vetch that were wrongly labeled as lentil were imported into India, sold as lentil and caused sickness in certain villages. In a more recent incident, consignments of vetch were deliberately wrongly labeled as lentil and imported into Sri Lanka. In Sri Lanka, lentils attract a concessionary duty whereas animal feed does not. In this case, officials found that the import racket had been carried out over a 12 months period that encompassed four shipments. Therefore, unscrupulous exports and or mixing of seed at point of delivery have caused massive damage to the Australian Pulse Industry's reputation on the international lentil markets.

Vetch is also used as a rotational crop for the re-nitrification of the soil and is highly likely to have been planted previously in a site where red lentil is grown. Different species of wild vetch can also grow in fields planted to lentil. Thus, there will inevitably be a degree of vetch regrowth and subsequent contamination in the following lentil crop. In order for the grains industry to regain an international reputation, an assurance to importers regarding the quality and purity of the product will need to be made.

DNA fingerprinting through the use of molecular markers will provide quality assurance of export lentil seed by certifying the purity of each export, which will ultimately result in a greater market share in these countries and provide a marketing-edge over lentil export competitors.

In order to provide a DNA fingerprint that can definitively differentiate red lentil cultivars from vetch (*V. sativa*) and wild vetch (*Vicia* spp.) a molecular marker test has been developed, based on nucleic acid sequence from the smallest ribosomal 5S rDNA gene. The marker was sufficiently sensitive to detect 0.1 percent vetch seed contamination in a lentil seed admixture by PCR (one vetch seed in 999 lentil seeds) (Figure 3). Commercial samples of split, red lentil collected from various retail outlets around Melbourne, Australia were DNA fingerprinted and analyzed for contamination by vetch. All samples analyzed were found to be lentil with a purity of at least 99%. No vetch was detected in any of these samples labeled as lentil.

FIGURE 3. A vetch-specific marker profile in a PCR titration of vetch:lentil seed admixtures. M, Molecular weight markers; 1, 100% lentil seed; 2, 100% vetch seed; 3, 20% vetch, 80% lentil; 4, 5% vetch, 95% lentil; 5, 2% vetch, 98% lentil; 6, 1% vetch, 99% lentil; 7, 0.1 % vetch, 99.9% lentil.

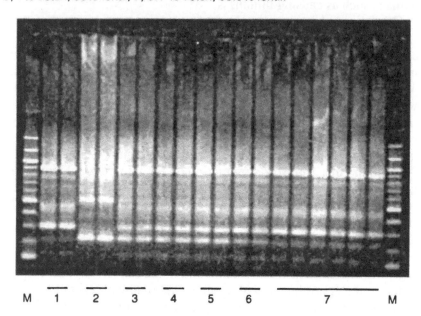

M   1   2   3   4   5   6   7   M

REFERENCES

Abo-elwafa, A., Murai, K., and T. Shimada (1995). Intra- and inter-specific variations in *Lens* revealed by RAPD markers. *Theoretical and Applied Genetics* 90, 335-340.

Ahmad, M., McNeil, D.L., Fautrier, A.G., Armstrong, K.F., and AM. Paterson (1996). Genetic relationships in *Lens* species and parentage determination of their interspecific hybrids using RAPD markers. *Theoretical and Applied Genetics* 92, 1091-1098.

Ahmad, M., Fautrier, A.G., Burritt, D.J., and D.L. McNeil (1997a). Genetic diversity and relationships in *Lens* species and their F_1 interspecific hybrids as determined by SDS-PAGE. *New Zealand Journal of Crop and Horticultural Science* 25, 99-108.

Ahmed, S. and R.A.A. Morrall (1996). Field reactions of lentil lines and cultivars to isolates of *Ascochyta fabae* f. sp. *lentis. Canadian Journal of Plant Pathology* 18: 362-369.

Ahmed, S.K., Morrall, R.A.A., and J.W. Sheard (1996). Virulence of *Ascochyta fabae* f.sp. *lentis* on lentils. *Canadian Journal of Plant Pathology* 18, 354-361.

Andrahennadi, C.P. (1994). Genetics and linkages of isozyme markers and resistance to seed-borne ascochyta infection in lentil. MSc. Thesis, University of Saskatchewan, Saskatoon, Canada.

Andrahennadi, C.P. (1997). RAPD markers for Ascochyta blight resistance, phylogenetic studies and cultivar identification in lentil. PhD. Thesis, University of Saskatchewan, Saskatoon, Canada.

Erskine, W., Adham, Y., and L. Holly (1989). Geographic distribution of variation in quantitative traits in a world lentil collection. *Euphytica* 43, 97-103.

Ferguson, M.E. and L.D. Robertson (1996). Genetic diversity and taxinomic relationships within the genus *Lens* as revealed by allozyme polymorphism. *Euphytica* 91, 163-172.

Ford, R., Garnier-Géré, P., Nasir, M., and P.W.J. Taylor (2000a). The structure of *Ascochyta lentis* in Australia revealed with RAPD markers. *Australasian Plant Pathology* 29, 36-45.

Ford, R., Nguyen, T., and P.W.J. Taylor (2000b). Development of SCARs linked to the AbR_1 gene and characterisation of other novel Ascochyta blight resistance sources in lentil. *Plant and Animal Genome VIII Conference*, San Diego, USA.

Ford, R., Pang E.C.K., and P.W.J. Taylor (1997). Diversity analysis and species identification in *Lens* using PCR generated markers. *Euphytica* 96, 247-255.

Ford, R., Pang, E.C.K., and P.W.J. Taylor (1999). Genetics of resistance to ascochyta blight (*Ascochyta lentis*) of lentil and identification of closely linked RAPD markers. *Theoretical and Applied Genetics* 98, 93-98.

Ford, R. (1999). Development of molecular markers for the genetic improvement of lentil. Ph. D. thesis, The University of Melbourne, Victoria, Australia.

Graham, G.C., Henry, R.J., and R.J. Redden (1994). Identification of navy bean varieties using random amplification of polymorphic DNA. *Australian Journal of Experimental Agriculture* 34, 1173-1176.

Havey, M.H. and F.J. Meuhlbauer (1989). Linkages between restriction fragment length, isozyme and morphological markers in lentil. *Theoretical and Applied Genetics* 77, 839-843.

Hoffman, D.L., Soltis, D.E., Muehlbauer, F.J., and G. Ladizinsky (1986). Isozyme polymorphism in *Lens* (Leguminosae). *Systematic Botany* 11, 392-402.

Jones, N., Ougham, H., and H. Thomas (1997). Markers and mapping: we are all geneticists now. *New Phytologist* 137: 165-177.

Kaiser, W.J. and B.C., Hellier (1993). *Didymella* sp. The teleomorph of *Ascochyta lentis* on lentil straw. *Phytopathology* 83: 692. (*Abstr.*).

Kaiser. W.J., Wang, B.C., and J.D. Rogers (1997). *Ascochyta fabae* and *A. lentis*: Host specificity, teleomorphs (*Didymella*), Hybrid analysis, and taxonomic status. *Plant Disease* 81: 809-816.

Kemal, S.A. and R.A.A. Morrall (1995). Virulence patterns in mating type 1 of *Ascochyta fabae* f.sp. *lentis* in Saskatchewan. *Canadian Journal of Plant Pathology* 17: 358 (*Abstr.*).

Ko, H.-L. and R.J. Henry (1994). Identification of barley varieties using the polymerase chain reaction. *Journal of the Institute of Brewing* 100, 405-407.

Ladizinsky, G. (1979). Species relationships in the genus *Lens* as indicated by seed-protein electrophoresis. *Botanical Gazette* 140, 449-451.

Ladizinsky, G. (1993). Wild lentil. *Critical Review in Plant Sciences* 12, 169-184.

Mayer, M.S. and P.S. Soltis (1994). Chloroplast DNA phylogeny of *Lens* (Leguminosae): origin and diversity of cultivated lentil. *Theoretical and Applied Genetics* 87, 773-781.

McDonald, B.A. and J.M. McDermott (1993). Population genetics of plant pathogenic fungi. *BioScience* 43, 311-319.

Morell, M.K., Peakall, R., Appels, R., Preston, L.R., and H.L. Lloyd (1995). DNA profiling techniques for plant variety identification. *Australian Journal of Experimental Agriculture* 35, 807-819.

Muench, D.G., Slinkard, A.E., and G.J. Scoles (1991). Determination of genetic variation and taxonomy in lentil (*Lens* Miller) species by chloroplast DNA polymorphism. *Euphytica* 56, 213-218.

Nasir, M. (1998). Improvement of drought and disease resistance in lentils in Nepal, Pakistan and Australia. Mid-Term Report, Australian Centre for International Agricultural Research project PN-9436.

Nasir M. and T.W. Bretag (1997a). Resistance to *Ascochyta fabae* f. sp. *lentis* in lentil accessions. Proc. International Food Legume Research Conference III, Adelaide, Australia. p.161

Nasir, M. and T.W. Bretag (1997b). Pathogenic variability in Australian isolates of *Ascochyta lentis. Australasian Plant Pathology* 26, 217-220.

Newbury, H.J. and B.V. Ford-Lloyd (1993). The use of RAPD for assessing variation in plants. *Plant Growth Regulation* 12, 43-51.

Nguyen, T., Brouwer, J.B., Taylor, P.W.J., and R. Ford (*In Press*). Resistance to ascochyta blight (*Ascochyta lentis*) in lentil accession ILL7537. *Australasian Plant Pathology.*

Pinkas, R., Zamir, D., and G. Ladizinsky (1985). Allozyme divergence and evolution in the genus *Lens. Plant Systematic Evolution* 151, 131-140.

Porta-Puglia, A., Bernier, C.C., Jellis, G.J., Kaiser, W.J., and M.V. Reddy. (1994). Screening techniques and sources of resistance to foliar diseases caused by fungi and bacteria in cool season food legumes. *Euphitica* 73: 11-25.

Porta-Puglia, A., Crino, P., and C. Mosconi (1996). Variability in virulence to chickpea of an Italian population of *Ascochyta rabiei. Plant Disease* 80, 39-41.

Sakr B. (1994). Inheritance and linkage of genetic markers and resistance to ascochyta blight in lentil. PhD. Thesis, Washington State University, Pullman, USA.

Sharma, S.K., Dawson, I.K. and R. Waugh (1995). Relationships among cultivated and wild lentils revealed by RAPD analysis. *Theoretical and Applied Genetics* 91, 647-654.

Sharma, S.K., Knox, M.R., and T.H.N. Ellis (1996). AFLP analysis of the diversity and phylogeny of *Lens* and its comparison with RAPD analysis. *Theoretical and Applied Genetics* 93, 751-758.

Taylor, P.W.J., Geijskes, J.R., Ko, H.-L., Fraser, T.A., Henry, R.J., and R.G. Birch (1995). Sensitivity of random amplified polymorphic DNA analysis to detect genetic changes in sugarcane during tissue culture. *Theoretical and Applied Genetics* 90, 1168-1173.

Vakulabharanam, V.R., Slinkard, A.E., and A. Vandenberg (1997). Inheritance of resistance to Ascochyta blight in lentil and linkage between isozyme. Proc. International Food Legume Research Conference III, Adelaide, Australia. p. 162.

Pigeonpea Nutrition
and Its Improvement

K. B. Saxena
R. V. Kumar
P. V. Rao

SUMMARY. Pigeonpea (*Cajanus cajan* [L.] Millsp.), known by several vernacular and trade names such as red gram, tuar, Angola pea, Congo pea, yellow dhal and oil dhal, is one of the major grain legume crops of the tropics and sub-tropics. It is a favorite crop of small holder dryland farmers because it can grow well under subsistence level of agriculture and provides nutritive food, fodder, and fuel wood. It also improves soil by fixing atmospheric nitrogen. India by far is the largest pigeonpea producer where it is consumed as decorticated split peas, popularly called as 'dhal.' In other countries, its consumption as whole dry seed and green vegetable is popular. Its foliage is used as fodder and milling by-products form an excellent feed for domestic animals. Pigeonpea seeds contain about 20-22% protein and appreciable amounts of essential amino acids and minerals. Dehulling and boiling treatments of seeds get rid of the most antinutritional factors such as tannins and enzyme inhibitors. Seed storage causes considerable losses in the quality of this legume. The seed protein of pigeonpea has been successfully enhanced by breeding from 20-22% to 28-30%. Such lines also agronomically performed well and have acceptable seed size and color. The high-protein lines were found

---

K. B. Saxena is Senior Scientist (Breeding), R. V. Kumar is Scientific Officer (Breeding), and P. V. Rao is Scientific Officer (Biochemistry), International Crops Research Institute for the Semi-Arid Tropics, Patancheru 502 324, Andhra Pradesh, India.

[Haworth co-indexing entry note]: "Pigeonpea Nutrition and Its Improvement." Saxena, K. B., R. V. Kumar, and P. V. Rao. Co-published simultaneously in *Journal of Crop Production* (Food Products Press, an imprint of The Haworth Press, Inc.) Vol. 5, No. 1/2 (#9/10), 2002, pp. 227-260; and: *Quality Improvement in Field Crops* (ed: A. S. Basra, and L. S. Randhawa) Food Products Press, an imprint of The Haworth Press, Inc., 2002, pp. 227-260. Single or multiple copies of this article are available for a fee from The Haworth Document Delivery Service [1-800-HAWORTH, 9:00 a.m. - 5:00 p.m. (EST). E-mail address: getinfo@haworthpressinc.com].

227

nutritionally superior to the cultivars because they would provide more quantities of utilizable protein and sulfur-containing amino acids. *[Article copies available for a fee from The Haworth Document Delivery Service: 1-800-HAWORTH. E-mail address: <getinfo@haworthpressinc.com> Website: <http://www.HaworthPress.com> © 2002 by The Haworth Press, Inc. All rights reserved.]*

KEYWORDS. Pigeonpea, *Cajanus cajan* (L.) Millsp., milling, *dhal*, dehusking, cooking time, protein, breeding, amino acid, storage, fodder

INTRODUCTION

In Asia and Africa population growth is the prime development constraint. Recently, India has crossed the alarming 'one billion' population mark and it is catching up fast the most populous country China. Providing significant quantity of quality food to the growing population with limited resources is a big challenge. Due to increase in rural family size, where the population growth is over 2% per year, the land holdings are dividing by each passing generation leading to increased pressure on unit land and reduction in the capacity of farmers to produce sufficient quantity of nutritive food for their families. In producing food crops, such farmers try to keep a balance between cereals and legumes, besides aiming to obtain some fodder for cattle. However, their small holdings and lack of basic resources like water, fertilizer, and pesticides restrict the food production. In general, the food production from small holdings is short both in quantity and quality and under such circumstances, the expectant and nursing mothers and young children are the most vulnerable lots. Recently, a number of weaning and supplementary foods have come in market but due to high prices they remain a luxury of urban middle and upper classes. Since the animal protein is also out of their reach, the problem of malnutrition among poverty-ridden masses is achieving a serious dimension in the country. To meet this challenge, a concerted effort is needed to increase the production of protein-rich legume crops which can be grown under subsistence level of farming and no crop other than pigeonpea suits most because it is drought tolerant, need minimum inputs, and can produce reasonable amounts of food, fodder, and fuel wood. Pigeonpea seeds contain about 20-22% protein and reasonable amounts of essential amino acids. India is the largest producer of pigeonpea with annual production of three million tons harvested from about four million hectares. In the past 10 years the area under pigeonpea is consistently increasing at the rate of 2% each per year (Ryan, 1997) but still its demand is out-scoring the supply and serious efforts are needed at every level to boost the production of this important legume. In India pigeonpea is predominantly consumed as *dhal* (decorticated dry split peas) but its whole seeds are also consumed in Africa and the Caribbean islands. In this chapter various aspects of pigeonpea nutrition are reviewed.

GRAIN QUALITY OF PIGEONPEA

Pal (1939) published perhaps the first review on the nutrition value of pulses in India. He compared different pulse crops for their digestibility coefficient, biological value, net protein value, and four essential amino acids. For biological value he judged pigeonpea as the best pulse crop and concluded that it makes the most nutritive food when eaten with rice. However, using an arbitrary scale for the overall nutrition value of the pulses, chickpea (*Cicer arietinum*) and black gram (*Phaseolus mungo*) were considered superior to pigeonpea. The nutritional value of food is determined by its chemical constituents and in pigeonpea a wide range is reported for these vital elements (Tripathi et al., 1975; Sharma et al., 1977; Narsimha and Desikachar, 1978; Manimekalai et al., 1979; Singh et al., 1984 a & b). Besides inherent genotypic differences, such variation can be attributed to environment where the crop was grown, methods of sampling and analyses, and method and length of seed storage periods.

Chemical Composition of Dry Seeds

The distribution of some dietary nutrients in different parts of dry pigeonpea seed as reported by Faris and Singh (1990) is given in Table 1. Broadly, pigeonpea seed contains 85% cotyledons, 14% seed coat, and less than 1% embryo. Carbohydrates and proteins are major constituents of cotyledons, embryo, and seed coat. Quantitatively, the cotyledons (66.7%) and seed coat (58.7%) are rich in carbohydrates while protein (49.6%) constitutes a major portion of embryo. Carbohydrates and fat are also present in significant quantities in embryo. About one-third of seed coat is made of fibers. The seed also contains amino acids, calcium, fiber, and iron. The contents of methionine and cystine, the sulfur-containing amino acids, range around 1% and they predominantly reside in cotyledons and embryo. Calcium is predominantly

TABLE 1. Distribution of nutrients in mature pigeonpea seed.

Constituent	Whole seed	Cotyledons	Embryo	Seed coat
Carbohydrates (%)	64.2	66.7	31.0	58.7
Protein (%)	20.5	22.2	49.6	4.9
Fat (%)	3.8	4.4	13.5	0.3
Fiber (%)	5.0	0.4	1.4	31.9
Ash (%)	4.2	4.2	6.0	3.5
Lysine[1]	6.8	7.1	7.0	3.9
Threonine[1]	3.8	4.3	4.7	2.5
Methionine[1]	1.0	1.2	1.4	0.7
Cystine[1]	1.2	1.3	1.7	-
Calcium[2]	296	176	400	917
Iron[2]	6.7	6.1	13.0	9.5
Thiamine[2]	0.63	0.40	-	-
Riboflavin[2]	0.16	0.25	-	-
Niacin[2]	3.1	2.2	-	-

Adapted from Faris and Singh (1990)
1: g 100^{-1} g protein
2: mg 100^{-1} g dry matter

found in seed coat and embryo. Singh and Jambunathan (1982) studied distribution of major protein fractions in different components of pigeonpea seed and found that globulins constitute about 65% of total protein (Table 2). In comparison to other protein fractions globulin is inferior in sulfur-containing amino acids. Albumin, though in relatively small quantity, is rich in sulfur amino acids. The portion of prolamin in seed is low. According to Eggum and Beames (1983) pigeonpea is rated inferior to most other legumes as far as sulfur-containing amino acids is concerned but, unlike other legumes, the high protein content of pigeonpea seed is not tightly linked to its low methionine content (Singh and Eggum, 1984). Nigam and Giri (1961) observed that stachyose and verbascose constitute major component among sugars of pigeonpea. The pigeonpea starch has been found to be stable to heat up to 90°C (Modi and Kulkarni, 1976).

TABLE 2. Major protein fractions (%) in dry pigeonpea seed.

Component	Albumin	Globulin	Glutelin	Prolamin
Whole seed	10.2	59.9	17.4	3.0
Embryo	17.0	52.7	21.3	2.7
Cotyledons	11.4	64.5	18.2	3.5
Seed coat	2.6	26.3	32.8	4.2

Adapted from Singh and Jambunathan (1982)

Chemical Composition of Immature Seeds

Physiologically mature green seeds of pigeonpea are consumed as vegetable in Dominican Republic, the Caribbean islands and some parts of India. In Dominican Republic 80% of the crop is exported as canned or frozen vegetable. Singh et al. (1977) compared vegetable pigeonpea (Figure 1) with that of pea (*Pisum sativum*) and found that pea seed had higher protein than pigeonpea but it was similar in crude fibre content. The trypsin inhibitors were more in pigeonpea when compared to pea but far less when compared with soybean. Nutritionally, green pigeonpea seed is considered superior to the *dhal* (Table 3). According to Faris et al. (1987) green pigeonpea seed is a rich source of iron, calcium, and magnesium when compared with its *dhal*. The green seed contains lower quantities of trypsin and amylase inhibitors and flatulence-causing sugars. The green seed cooks quickly and is also a better source of vitamin A. The protein and starch digestibility of green seed are higher than mature seed (Singh et al., 1984a). Singh et al. (1991) while studying chemical changes in the developing pigeonpea seeds, found that dry matter accumulation increased up to 28-32 days after flowering in different cultivars and recommended that for the best green pea yield, the crop should be harvested at nearly 30 days after flowering. The protein content and soluble sugars showed a gradual decrease with advancing maturation of seeds while its starch continued to increase. ICP 7035, a known vegetable type Indian pigeonpea landrace, was found to be more biochemically active in accumulating soluble sugars. This landrace was marginally low in calcium and magnesium at all the stages of seed development. The iron content of seeds also decreased as they approached maturity. In pigeonpea a range of pod color is found but it is not related to any quality parameter (Saxena et al., 1983).

FIGURE 1. Vegetable pigeonpea (left) and green pea pods (right).

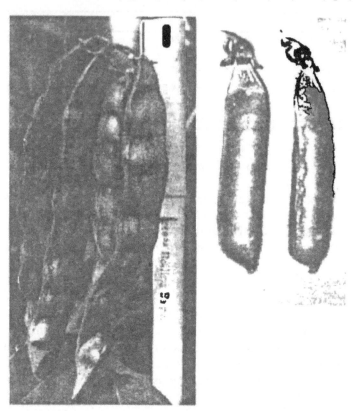

Antinutritional Factors

Like most legumes, pigeonpea also contains certain amounts of antinutritional factors. These include oligosaccharides (raffinose, stachyose, verbascose), polyphenols (phenols, tannins), phytolectins, and enzyme (trypsin, chymotrypsin, amylase) inhibitors. Singh (1988) studied a number of genotypes for quantifying important antinutritional factors and toxic substances present in pigeonpea seeds and found a large variation among genotypes for these traits (Table 4). Amylase and trypsin inhibitors and phenols were found in significant quantities. In addition, the flatulence causing sugars were also present in appreciable quantities. According to Kamath and Belavady (1980) pigeonpea seeds also contain certain unavailable carbohydrates which characteristically reduce the bioavailability of important nutrients.

TABLE 3. Comparison of pigeonpea green seed and *dhal* for various nutritional consituents.

Constituent	*Dhal*	Green seed
Starch content (%)	57.6	48.4
Protein (%)	24.6	21.0
Protein digestibility (%)	60.5	66.8
Trypsin inhibitor (unit mg^{-1})	13.5	2.8
Soluble sugars (%)	5.2	5.1
Crude fiber (%)	1.2	8.2
Fat (%)	1.6	2.3
Calcium[1]	16.3	94.6
Magnesium[1]	78.9	113.7
Copper[1]	1.3	1.4
Iron[1]	2.9	4.6
Zinc[1]	3.0	2.5

Adapted from Faris et al. (1987)
1 = mg 100 g $^{-1}$ sample

TABLE 4. Genotypic variation for major antinutritional factors in pigeonpea.

Factor	Genotypes	Range
Total phenols (mg g^{-1})	14	3.0-18.3
Tannins (mg g^{-1})	10	0.0-0.2
Trypsin inhibitor (units mg^{-1})	9	8.1-12.1
Chymotrypsin inhibitor (units mg^{-1})	9	2.1-3.6
Amylase inhibitor (units g^{-1})	9	22.5-34.2
Raffinose (g 100 g^{-1})	10	0.24-1.05
Stachyose (g 100 g^{-1})	9	0.35-0.86
Stachyose + verbascose (g 100 g^{-1})	4	1.60-2.30

Adapted from Singh (1988)

Godbole et al. (1994) reported the presence of protease inhibitor in seven-day old seed of variety TAT-10 while Ambekar et al. (1996) found that the inhibitors are either not synthesized or inactive up to 28 days of the seed development. No other plant part except seed recorded the presence of trypsin and chymotrypsin inhibitors (Mutimani and Paramjyothi, 1995). Since seed coat is rich in antinutritional factor it assumes greater importance where whole pigeonpea seeds are consumed. In many African, South American and Carribbean countries where dehulling facilities are not available and whole seed is consumed, predominantly white seeded cultivars are grown which contain relatively less quantity of polyphenols. Singh (1984) compared white, brown, and light brown seeded pigeonpea types for antinutritional factors (Table 5) and established strong relationship between seed coat color and antinutritional factors. He found three times greater quantity of polyphenols in the red seeded lines in comparison to the white seeded types. Similarly, the enzyme inhibition activity was much larger in the colored pigeonpea. In India almost entire pigeonpea production is converted into *dhal*. In this process the seed coat is removed and therefore the large amounts of tannins present in the colored pigeonpea pose no problem in its consumption. The amounts of polyphenols in *dhal* made

TABLE 5. Polyphenol contents and varietal differences in the enzyme-inhibitory property of pigeonpea polyphenols.

Cultivar	Testa color	Ployphenols (mg g^{-1} sample)	Enzyme inhibition[1] (%)			
			Trypsin	Chyco-trypsin	Human saliva	Hog pancreas
Hy 3C	White	3.7	37.9	36.0	34.5	21.8
NP (WR) 15	White	6.0	40.5	38.6	32.7	19.7
C 11	Light brown	14.2	91.5	90.3	86.0	80.9
BDN 1	Brown	15.2	90.3	91.6	79.4	69.3
No. 148	Brown	14.9	88.0	85.9	75.8	68.5
Mean		10.8	69.7	68.5	61.7	52.0
SE		±0.2	±2.1	±1.7	±1.4	±1.3

[1] Based on assay using 200 mg polyphenols for trypsin and chymoptrypsin, and 250 µg polyphenols for amylase inhibitions.
Adapted from Singh (1984)

either from red or white grain were found to be similar and ranged between 1.4 to 1.9 mg g^{-1} samples. Pichare and Kachole (1994) found no association between trypsin inhibitor activity and pod borer resistance in pigeonpea.

HUMAN NUTRITION

Role of Pigeonpea in Rural Diets

Methionine and cystine followed by tryptophan and threonine are the limiting essential amino acids in pigeonpea whereas lysine is the limiting amino acid in rice and wheat (Faris and Singh, 1990). A food combining cereals and pulses provide a balanced diet because they complement the amino acid profiles of each other. According to Hulse (1977), the mutual compensation is closest to the ideal value when the ratio by weight of cereals to legume is roughly 70:30. In southern and eastern Africa this ratio is 90:10 reflecting shortage of legume protein in the diet. Daniel et al. (1970) studied supplementation of cereal diets with various proportions of pigeonpea in rats and reported that supplementation of ration with pigeonpea significantly enhanced the nutritive value of the diet. Supplementation of rice diet with 8.5% and 16.7% pigeonpea *dhal* markedly improved the quality of diet (Table 6). Similarly, Kurien et al. (1971) demonstrated that a supplement of pigeonpea in maize diet significantly improved the quality of food.

TABLE 6. Effect of supplementary rice diets with varying levels of pigeonpea on the growth of young rats.[1]

Diet	Protein (%)	Gain in mass (%)	Protein intake (%)	Protein efficiency ratio
Rice	7.2	25.5	11.8	1.78
Rice + 8.5% pigeonpea	8.7	32.8	15.5	2.13
Rice + 16.7% pigeonpea	10.0	45.2	19.6	2.32
Rice + 25.0% pigeonpea	11.4	48.9	21.8	2.25

[1] Based on an experimental period of 4 weeks.
Adapted from Daniel et al. (1970)

Bidinger and Nag (1981) conducted a village level study including 240 families of different resource groups, representing six villages located in three agro-climatic zones of India. They observed that pigeonpea was by far the most preferred pulse crop and its consumption patterns differed widely by age group, farm size, and the village. The consumption rate was found linear with small farmers consuming the least amount and the large farmers the most. The consumption of pigeonpea was also found to be related to the production. National Institute of Nutrition in India recommends cereal:pulse ratio of 3:1 for very young children, 5:1 for women, and 6:1 for men. In most cases in villages, these standards could not be met (Table 7). Bidinger and Nag (1981) also reported that 10% of the protein and 5% of energy in the village diets came from pigeonpea. The maximum lysine provided from the diet was 21.7%. These values are low and reflect the low consumption of legumes. Prema and Kurup (1973) reported that pigeonpea contains cholesterol and phospholipid lowering effect. Globulin fraction of pigeonpea protein was found to have a significant hypolipidaemic action in rats fed with a high-fat and high-cholesterol diet. They reported marked reduction in the total and free cholesterol, phospholipids, and triglycerides contents in serum, liver, and aorta tissues of rat.

Nutrition Losses in Dehulling

In the Indian subcontinent pigeonpea is predominantly consumed in the form of *dhal* and conversion whole seed into *dhal* is a big industry in

TABLE 7. Relative cereal:pulse consumption by different age groups in six villages of central India.

Village	Age group		
	1 to 6	7 to 18	Adults
Aurepalle	31:1	35:1	37:1
Dokur	23:1	31:1	42:1
Shirapur	15:1	14:1	17:1
Kalman	14:1	18:1	20:1
Kanzara	7:1	9:1	10:1
Kinkheda	9:1	10:1	10:1

Adapted from Bidinger and Nag (1981)

the country. For commercial purposes, big machines are used for dehulling while in rural areas, dehulling is done by using traditional grinding stones called *chakki* or quern. Since the cotyledons of pigeonpea are attached tightly with seed coat by gums, the processing primarily involves loosening of husk followed by dehusking and splitting of the two cotyledons. Therefore, pigeonpea dehulling is not only difficult but also a specialized function when compared with other legumes. Losses of seed mass during the process of dehulling is a common event. Excluding the husk which accounts for about 15%, the *dhal* recovery in pigeonpea is around 60% by *chakki* and around 70% by machines (Singh and Jambunathan, 1981). This means even by using advanced technology about 15-17% of grain mass is lost. By using *chakki* such losses shoot up to 20-25%.

Reddy et al. (1979) studied the protein deposition pattern in pigeonpea seed and reported that the outer layers of the cotyledons are richer in protein in comparison to inner layers of seed. From nutrition point of view, this is a matter of concern since dehulling not only removes protein-rich germ but also the outer layers of the colytedons where relatively more protein constituents are housed. Fortunately, the protein quality in terms of amino acids is not adversely affected by dehulling. Singh and Jambunathan (1990) further reported that dehulling also removes about 20% calcium and 30% iron. To preserve the nutritive value of pigeonpea seed and minimizing the nutrient losses during dehulling it is essential that more efficient dehulling technology is developed and transferred to rural areas where by and large milling is still carried out by inefficient old-age techniques. According to Kurien (1981) under controlled conditions the *dhal* yield achieves the maximum efficiency of 80-84% but at commercial level the recovery remains around 70%. He also reported large varietal differences (72 to 82%) for *dhal* yield. Therefore, it can be assumed that with a combination of a superior variety and an efficient pigeonpea processing technology, the nutrient losses can be minimized.

Cooking Quality

Pigeonpea seeds dry, green, or processed in the form of *dhal* and other products are consumed after cooking. Therefore, besides various nutritional aspects the cooking time and other related parameters assume significant importance. Consumers always prefer a *dhal* that cooks fast and produces more volume upon cooking with high consistency and flavor. Cooking time recorded between 22 and 44 minutes for *dhal* and between 45 and 67 minutes for whole seed by Sharma et al. (1977) indicate the extent of genotypic variation present for this trait.

Cooking time of *dhal* was found to be independent of taste and flavor (Maninekalai et al., 1979). Jambunathan and Singh (1981) studied various physico-chemical characters in 25 pigeonpea cultivars and reported a considerable range (Table 8) for various quality parameters. The cooking time ranged between 24-68 minutes. They also found that quick cooking trait was associated with large seed size, high solid dispersal, water absorption, nitrogen solubility indices, and nitrogen content of the solids dispersed. Lines with high protein and small seeds in general take more time to cook. Narsimha and Desikachar (1978), Singh et al. (1984c), and Sharma et al. (1977) reported positive association of cooking time with its calcium and magnisium contents. The issue of pre-cooking, soaking, and cooking time needs to be resolved. In

TABLE 8. Variation for various physico-chemical characteristics in 25 pigeon-pea cultivars.

Constituent	Range	Mean
Solids dispersed (%)	20.8-54.7	37.4
Water absorption (g g^{-1} sample)		
dhal	1.69-2.65	2.25
whole grain	0.63-1.34	1.02
Increase in volume (v/v)		
dhal	1.18-1.86	1.51
whole grain	0.91-1.54	1.13
Starch (%)	51.5-63.4	58.6
Soluble sugars (%)	3.6-5.3	4.8
Nitrogen solubility index (%)	28.7-42.5	36.4
Nitrogen content in solids dispersed (%)	19.6-31.8	27.3
Protein (%)	19.7-25.2	22.1
100-seed mass (g)	6.2-20.7	9.6
Cooking time (min)	24-68	38

Adapted from Jambunathan and Singh (1981)

some experiments pre-soaking in water reduced cooking time (ICRISAT, 1987) while in others (Saxena et al., 1992) this treatment increased the cooking time significantly. Soaking in sodium bicarbonate helped in reducing cooking time in pigeonpea but it increased pH and thereby adversely affected the organoleptic quality of *dhal*.

According to Salunkhe (1982) cooking of pigeonpea improved the bioavailability of nutrients and also destroyed some antinutritional factors. Heat treatment also enhanced the starch digestibility. Lines, which take long time to cook also face the danger of loosing vital vitamins from the food. Cooking of seed after germination not only enhances the digestibility of starch (Jyothi and Reddy, 1981) but also reduces the levels of oligosaccharides (Iyenger and Kulkarni, 1977). The fermentation of seeds also helps in reducing inhibitory activity of the digestive enzymes (Rajalakshmi and Vanaja, 1967). It appears that a little research has been conducted in the past in this critical aspect of pigeonpea quality. Studies in understanding the role of various chemical constituents on cooking time in diverse genetic materials will help in resolving this issue.

Geervani (1981) studied the effect of boiling, pressure-cooking, and roasting on the quality of pigeonpea and reported that thiamine and riboflavine were destroyed by heat but niacin content was unaltered during all the treatments. Availability of lysine and methionine decreased more on roasting but the available methionine increased on boiling and pressure-cooking.

Quality Losses in Storage

Throughout the world and particularly in India and Africa, pigeonpea is predominantly cultivated by small holder resource poor farmers for meeting their domestic protein needs and to generate some income. These farmers generally store the whole seeds for about 8-12 months for sowing and round-the-year consumption. They process small quantities of grain through hand-operated mills as and when needed. In rural areas, the seeds are generally stored in gunny bags or bins made of mud and husk and during the storage period a considerable damage is caused by storage pests. Among these, bruchid (*Callosabruchus* spp.) is the major pest. In most cases the ripening pigeonpea pods are infested on the plants in the field and this infestation is carried to the storage bins through seeds. Since the bruchids complete their life cycle in about four weeks, they multiply fast inside the bin and cause considerable seed damage. This damage not only reduces *dhal* recovery and deteriorates germination but also adversely affects the hygine and nutritive value of seeds (Parpia, 1973). In pigeonpea, although some genotypic variation for bruchid resistance is reported (Uma Reddy and Pushpamma, 1981) but these differences are inconsistent and are not large enough to ignore the issue of storage pests.

According to the standard set by Prevention of Food Adulteration Act of 1967, food grains containing more than 10 mg uric acid per 100 g of food, arising as a result of insect infestation, are unfit for human consumption. The pigeonpea seeds when stored for five months turned unfit for consumption as their total uric acid content crossed the prescribed limit of 200 mg per 100 g of sample (Daniel et al., 1977). Cooking time of food legumes in general increases with storage time and pigeonpea is no exception. Vimla and Pushpamma (1983) reported that by storing pigeonpea for eight months the safety level of uric acid was crossed. They also found that cooking time of both undamaged and damaged pigeonpea seeds increased significantly after storing them for about 12 months, indicating that even improved storage methods failed to retain the quality traits of stored seeds (Vimla and Pushpamma, 1985). Srivastava et al. (1988) reported that with the increase in insect infestation and the advancement of storage period the parameters such as seed moisture, total ash, crude fibre, protein, and reducing sugar contents increased while fat, carbohydrates, and non-reducing sugars decreased. Daniel et al. (1977) found that lysine, threonine, and protein efficiency ratios were significantly and adversely affected in pigeonpea when the seeds were stored in jute bags. Uma Reddy and Pushpamma (1981) reported significant reduction in the amino acid contents in the infested seed samples and the decline in lysine was greater than those of methionine and tryptophan. Daniel et al. (1977) observed significant decrease in the protein efficiency ratio due to storage. The storage of pigeonpea seeds also resulted in the loss of vitamins. Such losses were less (10-26%) in the protected seed and high (32-49%) in the unprotected infested seeds (Uma Reddy and Pushpamma, 1981). Thiamine and niacin contents also registered decline during storage. A number of factors have been identified which determine the extent of quality loss during storage. These include moisture, temperature, relative humidity, corneus thickness, hardness of grain, and ovipositional differences of storage pests (Squire, 1933; Singh et al., 1977). The storage losses can be reduced to some extent by improving the storage conditions but the decline in some quality parameters is inevitable. A well directed research is needed to provide pigeonpea farmers a cost-effective and efficient seed storing technology.

ANIMAL NUTRITION

Pigeonpea is a wonderful plant because besides providing nutritious food for human beings it is a preferred animal fodder and feed also. Its fodder is relished by cattle, goats, and sheep while its harvest-trash, grain, and milling by-products form an excellent feed for various domestic animals.

Fodder

The perennial nature of pigeonpea plant allows it to produce tender leaves shortly after cutting the plants during its vegetative growth period and also after the harvest of seed crop. The fodder yields of pure stand cuts depend on both genotype and management practices, which include height and frequency of cuttings, availability of soil moisture and nutrition. The genotypic differences for vegetative growth have also been observed at ICRISAT. The long-duration pigeonpea cultivars that are photo-thermal sensitive produce large biomass when planted around the longest day of the year. The same genotype if planted later in the reducing daylengths produces less quantity of biomass. The selection of a suitable cultivar and appropriate agronomic management practice can produce plenty of quality forage from this crop.

Singh and Kush (1981) in India, Herrera et al. (1966) in Columbia, Parbery (1967) in Australia, and Shiying et al. (1999) in China have reported around 50 t ha^{-1} fodder yields in multiple cuttings from pigeonpea. The actual edible forage, however, is about 50% of the total yield because of the woody stem of the plant (Whiteman and Norton, 1981). Pigeonpea stands are also used for grazing purposes. In Hawaii the live-weight gain of over 1,120 kg ha^{-1} year^{-1} have been reported by Krauss (1932). Whiteman and Norton (1981) concluded that pigeonpea forage was superior to grass in gain head^{-1} indicating that the crop had a higher nutritive value and could carry a higher stocking rate than those of the grasses.

Generally, pigeonpea is used as forage for supplementing protein when the pasture quality is sub-standard. The young tender leaves and fresh flowers and pods form nutritive fodder for all grades of livestock. Leaves are the major forage component during the vegetative growth. As the plant approaches reproductive stage the fodder quality is enhanced due to the development of high-protein seeds. Therefore, the forage quality at a particular time will depend on the proportion of different plant parts. A proximate analysis of different pigeonpea plant parts as reported by various workers is summarized in Table 9. According to Krauss (1921) the fresh green foliate contain 23.7% crude protein and 35.7% fibre and seed meal has 25.3% protein and 7.3% fiber. Henke et al. (1940) compared pigeonpea fodder with other better known forage crops and concluded that pigeonpea produced the highest economic value of digestible nutrients per unit area when compared with leucern (*Medicago sativa*) and other species.

Whiteman and Norton (1981) conducted sheep feeding trials using pigeonpea pods and pangola grass (*Digitaria decembens*) in Australia. They concluded that pigeonpea pods with 7.5% crude protein fed as a sole diet were of low nutritive value and sheep lost 2% body weight. However, the inclusion of 33% of high quality forage such pangola grasses in the diet considerably improved the nutritive value of feed. They also pointed out that the harvest trash, which contains a significant proportion of leaves, would be a more nutritive feed than pods alone.

In China, pigeonpea is being promoted to meet the growing need of fresh quality fodder in the country because it can be grown well in the eroded soils of southern hilly regions for providing quality fodder under dry conditions. The ability of pigeonpea to allow 3-5 fodder cuttings within a year also makes it a useful stall-feeding crop. Pigeonpea being a perennial drought tolerant crop has shown high adaptation in a range of soil types of mountain regions of Du Au, Dahua, Huan Jiang, and Feng Shan counties of Guangxi province of China. According to Fuji

TABLE 9. Major nutrition constituents in different pigeonpea plant parts.

Component	Crude protein (%)	Crude fiber (%)	N-free extract (%)	Fat (%)	Ash (%)	Reference
Fresh green forage	23.7	35.7	26.3	5.3	8.7	Krauss (1921)
Whole tops, mature	18.8	29.4	40.0	5.2	5.6	Work (1946)
Whole tops, young	15.8	31.2	37.7	4.6	5.6	Work (1946)
Seed meal	25.3	7.3	61.2	1.7	4.1	Krauss (1921)
Mature dry seed	21.3	-	63.7	1.7	4.2	Morton (1976)
Pod meal	10.1	40.7	45.0	1.6	3.1	Krauss (1921)
Pods intact	7.0	-	42.8	0.4	5.7	Morton (1976)

and Zhanghong (1995) the foliage of pigeonpea is a quality fodder and goats, buffalo, cattle, and pig relish it.

A preliminary evaluation of ICRISAT pigeonpea varieties at Guangxi Academy of Agricultural Sciences in China showed that with multiple cuttings within a year variety ICPL 93047 produced 54 t ha^{-1} of fresh and 29 t ha^{-1} of dry fodder (Shiying et al., 1999). This experiment also showed that pigeonpea could grow well during winter and can meet the fodder needs when normal fodder supply is limited. It was observed that the goats and cattle liked the dry forage of pigeonpea better than green matter. S.C. Rao (Personal communication) compared tall and dwarf pigeonpea lines for forage production and their nutrition value in the southern plains of USA and reported that the dwarf genotype (PBNA) produced tender branches resulting in relatively less stem dry matter. In comparison to tall types (23 g kg^{-1}) the dwarf line produced greater (28.6 g kg^{-1}) nitrogen. The digestibility of the forage harvested from the dwarf line was also greater than the tall genotypes.

Feed

Whole grain, threshing trash and milling by-products are used as feed for cattle, poultry, and pigs. These pigeonpea by-products provide protein-rich substitute for domestic animals at cheaper rates. In countries where climate is hot and dry and other legume crops are difficult to grow, pigeonpea is an attractive alternative. In the first quarter of 20th century pigeonpea was extensively used as poultry meal in Hawaii. According to Krauss (1932) an equal mixture of cracked pigeonpea and cracked maize seed was considered the best poultry ration. Draper (1944), Springhall et al. (1974), and Wallis et al. (1986) considered pigeonpea as an ideal protein substitute for all types of poultry rations. Since whole pigeonpea seeds contain some amount of antinutritional factors heat treatment of grains was introduced in the animal ration preparations. This resulted in a significant increase in the apparent metabolizable energy content of pigeonpea meal (Nowkolo and Oji, 1985). Wallis et al. (1986) reported little effect of heating on growth rates, feed intake, and in feed conversion efficiency. Falvey and Visitpanich (1980) conducted pig-feeding trials using 30% ground pigeonpea seed in Thailand. They reported live-weight gain from 25 g day^{-1} to 159 g day^{-1}. By using boiled pigeonpea the live-weight gain was further increased to 205 g day^{-1}. This increase was attributed to the reduction of trypsin inhibitor activity that in turn improved the feed conversion ratio. The recently bred high-protein pigeonpea lines at ICRISAT are likely to enhance the utility of pigeonpea in animal ration.

Use of pigeonpea seeds as feed is a common practice in rural China and it is primarily fed to pigs and chickens and some times to cattle and goats also. For pigs, the boiled seeds of pigeonpea are used to prepare feed mixtures with other ingredients while raw seeds are fed to chickens. In 1992, Research Institute of Resource Insects in China studied the nutritional value of pigeonpea feed. In this experiment pigs were fed with feed mixtures prepared with different concentrations of pigeonpea (Fuji et al., 1995). They found that a mixture with 6-12% pigeonpea in the meal mixture, the gain in the meat-mass production was 78 g day^{-1} with a ratio of meat-mass to feed input of 3.54:1 and this efficiency-mark meets the Chinese National Standards. Based on this information, Fuji et al. (1995) developed various feed mixtures using pigeonpea seed (22% protein) and dry leaf powder (19% protein) as major source of protein.

As mentioned earlier that in India over 3 million tons of pigeonpea is converted into *dhal* annually by processing either at household level by *chakki* or at commercial mills. This conversion of whole dry seeds into *dhal* yields significant quantity (about 30%) of the by-products. These include approximately 3-8% brokens, 15% powder, and 10% husk. The powder and brokens are important source of protein for cattle feed (Pathak, 1970). Whiteman and Norton (1981) evaluated non-seed material collected from machine harvester and reported that it contains 13.9% crude protein, 0.35% phosphorus, 0.06% sulfur, and 7.3% ash. This ration when fed sole was inadequate for live-weight maintenance and they attributed it to its low sulfur content which is associated with nitrogen requirement of cattle and suggested that sulfur supplement is essential for utilization of this forage.

GENETIC ENHANCEMENT OF PROTEIN

Increasing yielding capacity of the food crops is the primary task of plant breeders and production of varieties resistant to diseases and pests is their perennial target. As for as demand for quality is concerned, it assumes importance after a certain quantitative level of food production has been achieved. In most third world countries food supplies have not kept pace with the rising population and therefore quality breeding never reached a priority level in any institution. Considering the state of malnutrition in most developing and under-developed countries and the role pigeonpea can play in subsistence farming, the genetic enhancement of protein content in pigeonpea is a logical approach for addressing this issue. A small increase in protein content of the adapted cultivars will lead to significant protein yield on a sustainable basis. For increased adoption of the enhanced-protein cultivars it is essential that they perform as good as normal cultivars in most agronomic traits such as seed size, disease resistance, and yield. This will ensure adequate returns to farmers. Since ICRISAT has global responsibility for pigeonpea improvement, it took this challenge and a project on breeding high-protein lines was implemented. To have an effective breeding program studies on genetic control were also conducted to develop efficient breeding methodology and selection and testing procedures. The results are discussed herein.

Genetic Control of Protein Content

Information on the genetic control of protein content in pigeonpea is limited. Saxena and Sharma (1990), while reviewing the subject, reported the presence of both additive and non-additive genetic variation in determining protein content in pigeonpea and this variation was found to be controlled by 3-4 genes (Dahiya et al., 1977). Reddy et al. (1979) reported that the magnitude of heterosis for protein was in the negative direction. Dahiya and Brar (1977) reported strong maternal influence in determining protein content of seed.

For better understanding of the nature of genetic parameters the parents should have a large variation for the traits. Since the genetic material used in earlier genetic studies had limited variability for protein, Durga (1989) conducted genetic analysis for protein in two high (30-31% protein), two medium (26-27% protein), and two low (22-23% protein) lines of pigeonpea to develop basic information on various genetic parameters. She concluded that (i) reciprocal differences in F_1 generation for protein were large, (ii) protein content was under additive and complementary gene effects, (iii) low-protein content was dominant or partially dominant over high-protein content, and (iv) protein content had moderately high (65.2%) narrow-sense heritability.

Breeding Methodology

As the first step in breeding for high protein, a search for high-protein trait was made in literature and in the pigeonpea germplasm available in ICRISAT gene bank. Swaminathan (1973) analyzed about 2000 pigeonpea germplasm and reported little variation for protein content. Although significant genotypic differences were reported for protein content in some studies (Yadav, 1984; Esh et al., 1959), the variation was not found large enough to allow selection of lines within the germplasm. Further, it was also noticed that the reported observations were based on single year data and perhaps had significant confounded influence of the sampling and environment. Considering all the factors together, it was decided to search for high-protein trait in the secondary gene pool and use them in breeding program. Since all the wild relatives of pigeonpea cannot be crossed with cultivated types, only the crossable species were examined. The results indicated that *Cajanus sericeus, C. lineatus,* and *C. scarabaeoides* had high-protein (Table 10). The *dhal* protein levels in this group ranged up to 31%. Therefore, these were selected as donor parents for hybridization. Breeding for high protein was carried out using pedigree method. Since the wild relatives of pigeonpea differ grossly from the cultivated types with respect to all agronomic traits and are unfit for cultivation and consumption, a breeding strategy was developed to select simultaneously for improved agronomic traits and high protein content.

TABLE 10. Protein content and seed size of high-protein lines and their parents.

Species/genotype	Protein (%)	100-seed mass (g)	Protein seed^{-1} (mg)
Cultivated species			
Baigani	23.7	11.2	26.5
Pant A-2	22.7	7.5	17.0
T.21	24.4	7.5	18.3
Wild species			
C. scarabaeoides	28.4	2.3	6.5
C. sericea	29.4	1.9	5.6
C. albicans	30.5	2.8	8.5
High-protein lines			
HPL 2	29.0	12.1	35.1
HPL 7	28.0	10.0	28.0
HPL 40	27.0	10.4	28.1
HPL 51	27.9	10.6	29.6

Adapted from Saxena et al. (1987a)

Hybridization and Selection

Crosses were made using the wild relative as female parent. In the subsequent segregating generation (F_2), as expected, a large variation was observed for plant type and seed characters. In each cross over 200 plants were examined for protein content and the segregants with desirable protein content were selected. In each subsequent generation the individuals with improved plant type were selected in field and the final selection of the single plants for generation advance was done after determining their protein content in the laboratory. Each plant sample was evaluated in duplicate and compared with a control cultivar grown in the same field. The selected plants were ratooned and selfed using muslin cloth bags to harvest genetically pure seed. After 10 generations of pedigree selection simultaneously for agronomic traits in field and seed characters in laboratory, several breeding lines were identified (Tables 10 and 11). In these selections the high-protein trait of the wild relative of pigeonpea and seed characters of cultivated type were recovered.

TABLE 11. Performance of some high-protein pigeonpea selections at Patancheru.

Year/line	Days to mature	100-seed mass (g)	Grain yield (t ha^{-1})	Protein (%)	Protein (kg ha^{-1})
1985					
HPL 40-5	169	9.6	2.10	26.9	452
HPL 40-17	169	8.5	2.07	26.5	440
BDN 1 (control)	168	9.6	2.02	23.2	373
SE	±0.9	±0.18	±0.18	±0.46	±37.3
CV %	0.9	3.4	17.3	3.0	17.0
1986					
HPL 8-10	163	10.5	1.66	26.5	353
HPL 8-16	162	10.5	1.57	27.4	344
ICPL 211 (control)	162	14.3	1.46	21.6	251
SE	±1.1	±0.15	±0.19	±0.21	±38.5
CV %	1.3	2.5	27.0	1.7	25.8

Relationship Between Seed Size and Protein

In pigeonpea, seed size is an important parameter from consumption and dehulling points of view. Negative correlations between seed size and protein have been reported in cereals such as pearl millet (Kumar et al., 1983) and sorghum (Wayne and Casady, 1974) and legumes (Blixt, 1979; Imam, 1979). In pigeonpea, Dahiya and Brar (1976) found no association between seed size and protein in 220 germplasm accessions, while Reddy et al. (1979) reported negative association between these two variables in inter-specific derivatives. Saxena et al. (1987b) while studying this relationship in 192 pigeonpea cultivars, found that the association between seed size and protein content was negative and partly controlled by genetic factors. Therefore, at the commencement of the project, aimed to develop high-protein lines in pigeonpea, there was some concern that the high-protein level of the wild species donors would remain associated with small seed size and antinutritional factors in the derived lines. According to Bahl et al. (1979) a negative relationship between seed size and protein was a result of the deposition of an increased amount of starch in seed which alters the starch:protein ratio. However, it has been found in *Vicia faba* (Abo-Hegazi, 1979), *Phaseolus vulgaris* (Gridley and Evans, 1979), and soybean (Hartwig and Hinson, 1972) that the negative correlation could be changed through breeding and selection of exceptional genotypes in segregating generations that appeared more efficient than expected in their protein synthesis and combine superior agronomic traits is possible.

The estimates of correlation between seed size and protein obtained in the breeding materials developed at ICRISAT indicated that in pigeonpea improved seed size and protein can be selected simultaneously (Saxena et al., 1987c). From the inter-specific crosses, some promising high-protein lines identified are HPL 2, HPL 7, HPL 40, and HPL 51 (Table 10). These lines combined high-protein and large seed size (Figure 2).

Agronomic Evaluation of High-Protein Lines

In pigeonpea, seed size is also associated with yield. Sharma and Saxena (1977) showed that these two variables are independent of each other in the 100-seed mass range of 9-12 g. This relationship, however, was positive in small seeded materials and negative in the materials larger than the seed size range indicated above. These observations in-

FIGURE 2. Seeds and pods of wild relative of pigeonpea (right), high-protein line (middle), and control cultivar (left)

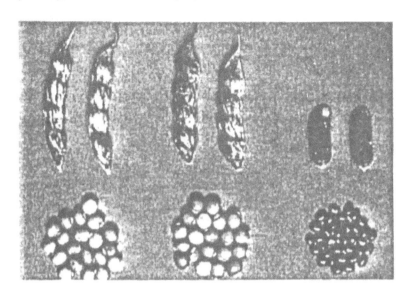

dicated that within the medium seed size range simultaneous improvement for seed protein, seed size as well as yield could be made.

The results of agronomic evaluation of the promising high-protein selections were very encouraging (Table 11). The high-protein selections were found similar to the control cultivars in important agronomic traits such as days to maturity, seed size, and dry seed yield. For protein content the selections were significantly superior to the controls and their protein content ranged between 26-27%. An estimate of total protein harvest in this trial revealed that by growing these high-protein lines in one hectare about 350-450 kg crude protein could be harvested. These values reflect the additional advantage of 80-100 kg protein ha^{-1}. Cultivation of these lines will markedly improve availability of the nutritive protein to the farmers without sacrificing the seed yield.

Nutritional Quality of High-Protein Lines

Since the high-protein trait in pigeonpea was transferred using traditional breeding methods from its wild relatives, known to possess various antinutritional factors, it was necessary to compare the derived lines with traditional cultivars for various nutritional quality parameters before releasing them for human or animal consumption. For this purpose, Singh et al. (1990) compared two high-protein lines (HPL 8 and HPL 24) with two control cultivars (C 11 and ICPL 211). The main findings are discussed below.

Chemical composition: There were large differences between the protein levels of high-protein lines (28.7 to 31.1%) and control cultivars (23.1 to 24.8%). As expected the starch component (54.3 to 55.6%) of the high-protein lines was relatively less than that of controls (58.7 to 59.3%). Also the high-protein lines (2.5 to 2.6%) were marginally lower in fat content when compared with control cultivars (2.9 to 3.1%). The differences in the major protein fractions of the high and normal-protein lines were large. In comparison to controls (60.3 to 60.5%), the globulin fraction was higher (63.5 to 66.2%) in the high-protein lines and the reverse was true for glutelin (Table 12). This variation in the storage proteins, however, was not large enough to influence the amino acid profiles of high and normal-protein lines (Singh et al., 1990). Also the activities of trypsin and chymotrypsin inhibitors were found more or less similar in the high-protein lines and the controls.

Biological evaluation: A series of rat feeding trials and laboratory

TABLE 12. Comparison of high-protein lines and control cultivars for starch, major protein fractions, and sulphur containing amino acids.

Constituent	High-protein lines		Controls		SE
	HPL 8	HPL 40	C 11	ICPL 211	
Starch[1]	54.3	55.6	58.7	59.3	±0.30
Protein[1]	28.7	31.1	24.8	23.1	±0.09
Albumin[2]	9.1	8.0	7.7	8.6	±0.34
Globulin[2]	63.5	66.2	60.5	60.3	±1.08
Prolami[2]	2.9	3.2	3.6	2.1	±0.06
Glutelin[2]	20.2	19.7	23.3	22.8	±0.75
Methionine[2]	1.0	1.0	1.1	1.1	±0.02
Cystine[2]	0.8	0.8	0.7	0.7	±0.01

Adapted from Singh et al. (1990)
[1]: (g 100−1 g *dhal*)
[2]: (g 100 1 g protein)

evaluations were conducted to assess various nutritional parameters for the high-protein lines. The raw seeds and cooked *dhal* samples from the high and normal-protein lines were found more or less similar in true protein digestibility, biological value, and net protein utilization. However the high-protein lines were found significantly superior in utilizable protein (Table 13). Singh et al. (1990) concluded that the high-protein lines are nutritionally superior to normal-protein cultivars as the former contain quantitatively more utilizable protein and sulfur containing amino acids. The whole seeds of the high-protein lines for animals and their *dhal* for human beings is nutritionally beneficial and its promotion will help in addressing the nutritional issues in rural areas.

TABLE 13. Biological evaluation (g 100^{-1} g) of raw whole pigeonpea seed and cooked *dhal* samples of high-protein selections and normal-protein control cultivars.

Parameter	High-protein lines		Control cultivars		SE
	HPL 8	HPL 40	C 11	ICPL 211	
Raw whole seed					
Protein	25.6	27.3	21.9	21.0	±0.48
TD	58.5	58.0	59.5	60.6	±1.08
BV	68.7	70.5	64.3	64.0	±1.13
NPU	40.2	40.9	38.3	38.8	±0.64
UP	10.3	11.2	8.4	8.1	±0.23
Cooked *dhal*					
Protein	27.6	30.8	23.9	22.8	±0.26
TD	83.7	82.9	84.3	85.7	±2.14
BV	67.0	65.3	66.7	62.9	±1.68
NPU	56.1	54.1	56.2	53.9	±1.06
UP	15.5	16.7	13.5	12.3	±0.25

TD : True protein digestibility
BV : Biological value
NPU: Net protein utilization
UP : Utilization protein
Adapted from Singh et al. (1990)

GENOTYPE-ENVIRONMENT INTERACTION
FOR SEED PROTEIN

Environment plays a significant role in the expression of both morphological and biochemical traits in crop plants. Some characters such as seed size show less environmentally induced variability while others exhibit relatively large variation and the seed protein belongs to the later group. The observed variation for protein over locations could be influenced by environment where it is grown including the soil type and its moisture and nutrient level. Some sites favor higher nitrogen accumulation in seed as compared to others and such marked differences have been demonstrated in almost all the cereals and legumes. Hamilton et al. (1951) observed a linear relationship between protein accumulation and the increase in the alcohol-soluble protein fraction of the total protein in maize. This resulted in reduced biological value of protein as neither lysine nor tryptophan is alcohol-soluble protein.

In crops like pigeonpea where flowering is determined by photoperiod and temperature, the rate of development and duration of vegetative and reproductive periods vary widely. The meteorological conditions of certain months may exert diverse effects on the nutritional metabolism in the varieties, thus influencing the crude protein content of seeds. Pietri et al. (1971) found no response of fertilization on protein content of pigeonpea while Sham (1976), Singh et al. (1974), and Esh et al. (1959) reported significant effects of location and fertilizer application on pigeonpea seed protein. Oke (1969) reported that incorporation of 20 ppm sulfur alone or in combination with phosphorus increased methionine content of pigeonpea. Jain et al. (1986) observed significant effect of location in the advanced breeding lines of short and medium maturity duration. Singh et al. (1984c) reported significant effects of growing season (rainy and post-rainy) on various quality parameters.

Saxena et al. (1984) reported the results of an extensive study conducted by ICRISAT in 1975 to characterize environmental variation for protein content in six pigeonpea cultivars at Hyderabad (17°N), Sehore (24°N), Mandasore (25°N), Pantnagar (29°), and Hisar (29°). At each location the plantings were done in different (4-12) months. They observed large and significant differences among locations and among months with in a location (Table 14). In general, the protein levels were high at higher latitudes. Within a variety grown at a particular location in different months, the variation for protein was also large. For example, in cv. 'Prabhat', planted at Hyderabad over 12 months, the seed protein content ranged from 21.6 to 25.2%; similarly at Pantnagar in 10

TABLE 14. Variation for *dhal* protein (%) in six pigeonpea cultivars planted in various months at different locations during 1975.

Cultivar (%)	Location	No. of months	Mean	Range	Variance	C.V.
Prabhat	Hyderabad	12	23.4	21.6-25.2	1.93	5.94
	Pantnagar	10	25.9	24.5-27.9	1.41	4.59
	Sehore	6	23.0	20.9-25.2	2.24	6.52
	Hisar	6	25.3	24.3-27.0	0.95	3.85
	Mandsore	7	24.5	23.2-26.2	0.90	3.88
Pusa Ageti	Hyderabad	12	23.8	21.0-26.4	2.34	6.45
	Pantnagar	7	26.7	24.7-29.3	2.73	6.19
	Sehore	8	23.9	22.1-25.4	1.29	4.76
	Hisar	6	24.9	24.2-25.4	0.18	1.72
	Mandsore	6	25.1	23.7-26.6	1.31	4.56
T. 21	Hyderabad	12	24.3	22.3-26.4	2.43	6.41
	Pantnagar	8	26.8	25.6-28.6	0.81	3.36
	Sehore	10	22.8	20.2-24.4	1.52	5.40
	Hisar	4	25.1	24.3-25.7	0.33	2.29
	Mandsore	8	26.2	24.5-28.2	1.79	5.10
No. 148	Hyderabad	12	24.0	20.7-26.8	2.79	6.96
	Pantnagar	4	26.4	25.3-28.1	1.43	4.53
	Sehore	9	23.9	22.1-26.2	1.58	5.25
	Hisar	6	25.2	24.0-26.3	0.79	3.52
	Mandsore	8	25.0	23.5-25.8	0.72	3.40
ST 1	Hyderabad	12	23.6	22.3-24.6	0.56	3.18
	Pantnagar	7	26.5	24.5-27.5	0.95	3.68
	Sehore	9	22.8	19.8-25.6	3.29	7.97
	Hisar	6	25.3	24.4-26.2	0.56	2.96
	Mandsore	7	24.8	21.4-27.4	4.45	8.50
PDM 1	Hyderabad	12	23.6	20.1-26.9	4.15	8.63
	Pantnagar	6	26.4	25.4-28.1	1.01	3.82
	Sehore	8	24.1	20.7-26.4	4.31	8.62
	Hisar	5	26.5	24.9-27.6	1.13	4.01
	Mandsore	7	23.9	22.1-26.1	3.44	7.74

plantings within a year the protein content of cv. Prabhat ranged from 24.5 to 27.9%. Besides environments, the pigeonpea lines used in this study differed considerably in their response to photoperiod and temperature and this caused a large variation in flowering and maturity. Therefore no attempt was made to identify the effect of any particular location, date of planting or any other specific factor on the protein values. The differences observed in maximum and minimum protein values within varieties amply demonstrated that the environment could have a significant role in determining the seed protein content in pigeonpea.

Genotype-environment interaction in the high-protein lines developed at ICRISAT was also studied (Saxena et al., 1987a). The replicated trials were conducted at six locations. Although statistically significant genotype-environment interactions were recorded, the high-protein lines recorded significantly superior protein content to controls. For example in HPL 24 the protein ranged between 30.9 to 32.3% while in control it ranged between 21.4 to 24.5 (Table 15). The data also showed that the extent of variation for protein-content was more or less similar in high-protein lines and control cultivar, but the high-protein trait was maintained at each location. A summary of performance of the

TABLE 15. Protein content of high-protein selections at different locations during 1985 and in different years at Patancheru.

Locations/year	HPL 24	HPL 25	HPL 26	HPL 28	Control	SE
Locations in 1985						
Patancheru	31.3	28.6	29.7	27.8	23.3	±0.26
Jalna	32.2	28.9	29.7	30.4	23.1	±0.69
SK Nagar	30.9	28.4	29.0	27.3	21.4	±0.36
Gulburga	32.1	29.9	-	27.6	23.0	±0.49
Gwalior	32.3	30.4	28.2	27.3	22.0	±0.71
Hisar	31.1	29.6	31.7	29.2	24.5	-
Years at Patancheru						
1981	28.3	28.3	27.6	27.6	20.9	-
1982	33.1	31.8	30.9	31.8	22.6	-
1983	29.3	29.8	29.7	29.5	23.6	-
1984	33.8	31.6	30.7	31.4	22.7	-
1985	31.4	29.1	29.0	27.9	22.9	-
1986	32.6	31.4	30.9	29.2	23.3	-
Mean	31.4	30.3	29.8	29.6	22.7	-

high-protein lines grown at Patancheru (Table 15) for six years indicated significant year-to-year variation. For example in 1982, 1984, and 1986 relatively high-protein estimates were recorded and in 1981 the estimates were relatively low. In spite of such variation over the years the superiority of high protein line was maintained.

REFERENCES

Abo-Hegazi, A.M.T. 1979. High protein lines in field beans *Vicia faba* from a breeding programme using gamma ray: 1. Seed yield and heritability of seed protein. pp. 33-36. *In* Proc. Symp. Seed Protein Impr. Cereals and Grain Legumes. Vol. II, IAEA, FAO and GSF. Neuherberg; Federal Republic of Germany.

Ambekar, S.S., S.C. Patil, A.P. Giri, and M.S. Kachole 1996. Trypsin and amylase inhibitors in pigeonpea seeds. Int. Chickpea and Pigeonpea Newsletter 3:106-107.

Bahl, P.N., S.P. Singh, H. Ram, D.B. Raju, and H.K. Jain. 1979. Breeding for improved plant architecture and high protein yields. pp. 297-306. *In* Proc. Symp. Seed Protein Impr. Cereals and Grain Legumes. Vol. I, IAEA, FAO and GSF. Neuherberg, Federal Republic of Germany.

Bidinger, P.D., and B. Nag. 1981. The role of pigeonpeas in village diets. pp. 357-364. *In* Proc. Int. Workshop Pigeonpeas, Vol. 1, 15-19 December 1980, ICRISAT Center, India. Patancheru, A.P., India.

Blixt, S. 1979. Natural and induced variability for seed protein in temperate legumes. pp. 3-21 *In* Proc. Symp. Seed Protein Impr. Cereals and Grain Legumes. Vol. II, IAEA, FAO and GSF. Neuherberg, Federal Republic of Germany.

Dahiya, B.S., and J.S. Brar. 1976. The relationship between seed size and protein content in pigeonpea (*Cajanus cajan* (L.) Millsp.). Trop. Grain Legume Bull. 3:18-19.

Dahiya, B.S., and J.S. Brar. 1977. Diallel analysis of genetic variation in pigeonpea (*Cajanus cajan*). Exptl. Agric. 13 (2):193-200.

Dahiya, B.S., J.S. Brar, and B.S. Bhullar. 1977. Inheritance of protein content and its correlation with grain yield in pigeonpea (*Cajanus cajan* (L.) Millsp.). Qual. Plant Pl. Food. Hum. Nutri. 27 (3-4):327-334.

Daniel, V.A., D. Narayanaswamy, B.L.M. Desai, S. Kurien, M. Swaminathan, and H.A.B. Parpia. 1970. Supplementary value of varying levels of red gram (*Cajanus cajan*) to poor diets based on rice and ragi. Ind. J. Nutr. Dietet. 7:358-362.

Daniel, V.A., P. Rajan, K.V. Sanjeevarayappa, K.S. Srinivasan, and M. Swaminathan. 1977. Effect of insect infestation on the chemical composition and the protein efficiency ratio of the proteins of Bengal gram and red gram. Ind. J. Nutr. Dietet. 14:70-74.

Draper, C.I. 1944. Algaroba beans, pigeonpeas and processed garbage in the layingmash. Hawaii Agrl. Exp. Sta. Prog. Rept. 44. Uni. of Hawaii, Honolulu, Hawaii, USA.

Durga, B.K. 1989. Genetic studies on protein content and nitrogen accumulation in pigeonpea. Ph.D. Thesis. Osmania Uni., Hyderabad, India. p.152.

Eggum, B.O., and R.M. Beames. 1983. The nutritive value of seed proteins. pp. 499-531. *In* Seed Protein: Biochemisry, Genetics, and Nutritive Value. (eds. W. Gottschalk, and H.P. Muller), Martinus Nijhoff/W. Junk Publishers, The Hague, The Netherlands.

Esh, G.C., T.S. De, and U.P. Basu. 1959. Influence of genetic strain and environment on the protein content of pulses. Science. 129:148.

Falvey, L., and T. Visitpanich. 1980. Nutrition of highland swine. 2. Preparation of pigeonpea seed fed in conjunction with rice bran and banana stalk. Thai J. Agrl. Sci. 13:29-34.

Faris, D.G., and U. Singh. 1990. Pigeonpea: Nutrition and Products. pp. 401-434. *In* The Pigeonpea (ed. Y.L. Nene, S.D. Hall, and V.K. Sheila), Wallingford, U.K. CAB International.

Faris, D.G., K.B. Saxena, S. Mazumdar, and U. Singh. 1987. Vegetable pigeonpea: a promising crop for India. Patancheru, A.P., India: ICRISAT. p. 13.

Fuji, L., and L. Zhenghong. 1995. Research on feeding pigeonpea dry grains to pigs. Feed Industry. 7: 29-31.

Fuji, L., L. Zhenghong, Y. Jie, and C. Kunxian. 1995. Study on feeding pig by pigeonpea seed. Fodder Industry. 16 (7):29-31.

Geervani, P. 1981. Nutritional Evaluation of Pigeonpea (Variety Hyderabad 3A) processed by traditional methods. pp. 427-434. *In* Proc. Int. Workshop Pigeonpeas. Vol. 2, 15-19 December, 1980, ICRISAT, Patancheru, A.P. India.

Godbole, S.A., T.G. Krishna, and C.R. Bhatia. 1994. Changes in protease inhibitory activity from pigeonpea (*Cajanus cajan* (L.) Millsp.) during seed development and germination. J. Sci. Food and Agri. 66: 497-501.

Gridley, H.E., and A.M. Evans. 1979. Prospects of combining high yield with increased protein production in *Phaseolus vulgaris* L. pp. 47-58. *In* Proc. Symp. Seed Protein Impr. Cereals and Grain Legumes. Vol. II, IAEA, FAO and GSF. Neuherberg, Federal Republic of Germany.

Hamilton, T.S., B.C. Johnson, and H.H. Mitchell 1951. The dependence of the physical and chemical composition of the corn kernel on soil fertility and cropping system. Cereal Chem. 28:163-177.

Hartwig, E.E., and K. Hinson. 1972. Association between chemical composition of seed and yield of soybeans. Crop Sci. 12:829-830.

Henke, L.A., S.H. Work, and A.W. Burt. 1940. Beef cattle feeding trials in Hawaii. Hawaii Agrl. Exp. Sta. Bull. 85. Uni. Hawaii, Honolulu, Hawaii, USA.

Herrera, P.G., C.J. Lotero, and L.V. Crowder. 1966. Cutting frequency with tropical forage legumes. Agrl. Tropicale. 22: 473-483.

Hulse, J.H. 1977. Problems of nutritional quality of pigeonpea and chickpea and prospects of research. pp. 88-100. *In* Nutritional Standards and Methods of Evaluation for Food Legume Breeders. (eds. Hulse, J.H., K.O. Rachie, and L.W. Billingsley), Intl. Dev. Research Center, Ottawa, Canada.

ICRISAT. 1987. Annual Report 1986. Patancheru, A.P., India. ICRISAT. pp. 195-196.

Imam, M.M. 1979. Variability in protein content of locally cultivated *Phaseolus* and *Vigna* spp. pp. 119-126. *In* Proc. Symp. Seed Protein Impr. Cereals and Grain Legumes. Vol. II, IAEA, FAO and GSF. Neuherberg, Federal Republic of Germany.

Iyengar, A.K., and P.R. Kulkarni. 1977. Oligosaccharide levels of processed legumes. J. Food. Sci. Tech. 14:222-223.

Jain, K.C., K.B. Saxena, U. Singh, L.J. Reddy, and D.G. Faris. 1986. Genotype environment interaction for protein content in pigeonpea. Int. Pigeonpea Newsletter. 5:18-19.

Jambunathan, R., and U. Singh. 1981. Grain quality of pigeonpea. pp. 351-356. *In* Proc. Int. Workshop Pigeonpeas, Vol. 1, 15-19 December 1980, ICRISAT, Patancheru, A.P., India.

Jyothi, E., and P.R. Reddy. 1981. The effect of germination and cooking on the in *vitro* digestibility of starch in some legumes. Nutr. Rep. Int. 23:799-804.

Kamath, M.V., and B. Belavady. 1980. Unavailable carbohydrates of commonly consumed Indian foods. J. of Sci. of Food and Agric. 31, 194-202.

Krauss, F.G. 1921. The pigeonpea (*Cajanus indicus*): its culture and utilization in Hawaii. Uni. of Hawaii Agril. Exp. Sta. Bull 46, Honolulu, Hawaii, USA.

Krauss, F.G. 1932. The pigeonpea (*Cajanus indicus*) its improvement, culture and utilization in Hawaii. Hawaii Agril. Exp. Sta. Bull. 64, Honolulu, Hawaii, USA.

Kumar, K.A., S.C. Gupta, and D.J. Andrews. 1983. Relationship between nutritional quality characters and grain yield in pearl millet. Crop Sci. 23:232-234.

Kurien, P.P. 1981. Advances in milling technology of pigeonpea. pp. 321-328. *In* Proc. Int. Workshop Pigeonpeas, Vol. 1, 15-19 December 1980, ICRISAT Center, Patancheru, A.P., India.

Kurien, S., D. Narayanaswamy, V.A. Daniel, M. Swaminathan, and H.A.B. Parpia. 1971. Supplementary value of pigeonpea (*Cajanus cajan*) and chickpea to poor diets based on kaffir corn and wheat. Nutri. Repts. Int. 4:227-232.

Manimekalai, G., S. Neelakantan, and R.S. Annapan. 1979. Chemical composition and cooking quality of some improved varieties of red gram dhal. Madras Agric. J. 66: 812-816.

Modi, J.D., and P.R. Kulkarni. 1976. Studies on starches of ragi and red gram. J. Food. Sci. Tech. 13 (1): 9-10.

Morton, J.F. 1976. The pigeonpea (*Cajanus cajan* (L.) Millsp.), a high protein tropical bush legume. Hort. Sci. 11(1): 11-19.

Mutimani, V.H., and S. Paramjyothi. 1995. Protease inhibitors in some pigeonpea lines. Int. Chickpea and Pigeonpea Newsletter. 2:79-81.

Narasimha, H.V., and H.S.R. Desikachar. 1978. Objective methods for studying cookability of tur pulse (*Cajanus cajan*) and factors affecting varietal differences in cooking. J. Food. Sci. Tech. 15: 47-50.

Nigam, V.N., and K.V. Giri. 1961. Sugar in pulses. Can. J. Biochem. Physiol. 39: 1847-1853.

Nwokolo, E., and U.T. Oji. 1985. Variation in metabolizable energy of raw and autoclaved white and brown varieties of three tropical grain legumes. J. Animal Feed Sci. Tech. 13:141-146.

Oke, O.L. 1969. Sulphur nutrition of legumes. Expl. Agric. 5:111-116.

Pal, R.K. 1939. A review of the literature on the nutritive value of pulses. Ind. J. Agric. Sci. 9 (1):133-137.

Parbery, D.B. 1967. Pasture and fodder crop plant introductions at Kimberley Research Station, W. Australia 1963-1964. 1. Perennial Legumes. CSIRO Division of Land Research and Technology Memoir 6716. CSIRO, Melbourne, Australia.

Parpia, H.A.B. 1973. Utilization poblems in grain legumes. pp. 281-295. *In* Nutritional Impr. Food Legumes by Breeding. Protein Advisory Group, United Nations, New York.

Pathak, G.N. 1970. Red gram. pp. 14-53. *In* Pulse Crops of India. Ind. Council of Agric. Research, New Delhi, India.

Pichare, M.M., and M.S. Kachole. 1994. Protease inhibitors of pigeonpea and its wild relatives. Int. Chickpea and Pigeonpea Newsletter. 1:44-46.

Pietri, R., R. Abrams, and F.J. Julie. 1971. Influence of fertility level on the protein content and agronomic characters of pigeonpea in an oxysol. J. Agric. Univ. Puerto Rico. 55(4): 474-477.

Prema, L., and P.A. Kurup. 1973. Hypolipidaemic activity of the protein isolated from *Cajanus cajan* in high fat-cholesterol diet fed rats. Ind. J. Biochem. Biophy. 10:293-296.

Rajalakshmi, R., and K. Vanaja. 1967. Chemical and biological evaluation of the effects of fermentation on the nutritive value of foods prepared from rice and legumes. British J. Nutr. 21:467-473.

Reddy, L.J., J.M. Green, U. Singh, S.S. Bisen, and R. Jambunathan. 1979. Seed protein studies on *Cajanus cajan, Atylosia* spp. and some hybrid derivatives. pp. 105-117. *In* Seed Protein Impr. Cereals and Grain Legumes, volume II. Intl. Atomic Energy Agency, Vienna.

Ryan, J.G. 1997. A global perspective on pigeonpea and chickpea sustainable production system: present status and future potential. pp. 1-32. *In* Recent Adv. in Pulses Res. (ed. A.N. Asthana and Masood Ali) Ind. Soc. Pulses Res. Dev. Ind. Inst. of Pulses Research, Kanpur, India.

Salunkhe, D.K. 1982. Legumes in human nutrition: current status and future research needs. Current Sci. 51:387-394.

Saxena, K.B., and D. Sharma. 1990. Pigeonpea genetics. pp. 137-157. *In* The Pigeonpea (ed. Y.L. Nene, S.D. Hall, and V.K. Shiela) CAB International, UK.

Saxena, K.B, S.J.B.A. Jayasekera, and H.P. Ariyaratne. 1992. Pigeonpea varietal adaptation and production studies in Sri Lanka. ICRISAT, Patancheru. p. 173.

Saxena, K.B., U. Singh, and D.G. Faris. 1983. Does pod color affect the organoleptic qualities of vegetable pigeonpeas? Intl. Pigeonpea Newsletter. 2:74-75.

Saxena, K.B., R.V. Kumar, D.G. Faris, and L. Singh. 1987a. Breeding for special traits. Pigeonpea Breeding Annual Report, International Crops Research Institute for the Semi-Arid Tropics, Patancheru, A.P., India. p. 72.

Saxena, K.B., M.D. Gupta, and U. Singh. 1987b. Can seed size and protein content in pigeonpea be increased simultaneously? Int. Pigeonpea Newsletter. 6:29-31.

Saxena, K.B., D.G. Faris, U. Singh, and R.V. Kumar. 1987c. Relationship between seed size and protein content in newly developed high protein lines of pigeonpea. Plant Foods Hum. Nutr. 36: 335-340.

Saxena, K.B., D.G. Faris, and R.V. Kumar. 1984. Breeding for special traits. Pigeonpea Breeding Annual Report, International Crops Research Institute for the Semi-Arid Tropics, Patancheru, India. p. 99.

Sham, N.L. 1976. Effect of nitrogen, phosphorus and sulphur on protein content of arhar *Cajanus cajan* (L.). Seed Farming 2(5): 37-39.

Sharma, D., and K.B. Saxena. 1977. Relationship between seed size and yield in pigeonpea. Pigeonpea Breeding Annual Report, 1976-77. International Crops Research Institute for the Semi-Arid Tropics, Patancheru, India. p. 199.

Sharma, Y.K., A.S. Tiwari, K.C. Rao, and A. Mishra. 1977. Studies on chemical constituents and their influence on cookability in pigeonpea. J. Food Sci. Tech. 14:38-40.

Shiying, Y., L. Yuxi, C. Chenbin, L. Shichun, and L. Hunchao. 1999. Observations on introduced pigeonpea. Guangxi Agric. Sci. 6: 309-310.

Singh, D.N., and A.K. Kush. 1981. Effect of population density on growth pattern and yielding ability in pigeonpea. pp. 165-174. *In* Proc. Int. Workshop Pigeonpeas, Vol. 1, 15-19 December 1980, ICRISAT Center, Patancheru, A.P., India.

Singh, L., and D. Sharma, and A.D. Deodhar. 1974. Effect of environment on protein content of seeds and implication in pulse improvement. pp. 731-808. *In* Proc. 2nd General Cong. SABRAO, New Delhi, India.

Singh, L., N. Singh, M.P. Shrivastava, and A.K. Gupta. 1977. Characteristics and utilization of vegetable types of pigeonpeas (*Cajanus cajan* (L.) Millsp.). Ind. J. Nutri. Diet. 14:8-10

Singh, U. 1984. The inhibition of digestive enzymes by polyphenols of chickpea (*Cicer arietinum* L.) and pigeonpea (*Cajanus cajan* (L.) Millsp.). Nutr. Rep. Int. 29:745-753.

Singh, U. 1988. Antinutritional factors of chickpea and pigeonpea and their removal by processing. Pl. Foods Human Nutri. 38, 251-261.

Singh, U., and B.O. Eggum. 1984. Factors affecting the protein quality of pigeonpea (*Cajanus cajan* L.). Qual. Plant. Plant Foods Human Nutr. 34:273-283.

Singh, U., and R Jambunathan. 1990. Pigeonpea: post-harvest technology. pp. 435-455. *In* The Pigeonpea (ed. Y.L. Nene, S.D. Hall and V.K. Sheila), Wallingford, UK. CAB International.

Singh, U., and R. Jambunathan. 1981. A survey of the methods of milling and consumer acceptance of pigeonpeas in India. pp. 419-425. *In* Proc. Intl. Workshop Pigeonpeas, Vol. 2:15-19 December 1980. ICRISAT, Patancheru, A.P. India.

Singh, U., and R. Jambunathan. 1982. Distribution of seed protein fractions and amino acids in different anatomical parts of chickpea (*Cicer arietinum* L.) and pigeonpea (*Cajanus cajan* L.). Qual. Plant. Plant Foods Human Nutr. 31: 347-354.

Singh, U., K.C. Jain, R. Jambunathan, and D.G. Faris. 1984a. Nutritional quality of vegetable pigeonpeas [*Cajanus cajan* (L.) Millsp.]: dry matter accumulation, carbohydrates and proteins. J. Food Sci. 49:799-802.

Singh, U., K.C. Jain, R. Jambunathan, and D.G. Faris. 1984b. Nutritional quality of vegetable pigeonpeas [*Cajanus cajan* (L.) Millsp.]: mineral and trace elements. J. Food Sci. 49:645-646.

Singh, U., M.S. Khardekar, and K.C. Jain. 1984c. Effect of growing season on the cooking quality of pigeonpea. Int. Pigeonpea Newsletter 3:50-51.

Singh, U., P. Venkateshwara Rao, K.B. Saxena, and L. Singh. 1991. Chemical changes at different stages of seed development in vegetable pigeonpeas (*Cajanus cajan*). J. Sci. Food Agric. 57:49-54.

Singh, U., R. Jambunathan, K.B. Saxena, and N. Subramaniam. 1990. Nutritional quality evaluation of newly developed high protein genotypes of pigeonpea [*Cajanus cajan* L.) J. Sci. Food Agric. 50:201-209.

Springhall, J., J.O. Akinola, and P.C. Whiteman 1974. Evaluation of pigeonpea (*Cajanus cajan*) seed meal in chicken rations. pp. 117-119. *In* Proc., Austr. Poultry Sci. Convention, Sydney, Australia.

Squire, F.A. 1933. Guiva Department of Agriculture, Rice Bull. No. 1: 51-57.

Srivastava, S., D.P. Mishra, and B.P. Khare. 1988. Effect on insect infestation on biochemical composition of pigeonpea (*Cajanus cajan* (L) Millsp.) seeds stored in mud bin. Bull. Grain Tech. 26: 120-125.

Swaminathan, M.S. 1973. Basic research for further improvement of pulse crops in South East Asia. pp. 61-68. *In* Nutritional improvement of food legumes by breeding (ed. M.M. Milner). Protein Advisory Group, United Nations.

Tripathi, R.D., G.P. Srivastava, M.C. Misra, and S.C. Sinha. 1975. Comparative studies in the quality characteristics of early and late cultivars of red gram (*Cajanus cajan* L.). Ind. J. Agric. Chem. 8: 57-61.

Uma Reddy, M., and P. Pushpamma. 1981. Effect of insect infestation and storage on the nutritional quality of different varieties of pigeonpea. pp. 443-451. *In* Proc. Int. Workshop Pigeonpeas, Vol. 2:15-19 December 1980. ICRISAT, Patancheru, A.P. India.

Vimala, V., and P. Pushpamma. 1983. Storage quality of pulses stored in three agro-climatic regions of Andhra Pradesh-1-Quantitative changes. Bull. Grain Tech. 21:54-62.

Vimala, V., and P. Pushpamma. 1985. Effect of improved storage methods on cookability of pulses stored for one year in different containers. J. Food Sci. Tech. 22:327-329.

Wallis, E.S., D.G. Faris, R. Elliott, and D.E. Byth. 1986. Varietal improvement of pigeonpea for smallholder livestock production systems. pp. 536-553. *In* Proc. of the Crop-livestock Systems Research Workshop, 7-11 July 1986, Khon Kaen, Thailand. Farming Systems Research Institute, Department of Agriculture, Thailand and Asian Rice Farming Systems Network, International Rice Research Institute, Philippines.

Wayne, J.C., and A.J. Casady. 1974. Heritability and interrelationship of grain protein content with other agronomic traits in sorghum. Crop Sci. 14:622-624.

Whiteman, P.C., and B.W. Norton. 1981. Alternative uses of pigeonpeas. pp. 365-377. *In* Proc. Intl. Workshop Pigeonpeas, Vol. 1, 15-19 December 1980, ICRISAT Center, India. Patancheru, A.P., India.

Work, S.H. 1946. Digestible nutrient content of some Hawaiian feeds and forages. Uni. Hawaii. Agril. Exp. Stn. Tech. Bull. 4, Honolulu, Hawaii, USA.

Yadav, S.P. 1984. Physical and chemical analyses of pigeonpea (*Cajanus cajan*) cultivar and F_1 crosses. J. Sci. Food Agric. 35:833-836.

Legume Forage Quality

J. H. Cherney
D. J. R. Cherney

SUMMARY. Legume forage quality research is now concerned not only with nutritive value of forage for ruminant animals, but the impact of these nutrients on environmental quality. If we are to move to a more sustainable agriculture worldwide, more legumes must be incorporated into animal production systems. The goal of forage legume breeders is to tailor legume nutritive value to the needs of the consuming animals. A generalized priority list for legume nutritive value research includes high fiber digestibility, low anti-quality compounds, appropriate condensed tannins, reduced nonprotein nitrogen, and high sulfur amino acids. Although breeders have regularly found a wide range in forage quality within a given legume species, very few varieties with proven forage quality advantages have been released. Yield and persistence issues have dominated forage legume breeding. Improvement in forage quality often is linked to a reduction in yield and/or persistence, and also frequently results in complex genotype × environment interactions. Transgenic technology has almost limitless potential for improving legume forage quality and environmental quality, but only if the public can be convinced that transgenics are an acceptable risk. Structural and functional roles of legume plant cell walls and their relationships to forage quality are poorly understood. Research on the basic processes of forage legume growth and the relationships between growth and forage composition should result in the development of more accurate simulation models, and trans-

---

J. H. Cherney is E.V. Baker Professor of Agriculture, Department of Crop and Soil Sciences, and D. J. R. Cherney is Senior Research Associate, Department of Animal Science, Cornell University, Ithaca, NY 14853.

Contribution from the Cornell University Agricultural Experiment Station.

[Haworth co-indexing entry note]: "Legume Forage Quality." Cherney, J. H., and D. J. R. Cherney. Co-published simultaneously in *Journal of Crop Production* (Food Products Press, an imprint of The Haworth Press, Inc.) Vol. 5, No. 1/2 (#9/10), 2002, pp. 261-284; and: *Quality Improvement in Field Crops* (ed: A. S. Basra, and L. S. Randhawa) Food Products Press, an imprint of The Haworth Press, Inc., 2002, pp. 261-284. Single or multiple copies of this article are available for a fee from The Haworth Document Delivery Service [1-800-HAWORTH, 9:00 a.m. - 5:00 p.m. (EST). E-mail address: getinfo@haworthpressinc.com].

genic technology can provide us with the tools to understand these basic processes. *[Article copies available for a fee from The Haworth Document Delivery Service: 1-800-HAWORTH. E-mail address: <getinfo@haworthpressinc. com> Website: <http://www.HaworthPress.com> © 2002 by The Haworth Press, Inc. All rights reserved.]*

KEYWORDS. Anti-quality, digestibility, environmental quality, nutritive value, palatablility, transgenic

INTRODUCTION

Legumes (plant family Leguminosae) are widely distributed around the world and their forage quality attributes have received renewed attention. Forage quality research has tended to focus on grasses and this point is exemplified in a recent forage quality symposium (Fahey and Hussein, 1999), where a series of papers on forage quality improvements and challenges almost completely lack references to legumes.

If the price of fossil fuels continues to increase worldwide, economics will favor the increased use of legume forage crops over commercial N fertilization of grasses. Increased research on manipulating the content of legume forage for improved nutritive value and for increased environmental quality should yield high returns. Returns from agricultural research in general have been traditionally high, and returns from forage quality research have been equally high. Ford (1999) estimated that potential forage quality improvements on an average USA dairy farm might exceed $3,000 (US) annually. Hopkins et al. (1994) estimated that if the UK livestock industry switched from N-fertilized grass to legume-based systems the annual savings would be approximately £300 million.

Legumes have been recognized for centuries for their benefits as green manures, but the reasons for this benefit were not clarified until the 20th century. The benefits of biological dinitrogen fixation, along with other factors resulting in high nutritive value of legume forage, are now being promoted as key components of sustainable animal agriculture. The challenge of animal agriculture is to produce high quality forage for high producing animals, while at the same time minimizing pollution risk (Wilkins, 2000).

Forage Quality Defined

Earlier forage quality definitions focused on the plant itself, and this was followed by definitions focusing on the animal and on animal intake in particular (Fahey and Hussein, 1999; Reid, 1994). More recently, legume forage quality also was related to C and N cycling as well as other environmental concerns (Frame et al., 1997). For the purposes of this review, we define legume forage quality as the forage traits that affect nutritive value and subsequent animal performance, as well as those impacting environmental quality. The importance of forage quality for sustainable agriculture will equal or perhaps surpass the importance of forage quality for animal performance in the future.

Most of our attention in this review will not focus on forage yield or stand persistence issues. The authors readily admit that there must be quantity in order to have quality. Indeed, the greatest strides in forage quality improvement likely will be achieved by development or introduction of a high producing, persistent legume into a region where none existed previously. Most of the world's forage legumes are grown in association with grasses, and the dietary interaction of legumes and grasses undoubtedly influences the legume forage ideotype. This review discusses potential changes in legume forage quality only, although acknowledging the importance of the concept of legumes grown with grass as a nutritive supplement to the grass (Clark and Kanneganti, 1998). Introduction of new, persistent legumes, or improvement in stand persistence through better pest control, impact quality primarily by impacting quantity. There are, however, examples of pest control that directly affect forage quality.

Legume species discussed here reflect the body of literature available. Lucerne *(Medicago sativa)*, *Trifolium* species and *Lotus* species are the principal temperate legumes (Frame et al., 1997), while a wide variety of tropical legume species are utilized for forage around the world. Quality issues in tropical legumes differ significantly from temperate legumes and most tropical species are utilized exclusively for pasture purposes. This review includes a discussion of various methods of plant manipulation of forage quality for meeting animal needs and addressing environmental issues. Desirable quality traits from an animal perspective are considered first.

ANIMAL PERFORMANCE

Wide variation in quality among legume species exists. There are certainly many characteristics that could be selected for in a breeding program for improved animal performance. Historically, however, breeding legumes for improvements in forage quality has not been a high priority (Bray, 1982). Legumes are typically bred for yield, persistence, and disease and pest resistance (Vogel and Sleper, 1994). Although some references to strain selection for quality go back several centuries (Casler and Vogel, 1999), these were primarily attempts to protect forage quality or to improve quality by increasing quantity. If consumed by animals, these legumes must also be palatable, so some breeding efforts have focused on palatability. Selection for forage quality was handicapped by the lack of fast, accurate chemical analysis methods, and by the fact that legumes were considered high quality forage to begin with. When forage quality improvement was first seriously undertaken, efforts focused on grasses, not legumes (Burton, 1972), and this tendency continues today (Casler and Vogel, 1999).

A major task for legume breeders is to tailor forage quality to more precisely meet animal needs (Humphreys, 1999). As animal nutritionists have become more adept at determining nutrient constituents that are important to animal performance, and have developed the analytical techniques to measure these constituents (Cherney, 2000; Reed et al., 2000), their requests to legume breeders have become more specific. Development of forage quality analysis procedures that were both efficient and related to animal performance allowed plant breeders to seriously select for forage quality improvement in large populations. Major steps forward in laboratory analyses were *in vitro* digestibility (Tilley and Terry, 1963), the detergent fiber system (Goering and Van Soest, 1970), and near infrared reflectance spectroscopy (NIRS) (Shenk et al., 1979). Chemical forage analyses can be used to identify factors that limit animal performance and can be used to predict animal performance (Cherney, 2000).

Given the importance of protein and energy balance for maximizing rumen microbial protein output (Hoover and Stokes, 1991), some nutritionists have been urging breeders to focus on the balance between energy and protein, as well as overall digestibility (Beever and Reynolds, 1994). Viands (1995), in consultation with animal nutritionists, recommended three major areas where breeders could improve animal performance: (a) increase escape protein, (b) make more energy available by increasing carbohydrate concentration, and (3) improve rate of digestion during the first 24 hours. Van Soest (1995) suggested that this could be done by possibly lowering lignin levels or increasing pectin levels.

For many years breeders were urged to breed for higher protein values, and they were very successful at this. Statistics provided by the Northeast Dairy Herd Improvement Association (Northeast DHIA Forage Lab Tables of Feed Composition) for samples that they processed indicated that average crude protein (CP) for legume silages and hays (mostly lucerne) were 19.7% and 19.3% of dry matter (DM), respectively. These values are well above values recommended for most classes of livestock that consume forages (NRC, 1985, 1989, 1996), yet Dhiman and Satter (1993) concluded that protein limited milk production in cows fed a diet high in lucerne haylage (75%).

Hill et al. (1988) estimated cost reductions in dairy cow feed costs resulting from changes in forage quality traits and concluded that lucerne breeders could justify selection for increased protein content in lucerne. Cherney (1995) used the Cornell Net Carbohydrate and Protein System (CNCPS; Fox et al., 1992) to demonstrate changes in lucerne crude protein and protein solubility on milk production. In that study, Cherney (1995) demonstrated that protein solubility has a large impact on estimated milk. Protein content is unlikely to limit milk production if lucerne (or other legume) is the sole source of nutrient (Figure 1a). Protein also will be unlikely to limit production if lucerne is used as the primary fiber source in the diet and concentrates supply the additional 40 to 60% of the diet (Figure 1b). Protein may be limiting in situations where the legume (i.e., lucerne) is the primary diet constituent and a highly fermentable carbohydrate (i.e., high moisture corn) is the other diet constituent (Figure 1c), which accounts for the observations of Dhiman and Satter (1993). In situations where energy is limiting, rather than protein, increasing protein may decrease milk production due to the energy required to dispose of excess ruminal ammonia (Cherney, 1995). It would, therefore, be more advisable to concentrate on decreasing protein solubility rather than increasing protein content.

Improving forage quality by increasing CP and/or decreasing fiber can have negative consequences for yield or persistence (Casler, 1999; Humphreys, 1999). Selection for lower lignin often results in lower yield (Buxton and Casler, 1993). Lignin-like materials and phenolic compounds may also be associated with disease and pest resistance and cold hardiness (Buxton and Casler, 1993). Viands (1995) reported that selection for improved CP and fiber values in lucerne was correlated with increased lodging and slower regrowth. While this certainly seems a daunting problem to tackle, with increased awareness of the specific constituent that animal nutritionists are interested in, and new breeding techniques, it may be possible to make advances in this area.

Nutritive value characteristics of legumes in order of importance were ranked by Wheeler and Corbett (1989) (Table 1), and summarized

FIGURE 1. The influence of alfalfa crude protein and protein solubility on milk production estimated (CNCPS; Fox et al., 1992) from available metabolizable energy (ME milk) and metabolizable protein (MP milk). (a) Alfalfa is the only source of nutrients. ME limits milk production until protein solubilities exceed 50%. (b) Alfalfa is the primary forage source in the diet. ME limits milk production. (c) Corn and alfalfa are used as nutrients. MP limits milk production until protein solubilities are reduced to less than 40%. (Adapted from Cherney, 1995.)

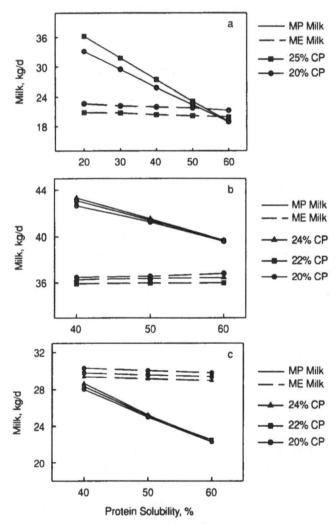

TABLE 1. Ranking of quality criteria for breeding legumes for liveweight gain and wool production (adapted from Wheeler and Corbett, 1989).

Rank	Liveweight Gain	Wool
1	High digestibility	High digestibility
2	Easy comminution	High S-amino acids
3	Appropriate tannins	Easy comminution
4	High NSC	Appropriate tannins
5	Adequate minerals	High CP
6	High CP	High NSC
7	High S-amino acids	High palatability
8	Low anti-quality constituents	Adequate minerals
9	High palatability	Low anti-quality constituents
10	High lipid	High lipid
11	Erect growth	Erect growth

by Poppi et al. (1999) for all forages. An updated list of nutritive value characteristics specifically related to legumes is as follows:

- high fiber digestibility
- low anti-quality compounds
- appropriate condensed tannins
- reduced nonprotein nitrogen
- high sulfur amino acids
- high nonstructural carbohydrates
- high lipid

Nutritive value traits suggested for targeting in temperate legumes include improvement of fiber digestibility and improved protein quality (Martin, 2000), as well as anti-quality components. Legume anti-quality compounds include phyto-estrogens and saponins, as well as cyanogens and mimosine in tropical browse (D'Mello and MacDonald, 1996). Condensed tannins are separated from anti-quality compounds because of their ability to either increase or decrease forage nutritive value (Waghorn et al., 1999). Temperate legumes can produce forage with optimum total nitrogen (crude protein) and total fiber (neutral detergent fiber) content through appropriate harvest management. Legumes already possess a fast rate of particle breakdown in the rumen (Beever and Thorp, 1996).

The largest single predictor of animal performance is intake. Increasing intake of digestible dry matter should therefore, be a primary objective of the breeder. In many parts of the world, the animal consumes most of its forage by grazing, often in mixed swards. In these situations, forage quality in terms of protein and fiber may not be the issue, but rather improving the overall quality of the sward (Allen, 1993). Bray (1982) suggested that improvement of grazing legumes is really a matter of fine-tuning, because many of the used species are already selected for persistence and yield. He suggests selecting for compatibility, specifically (a) compatibility with grass, (b) quality for the grazing animal, (c) compatibility with Rhizobium, and (d) and compatibility with insect pollinators (Bray, 1982). Bray (1982) suggested that breeding efforts to improve animal performance should be focused on reducing anti-quality components, rather than selecting for factors such as digestibility. His ideas on the prospects of improving forage quality for grazing are summarized in Table 2. Twenty years later, scientists are still commenting on the importance of developing lucerne and white clover (*Trifolium repens*) cultivars with the appropriate amount of tannins (Waghorn et al., 1999).

Although management practices exist to deal with bloat (Mayland and Cheeke, 1995), it continues to be a problem with high quality legumes. Plant pectins, lipids, lipoproteins, saponins, soluble proteins, tannins, and cell wall structures are all factors that have been implicated in the etiology of bloat (Allen, 1993). Ruminal microorganisms converted much less of the protein in birdsfoot trefoil (*Lotus corniculatus*) to ammonia than did lucerne (Cherney et al., 1994), clearly demonstrating the advantage of small amounts of tannins in legumes. On the other end of the spectrum are legumes such as Sercia lespedeza (*Lespedeza cuneata*), with poor animal performance due to high tannin content (Allen, 1993). A low tannin variety had higher neutral detergent fiber (NDF), acid detergent fiber (ADF), and DM digestibility, as well as a higher intake when fed to sheep versus a high tannin variety (Terrill et al., 1989).

Although there is considerable literature on the potential of modifying condensed tannins in legumes (Morris and Robbins, 1997; Morris et al., 1999), there are still analytical hurdles to overcome in appropriately and quantitatively characterizing tannins (Larkin et al., 1999). Birdsfoot trefoil has not been known to cause bloat, presumably due to the

TABLE 2. Prospects of improving grazing legume quality through selection (adapted from R. A. Bray, 1981).

Species	Character	Method Used	Priority	Chance of Success
Temperate				
M. sativa	bloat	reduce foaming proteins	medium	low
		increase tannins	medium	low
	digestibility	more digestible stems	medium	low
T. pratense	bloat	tannins	medium	low
	digestibility	more digestible stems	medium	low
T. repens	bloat	tannins	medium	low
	oestrogens		low	high
	cyanogenesis		nil	high
T. subterraneum	bloat	tannins	low	low
	oestrogens		high	certain
Tropical				
Stylosanthes spp. (perennial)	increased leaf retention		high	high
Leucaena leucocephalia		mimosine	high	medium
Macroptilium atropupureum	maintenance of high quality	rust resistance	high	high

presence of tannins (Beuselinck and Grant, 1995). Mupangwa et al. (2000) found that condensed tannin content of cassia (*Cassia rotundifolia*) was high enough to result in anti-quality effects, while condensed tannin content of lablab (*Lablab purpureus*) and sirato (*Macroptilium atropurpureum*) was too low to provide beneficial N utilization effects.

There are over 24,000 secondary plant compounds that are involved in plant adaptation to environment, but that are not part of the primary biochemical pathways of cell growth and reproduction (Reed et al., 2000). Many of these compounds, which include alkaloids, nonprotein amino acids, cyanogenic glycosides, volatile terpenoids, saponins, phenolic acids, hydrolysable tannins, proanthocyanidins and eostrogenic isoflavones, have antinutritional or toxic effects on mammals (Harborne, 1993). Some of these secondary plant compounds, particularly condensed tannins, are also associated with improved protein digestion and metabolism, and in protection against bloat (Waghorn et al., 1999). Condensed tannins may also have antihelminthic properties. Niezen et al. (1995) observed lower fecal *Trichostongylus colubriformis* egg counts and worm burdens and higher daily live-weight gains in sheep grazing sulla (*Hedysarum coronarium*) than those grazing lucerne. Sulla contains condensed tannins, while lucerne does not. The beneficial or toxic effects of any particular secondary plant compound is likely dose-dependent, as the biochemical pathways and mechanisms for the beneficial or toxic effect are likely to be the same (Shahidi, 1995). A full understanding of the biochemical nature and mechanism of action of these secondary compounds is necessary for breeders to be able to use new or conventional techniques to overcome the harmful effects of these compounds or maximize their beneficial effects.

Reproductive problems and low production in grazing ruminants in the tropics has been associated with mineral excesses or deficiencies (McDowell and Valle, 2000). A complete analysis of minerals was Minson's (1981) first priority where animal production was low in the tropics. Livestock producers in tropical countries typically do not supplement grazing livestock with minerals, with the possible exception of common salt, and forages are unlikely to meet animal's needs (McDowell and Valle, 2000). Soils in the tropics are generally poor, high in acidity and low in fertility. Legumes are selected for their ability to withstand these environments. Soil has a major impact on plant mineral status, but there are forage species and variety differences. Legumes are generally richer in a number of minerals (Fleming, 1973), but this is not always the case. For example, *Stylosanthes* was selected to grow in environments with low available soil P. This results in a forage with exceedingly low P (McDowell and Valle, 2000). McDowell and Valle (2000) recommended that new forage strains and varieties should include mineral concentrations as quality criteria. Until significant advances are made in this area, the most practical, cost-effective solution will continue to be mineral supplementation.

PLANT MANIPULATION TO MEET ANIMAL NEEDS

Classical Breeding

There have been relatively few released forage cultivars with proven improvements in forage quality through selection, and most of these have been grasses (Casler and Vogel, 1999). Popular legume forage crops (lucerne and clovers) are inherently high in forage quality. Frame et al. (1997) listed white clover breeding objectives in New Zealand and in the UK, and none of the objectives mention forage quality. Minor legume crops, particularly tropical legumes, may contain anti-quality components (Minson, 1988), but there is less interest in investing in breeding programs for such legumes. Therefore, few advances have been made in legume forage quality through classical breeding to-date, in spite of the promising variation frequently reported in the literature (McGraw et al., 1989; Papadopoulos and Kelman, 1999; Van Wijk et al., 1993). Various legume breeding techniques are described by Rumbaugh et al. (1988).

There certainly is potential for a variety of forage quality improvements in legumes. Moutray (1996) noted that a wide range in composition of lucerne forage could be found in clonal nurseries and small improvements in digestibility and protein content of lucerne have been realized. Sufficient variation in N concentration, digestibility, and fiber content for breeding and selection programs also were found in birdsfoot trefoil (McGraw et al., 1989; Papadopoulos and Kelman, 1999). While there is a need to select for reduced condensed concentration in *Lotus uliginosus*, it appears that *L. corniculatus* forage is not high enough in condensed tannins to require a breeding effort (Papadopoulos and Kelman, 1999). Persistence problems with red clover (*Trifolium pratense*) overshadow any concerns about bloat or anti-quality isoflavones as a breeding objective (Taylor and Quensenberry, 1996).

Legume breeding related to environmental quality is likely to emerge as a top priority. Environmental quality can be positively influenced by transferring phytase expression (Austin-Phillips et al., 1999), herbicide-degrading ability, or by developing plants that clean up soils contaminated with nitrates or other pollutants (Martin, 2000). Blumenthal et al. (1999) determined that lucerne cultivars selected for improved forage quality or different root system architecture were no more effective in removing sub-soil nitrate than standard cultivars. Non-N_2-fixing germplasm, however, was much more effective in nitrate phytoremediation than N_2-fixing germplasms (Blumenthal et al., 1999). Development of aluminum tolerance in lucerne (Tesfaye et al., 2000) and other legumes would allow environmentally friendly legume forage production on the vast world-wide area containing acid soils.

Management and environment have a significant influence on legume forage quality (Viands, 1995). Genetic improvements in legume forage quality may not be realized due to the likelihood of environment × genotype interactions (Casler, 1999; Julier et al., 2000). Sheaffer et al. (1998) found a complex location × entry × stand age interaction for forage quality of lucerne cultivars. Some varieties were stable for forage quality attributes over locations, while others were not. Another muli-location study comparing forage quality of a different set of lucerne cultivars found consistent forage quality differences among cultivars across harvest regimes and locations (Sheaffer et al., 2000). Selection for high and low lignin concentration in lucerne resulted in morphological changes, as well as complex changes in stem fiber components (Kephart et al., 1990). Significant phenotypic expression of the multifoliolate trait in legumes can result in changes in forage quality (Volenec et al., 1987). Stage of maturity, temperature and photoperiod, however, all influenced multifoliolate expression in lucerne (Juan et al., 1993).

While there are numerous woody tropical leguminous species used as browse (Skerman et al., 1988) our focus is on herbaceous legumes. Research in tropical and subtropical legume forage quality has been holistic due to the fact that most of these legumes are found in mixed species pastures. Hutton (1988) listed eight characteristics of successful tropical pasture legumes, and most of these related to yield and persistence. High nutritive value was assumed, although freedom from anti-quality compounds was considered important (Hutton, 1988). Cameron et al. (1993) stated that the overriding significance of adaptation and persistence for *Stylosanthes* species and other tropical legumes meant that quality should not receive much weight as a selection criterion. Adaptation and persistence also were the exclusive focus of a review of sub-Sahara African forage legumes (Dzowela, 1993). Minson (1985) concluded that chemical composition, energy value and voluntary intake were the three primary areas of concern with tropical legumes. Specific concerns with tropical legume quality were low digestibility and low sodium, but it was stressed that modifications of legume quality must maximize the nutritional benefits of legume-grass mixtures (Minson, 1985). The potential for modifications in legume forage quality was tremendously expanded with the onset of transgenics technology.

Transgenics

No technology breakthrough in history has had more potential or more controversy than the transfer of genes through molecular biology. The public is primarily concerned with the prospect that they are being asked to take an unknown risk while large corporations get the benefit (Jordan, 2000). Specific potential risks of transgenics as viewed by the general public include introduction of inadvertent toxins or allergens, negative changes in levels of essential nutrients, or negative effects of antibiotic therapies (Kaeppler, 2000). The effects of transgenics on non-target organisms and the potential for gene escape to weeds also are concerns. Such concerns have led to restrictions on this technology in Europe, and proposed restrictive legislation in many states in the USA. Debate over the potential risks of biotechnology will continue. We provide here a fleeting, current assessment of the potential of transgenics for improving legume forage quality.

Specific transgenic methods applied to forage crops are discussed by Spangenberg et al. (1999). In general, transgenics technology is poorly developed in forage legumes, compared to many other crops (McKersie, 1997a). There is considerable interest in value-added products involving legume transgenics but most of these attempts do not impact forage legume quality. The potential of this technology applied to legumes to improve environmental quality (Barton and Dracup, 2000; McKersie, 1997a) may exceed its potential to improve forage nutritive value. Glyphosate herbicide tolerant lucerne currently is being tested with commercial release in the USA predicted by 2004 (McCaslin and Fitzpatrick, 2000).

Legume forage nutritive value improvement should focus on fiber digestibility, protein quality and/or anti-quality components, as mentioned earlier. The legume receiving the most attention at this time is lucerne. Lucerne is one of the easiest crops to manipulate from a transgenic standpoint, because of the availability of transformable and regenerable lines (Bingham, 2000). The rapid accumulation of basic transgenics knowledge from model systems can be directly applied to lucerne in many cases. While there is considerable interest in white clover transgenics in Europe, research has been stymied. Sulfur-rich zein protein concentration was increased in both lucerne and birdsfoot trefoil with transgenic technology (Bellucci et al., 1997). Improved digestibility of lucerne has been attempted by modifying plant lignin, following the example of brown-midrib mutants in grasses (Cherney et al., 1991). Baucher et al. (1999) down-regulated one of the lignin biosynthesis enzymes, cynnamil-alcohol dehydrogenase, and modified the syringyl/guaicyl ratio in lignin. Reduced lignin and increased digestibility were found in transgenic lucerne by incorporating a caffeic acid O-methyltransferase antisense gene (Ni et al., 1993). Similar success with this same lignin biosynthesis enzyme was achieved in stylo (*Stylosanthes humilis*) (Rae et al., 1996). Lignin modification must be followed by a scheme for assessing the management implications and other consequences of plant manipulation (Clark and Wilson, 1993).

The overall goals of transgenic technology are similar to conventional breeding, that is, to identify, create, preserve and transfer genetic variability (McKersie, 1997b). Transfer of transgenic variability will be time consuming and will require supplemental classical breeding to complete the process, such as the backcrossing necessary with lucerne (Bingham, 2000). There will undoubtedly be numerous attempts to improve nutritive value of legumes with transgenics in the near future, but this effort is likely to be confined to the major legume species in the developed countries. Since large corporations patent transgenic processes, it will be a challenge to apply this technology to minor crops, called *orphan crops* by Jordan (2000). The potential to eliminate anti-quality factors from tropical forage legumes via transgenics may not be realized unless corporations agree to allow public good research on such *orphan crops*.

Management

Most gains in legume forage quality in the past were due to improved harvest or pasture management schemes. As with most forage crops, legume quality tends to decline with plant age, and first-cut forage of the season tends to be lower in quality compared to regrowth cuts in both temperate and tropical legumes (Cameron et al., 1993; Sheaffer et al., 1998). Plant age is the primary quality variable in lucerne and birdsfoot trefoil affected by management, although potato leafhopper (*Empoasca fabae*) has a significant effect on lucerne forage quality in the USA (Barnes and Sheaffer, 1995). Primary management variables that affect red clover forage quality are plant age, cutting height and environmental stress (Taylor and Quesenberry, 1996). Use of both white clover and red clover has declined for several decades (Frame et al., 1997; Taylor and Quesenberry, 1996), although the potential for legumes in sustainable agriculture has resulted in a resurgence of clover research. There is a large number of both temperate and tropical, annual and perennial legumes available for forage use, but management concerns involve yield and persistence, not quality. Their use in forage management systems hinges on solving establishment, persistence or yield problems. A typical example of this is kura clover (*Trifolium ambiguum*) (Seguin et al., 1999).

Modeling Forage Quality

To improve our control of forage quality in legume crops, we must be able to improve forage quality predictions (Fick et al., 1994). Modeling attempts to understand and predict the dynamic nature of forage quality by integrating environmental, physiological and/or morphological attributes related to forage quality. The ideal legume quality estimation method for neutral detergent fiber should have the following characteristics (Cherney and Sulc, 1997):

- Accuracy to within 30 to 40 g NDF kg^{-1} dry matter.
- Fast.
- Inexpensive.
- Easy to use.
- Consistent across a wide range of environments.
- Accurate across all harvests during the season.
- Allow for prediction into the future.
- Require legume producers to visit and assess their fields.

Attempts to understand and predict quality based on chemical analyses of forage crops began in the 1950s (Reid et al., 1959). Current approaches to forage quality predictions for legumes include age and weather-based equations, mean stage, and leafiness (Fick et al., 1994).

There have been numerous attempts to improve upon the lucerne quality prediction standard set by Kalu and Fick (1981). A rapid method for estimating lucerne quality based on Wisconsin predictive equations for lucerne quality (Hintz and Albrecht, 1991) was successfully field tested over a wide range of environments (Sulc et al., 1997). Allen and Beck (1996) suggested using growing degree days in combination with plant height and maturity stage for estimating lucerne fiber content. A method for predicting first cutting lucerne fiber content based on early sampling and analysis, followed by use of predictive equations, was suggested by Cherney and Sulc (1997). The ability to accurately predict legume forage quality is hindered most by our ability to predict and model environmental variation.

ENVIRONMENTAL QUALITY CONSIDERATIONS

Carbon-Nitrogen Dynamics

The dynamics of carbon and nitrogen are tightly linked in forage ecosystems (Kephart et al., 1995) as well as in legume plant systems (Gregerson et al., 1999). Sustainability of crop ecosystems is of concern, as significant loss of soil organic matter has occurred in regions such as North America since the onset of intensive crop management practices (Buyanovsky et al., 1987). Carbon is conserved in forage legumes and utilized for plant structure and yield, and beneficially recycled to the atmosphere (Heichel et al., 1988). Inclusion of legume crops in the rotation resulted in increased labile carbon concentrations compared with long fallow or continuous wheat (Blair and Crocker, 2000).

Van der Meer and Van der Putten (1995) concluded that liberal use of inorganic N fertilizers and concentrates resulted in large N losses and P accumulation in the system. On the other hand, Titchen and Phillips (1996) concluded that environmental losses of N were more closely correlated with productivity per hectare than whether the N source was inorganic fertilizer or atmospheric N. Thus, the environmental benefit of legumes in pastures may be due to a reduction in carrying capacity, as well as a reduction in fossil fuels needed to produce inorganic N fertilizer.

THE FUTURE

Expansion of legume use around the world will improve environmental quality and sustainability, but only if such systems are carefully managed (Titchen and Phillips, 1996). The opportunities for significant advances in nutritive value and environmental quality of forage legumes through transgenic technology will be realized if public fears about the technology are resolved. These opportunities exist for all legume species, not just the few currently targeted (Webb, 1996). Increased transgenic activity in Europe is not likely in the near future due to unfavorable attitudes from the public and politicians (Veronesi and Rosellini, 2000). Careful monitoring of released transgenics, even if they are viewed as substantially equivalent to existing varieties, will increase public trust. Broad legal agreements between the private and public sectors to allow access to transgenic technology for public good research also would help to overcome negative public opinion (Jordan, 2000) and would allow for the possibility of improved legume forage quality in third-world legume species.

Removal of anti-quality components from forage legumes is the easiest path to increased use of legumes. Elimination of legume bloat and other negative legume traits is possible in the foreseeable future. Other transgenic modifications of legumes could result in an overall reduction in phosphorus and nitrogen in animal manure as well as other environmental improvements (Martin, 2000).

More research on the basic processes of forage legume growth and the relationships between growth and forage composition are necessary for the development of more accurate simulation models (Fick et al., 1994). Structural and functional roles of legume plant cell walls and their relationships to forage quality are poorly understood and require significant research in the future (Hatfield et al., 1999). Transgenics may provide us tools to aid our understanding of genetic and environmental influences on forage composition, thus improving models with mechanistic concepts. Increased legume production would have the most impact on both forage nutritive value and environmental quality in Europe (Beever and Thorp, 1996), as well as in tropical and subtropical regions.

REFERENCES

Allen, M. and J. Beck. (1996). Relationship between spring harvest alfalfa quality and growing degree days. In *Proceedings of the 26th National Alfalfa Symposium*, East lansing, MI, 4-5 March 1996. Davis, CA: Certified Alfalfa Seed Council. pp. 16-25.

Allen, V.G. (1993). Advances in forage breeding for improved animal performance. In *Proceedings of the Cornell Nutrition Conference for Feed Manufacturers*, October 19-21, 1993, Rochester, NY. pp. 63-73.

Austin-Phillips, S., R.G. Koegel, R.J. Straub, and M. Cook. (1999). Animal feed compositions containing phytase derived from transgenic alfalfa and methods of use thereof. *Official Gazette of the United States Patent and Trademark Office Patents.* 1222(1), 27 July, 1999.

Barnes, D.K. and C.C. Sheaffer. (1995). Alfalfa. In *Forages: An Introduction to Grassland Agriculture*, Vol. 1, Ames, IA: Iowa State University Press. pp. 205-216.

Barton, J.E. and M. Dracup. (2000). Genetically modified crops and the environment. *Agronomy Journal* 92:797-803.

Baucher, M., M.A. Bernard-Vailhe, B. Chabbert, J.M. Besle, C. Opsomer, J. Botterman, and M. Van Montagu. (1999). Down-regulation of cinnamyl alcohol dehydrogenase in transgenic alfalfa (*Medicago sativa* L.) and the effect on lignin composition and digestibility. *Plant Molecular Biology* 39:437-447.

Beever, D.E. and C.K. Reynolds. (1994). Forage quality, feeding value and animal performance. In Grassland and Society, *Proceedings of the 15th General Meeting of the European Grassland Federation*, ed. L. 't Mannetje and J. Frame, Wageningen, The Netherlands, pp. 48-60.

Beever, D.E. and C. Thorp. (1996). Advances in the understanding of factors influencing the nutritive value of legumes. In *Legumes in Sustainable Farming Systems*, Occasional Symposium No. 30, 2-4 September, 1996, ed. D. Younie, British Grassland Society. pp. 194-207.

Bellucci, M., F. Paolocci, A. Alpini, and S. Arcioni. (1997). A biotechnological approach to increase the content of sulphur-rich proteins in forage leguminous species. In *Proceedings of the 17th General Meeting of the European Grassland Federation*, 18-21 May, 1997, ed. G. Nagy and K. Peto, Debrecen, Hungary: Debrecen Agricultural University. pp. 219-221.

Beuselinck, P.R. and W.F. Grant. (1995). Birdsfoot trefoil. In *Forages: An Introduction to Grassland Agriculture*, Vol. 1, Ames, IA: Iowa State University Press. pp. 237-248.

Bingham, E.T. (2000). Biotech traits in alfalfa–An overview. In *Proceedings of the 37th North American Alfalfa Improvement Conference*. 16-19 July 2000, Madison, WI. pp. 385-388.

Blair, N. and G.J. Crocker. (2000). Crop rotation effects on soil carbon and physical fertility of two Australian soils. *Australian Journal of Soil Research* 38:71-84.

Blumenthal, J.M., M.P. Russelle, and J.F.S. Lamb. (1999). Subsoil nitrate and bromide uptake by contrasting alfalfa entries. *Agronomy Journal* 91:269-275.

Bray, R.A. (1982). Selecting and breeding better legumes. In *Nutritional Limits to Animal Production from Pastures*, ed. J.B. Hacker, Farnham Royal, UK: Commonwealth Agricultural Bureaux, pp. 287-303.

Burton, G.W. (1972). Registration of Coastcross-1 bermudagrass. *Crop Science* 12:125.

Buxton, D.R. and M.D. Casler. (1993). Environmental and genetic effects on cell wall composition and digestibility In *Forage Cell Wall Structure and Digestibility*, ed. H.G. Jung, et al. ASA, CSSA, SSSA, Madison, WI, pp. 685-714.

Buyanovsky, G.A., C.L. Kucera, and G.H. Wagner. (1987). Comparative analyses of carbon dynamics in native and cultivated ecosystems. *Ecology* 68:2023-2031.

Cameron, D.F., C.P. Miller, L.A. Edye, and J.W. Miles. (1993). Advances in research and development with stylosanthes and other tropical pasture legumes. In *Grasslands for our World*, ed. M.J. Baker, Wellington, NZ: SIR Publishing. pp. 796-801.

Casler, M.D. (1999). Breeding for improved forage quality: Potentials and problems. In *Proceedings of the XVIII International Grassland Congress*, Volume 3, ed. J.G. Buchanon-Smith, L.D. Bailey, and P. McCaughey. 8-17 June, 1997. Winnipeg and Saskatoon. Calgary, Canada: Association Management Centre. pp. 323-332.

Casler, M.D. and K.P. Vogel. (1999). Accomplishments and impact from breeding for increased forage nutritional value. *Crop Science* 39:12-20.

Cherney, D.J.R. (1995). How much protein do we want in alfalfa forage? In *Proceedings of the 25th National Alfalfa Symposium*, Syracuse, NY. Davis, CA: Certified Alfalfa Seed Council. pp. 16-23.

Cherney, D.J.R. (2000). Characterization of forages by chemical analysis. In *Forage Evaluation in Ruminant Nutrition*, ed. D.I. Givens et al. Wallingford, UK: CABI Publishing, pp. 281-300.

Cherney, D.J.R., J.B. Russell, and J.H. Cherney. (1994). Factors affecting the deamination of forage proteins by ruminal microorganisms. *Journal of Applied Animal Research* 5:101-112.

Cherney, J.H., D.J.R. Cherney, D.E. Akin, and J.D. Axtell. (1991). Potential of brown-midrib, low-lignin mutants for improving forage quality. In *Advances in Agronomy.*, Vol. 46, ed. D.L Sparks, Orlando, FL: Academic Press, Inc. pp. 157-198.

Cherney, J.H. and R.M. Sulc. (1997). Predicting first cutting alfalfa quality. In *Silage: Field to Feedbunk*. NRAES-99. Ithaca, NY: Northeast Regional Agricultural Engineering Service. pp. 53-65.

Clark, D.A. and V.R. Kanneganti. (1998). Grazing management systems for dairy cattle. In *Grass for Dairy Cattle*, ed. J.H. Cherney and D.J.R. Cherney, Wallingford, UK: CABI Publishing. pp. 311-334.

Clark, D.A. and J.R. Wilson. (1993). Implications of improvements in nutritive value for plant performance and grassland management. In *Grasslands for Our World*, ed. M.J. Baker, Wellington, NZ: SIR Publishing. pp. 165-171.

Dhiman, T.R. and L.D. Satter. (1993). Protein as the first limiting nutrient for lactating dairy cows fed high proportions of good quality alfalfa silage. *Journal of Dairy Science* 76:1960-1971.

D'Mello, J.P.F. and A.M.C. MacDonald. (1996). Anti-nutrient factors and mycotoxins in legumes. In *Legumes in Sustainable Farming Systems*, Occasional Symposium No. 30, 2-4 September, 1996, ed. D. Younie, British Grassland Society. pp. 208-216.

Dzowela, B.H. (1993). Advances in forage legumes: A subSahara African perspective. In *Grasslands for Our World*, ed. M.J. Baker, Wellington, NZ: SIR Publishing. pp. 802-806.

Fahey, Jr., G.C. and H.S. Hussein. (1999). Forty years of forage quality research: Accomplishments and impact from an animal nutrition perspective. *Crop Science* 39:4-12.

Fick, G.W., P.W. Wilkens, and J.H. Cherney. (1994). Modeling forage quality changes in the growing crop. In *Forage Quality, Evaluation and Utilization*, ed. G.C. Fahey, Jr., Madison, WI: American Society of Agronomy, Inc. pp. 757-795.

Fleming, G.A. (1973). Mineral composition of herbage. In *Chemistry and Biochemistry of Herbage*, ed. G.W. Butler and R.W. Bailey. Academic Press, London, pp. 529-563.

Ford, S.A. (1999). Economic evaluation of forage quality gains–past and future. *Crop Science* 39:21-27.

Fox, D.G., C.J. Sniffen, J.D. O'Connor, J.B., Russell, and P.J. Van Soest. (1992). A net carbohydrate and protein system for evaluating cattle diets. III. Cattle requirements and diet adequacy. *Journal of Animal Science* 71:694-701.

Frame, J., J.F.L. Charlton, and A.S. Laidlaw. (1997). *Temperate Forage Legumes*, Wallingford, UK: CAB Intl.

Goering H.K. and P.J. Van Soest. (1970). Forage fiber anaylses (apparatus, reagents, procedures and some applications). Agricultural Handbook No. 379. ARS, USDA, Washington, DC.

Gregerson, R.G., D.L. Robinson, and C.P. Vance. (1999). Carbon and nitrogen metabolism in *Lotus*. In *Trefoil: The Science and Technology of the Lotus*, ed. P.R. Beuselinck, Madison, WI: ASA, CSSA, SSSA. pp. 167-185.

Harborne, J.B. (1993). *Introduction to Ecological Biochemistry*. Academic Press, London, p. 318.

Hatfield, R.D., J. Ralph, and J.H. Grabber. (1999). Cell wall structural foundations: Molecular basis for improving forage digestibilities. *Crop Science* 39:27-37.

Heichel, G.H., R.H. Delaney, and H.T. Cralle. (1988). Carbon assimilation, partitioning, and utilization. In *Alfalfa and Alfalfa Improvement*, Agronomy Monograph 29, ed. A.A. Hansen et al., Madison WI: ASA, CSSA, SSSA, pp. 195-228.

Hill, R.R., Jr., J.S. Shenk, and R.F. Barnes. (1988). Breeding for yield and quality. In *Alfalfa and Alfalfa Improvement*, Agronomy Monograph 29, ed. A.A. Hansen et al., Madison, WI: ASA, CSSA, SSSA, pp. 809-825.

Hintz, R.W. and K.A. Albrecht. (1991). Prediction of alfalfa chemical composition from maturity and plant morphology. *Crop Science* 31:1561-1565.

Hoover, W.H. and S.R. Stokes. (1991). Balancing carbohydrates and proteins for optimum rumen microbial yield. *Journal of Dairy Science* 74:3660-3644.

Hopkins, A., A. Davies, and C. Doyle. (1994). *Clovers and Other Grazed Legumes in UK Pasture Land*. IGER Technical Review No. 1.

Humphreys, M.O. (1999). The contribution of conventional plant breeding to forage crop improvement. In *Proceedings of the XVIII International Grassland Congress*, Volume 3, ed. J.G. Buchanon-Smith, L.D. Bailey, and P. McCaughey. 8-17 June, 1997. Winnipeg and Saskatoon. Calgary, Canada: Association Management Centre. pp. 71-78.

Hutton, E.M. (1988). Selection and breeding of tropical pasture legumes. In *Tropical Forage Legumes*, 2nd Edition, ed. Skerman et al., Rome, Italy: FAO, pp. 173-184.

Jordan, M.C. (2000). The privatization of food: Corporate control of biotechnology. *Agronomy Journal* 92:803-806.

Juan, N.A., C.C. Sheaffer, and D.K. Barnes. (1993). Temperature and photoperiod effects on multifolilolate expression and morphology of alfalfa. *Crop Science* 33: 753-578.

Julier, B., C. Huyghe, and C. Ecalle. (2000). Genetic variation and variety × environment interaction for digestibility, forage yield and protein content in alfalfa. *Proceedings of the 37th North American Alfalfa Improvement Conference*. 16-19 July 2000, Madison, WI. pp. 321.

Kaeppler, H.F. (2000). Food safety assessment of genetically modified crops. *Agronomy Journal* 92:793-797.

Kalu, B.A. and G.W. Fick. (1981). Quantifying morphological stage of development of alfalfa for studies of herbage quality. *Crop Science* 21:267-271.

Kephart, K.D., D.R. Buxton, and R.R. Hall, Jr. (1990). Digestibility and cell-wall components of alfalfa following selection for divergent herbage lignin concentration. *Crop Science* 30:207-212.

Kephart, K.D., C.P. West, and D.A. Wedin. (1995). Grassland ecosystems and their improvement. In *Forages: An Introduction to Grassland Agriculture*, Vol. 1, Ames, IA: Iowa State University Press. pp. 141-153.

Larkin, P.J., G.J. Tanner, R.G. Joseph, and W.M. Kelman. (1999). Modifying condensed tannin content in plants. In *Proceedings of the XVIII International Grassland Congress*, Volume 3, ed. J.G. Buchanon-Smith, L.D. Bailey, and P. McCaughey. 8-17 June, 1997. Winnipeg and Saskatoon. Calgary, Canada: Association Management Centre. pp. 167-178.

Martin, N.P. (2000). Opportunities for biotech traits in alfalfa down on the farm. In *Proceedings of the 37th North American Alfalfa Improvement Conference*. 16-19 July 2000, Madison, WI. pp. 401-405.

Mayland, H.F. and P.R. Cheeke. (1995). Forage-induced animal disorders. In *Forages, The Science of Grassland Agriculture*, 5th ed. Vol. 2., ed. R.F. Barnes et al. Ames, IA: Iowa State University Press, pp. 121-135.

McCaslin, M. and S. Fitzpatrick. (2000). Roundup ready alfalfa. In *Proceedings of the 37th North American Alfalfa Improvement Conference*. 16-19 July 2000, Madison, WI. pp. 396-400.

McDowell, L.R. and G. Valle. (2000). Major minerals in forages. In *Forage Evaluation in Ruminant Nutrition*, ed. D.I. Givens et al. Wallingford, UK: CABI Publishing, pp. 373-397.

McGraw, R.L., P.R. Beuselinck, and G.C. Marten. (1989). Agronomic and forage quality attributes of diverse entries of birdsfoot trefoil. *Crop Science* 29:1160-1164.

McKersie, B.D. (1997a). Improving forage production systems using biotechnology. In *Biotechnology and the Improvement of Forage Legumes*, ed. B.D. McKersie and D.C.W. Brown, Wallingford, UK: CAB International, pp. 3-21.

McKersie, B.D. (1997b). Summary and future prospects for the improvement of forage legumes using biotechnology. In *Biotechnology and the Improvement of Forage Legumes*, ed. B.D. McKersie and D.C.W. Brown, Wallingford, UK: CAB International, pp. 427-435.

Minson, D.J. (1981). An Australian view of laboratory techniques for forage evaluation. In *Forage Evaluation: Concepts and Techniques*, ed. J.L. Wheeler, and R.D. Mochrie, Griffen Press, Netley, South Australia, pp. 57-71.

Minson, D.J. (1985). Nutritional value of tropical legumes in grazing and feeding systems. In *Forage Legumes for Energy-Efficient Animal Production, Proceedings of a Trilateral Workshop*, 30 April to 4 May, 1984, Palmerston North, NZ, loc: USDA-ARS. pp. 192-196.

Minson, D.J. (1988). The chemical composition and nutritive value of tropical legumes. In *Tropical Forage Legumes*, 2nd Edition, ed. Skerman et al., Rome, Italy: FAO, pp. 185-193.

Morris, P. and M.P. Robbins. (1997). Manipulating condensed tannins in forage legumes. In *Biotechnology and the Improvement of Forage Legumes*, ed. B.D. McKersie and D.C.W. Brown, Wallingford, UK: CAB International, pp. 147-173.

Morris P., K.J. Webb, M.P. Robbins, L. Skot, and B. Jorgensen. (1999). Application of biotechnology to *Lotus* breeding. In *Trefoil: The Science and Technology of the Lotus*, ed. P.R. Beuselinck, Madison, WI: ASA, CSSA, SSSA. pp. 199-228.

Moutray, J. (1996). Alfalfa breeding efforts–Challenges for 2000 and beyond. In *26th National Alfalfa Symposium*, 4-5 March 1996, East Lansing, MI, Certified Alfalfa Seed Council. pp. 12-15.

Mupangwa, J.F., T. Acamovic, J.H. Topps, N.T. Ngongoni, and H. Hamudikuwanda. (2000). Content of soluble and bound condensed tannins of three tropical herbaceous forage legumes. *Animal Feed Science and Technology* 83:139-144.

National Research Council (1985). *Nutrient Requirements of Sheep*. National Academy Press, Washington, DC.

National Research Council (1989). *Nutrient Requirements of Dairy Cattle*, 6th revised ed. National Academy Press, Washington, DC.

National Research Council (1996). *Nutrient Requirements of Beef Cattle*. National Academy Press, Washington, DC.

Ni, W., N.L. Paiva, and R.A. Dixon. (1993). Reduced lignin in transgenic plants containing an engineered caffeic acid *O*-methyltransferase antisense gens. *Transgenic Research* 3:120-126.

Niezen, J.H., T.S. Waghorn, W.A.G. Charleston, and G.C. Waghorn. (1995). Growth and gastrointestinal nematode parasitism in lambs grazing either lucerne (*Medicago sativa*) or sulla (*Hedysarum coronarium*) which contains condensed tannins. *Journal of Agricultural Science*, Cambridge 125:281-289.

Papadopoulos, Y.A. and W.M. Kelman. (1999). Traditional breeding of *Lotus* species. In *Trefoil: The Science and Technology of the Lotus*, ed. P.R. Beuselinck, Madison, WI: ASA, CSSA, SSSA. pp. 187-198.

Poppi, D.P., S.R. McLennan, S. Bediye, A. de Vega, and J. Zorrila-Rios. (1999). Forage quality: Strategies for increasing nutritive value of forages. In *Proceedings of the XVIII International Grassland Congress*, Volume 3, ed. J.G. Buchanon-Smith, L.D. Bailey, and P. McCaughey. 8-17 June, 1997. Winnipeg and Saskatoon. Calgary, Canada: Association Management Centre. pp. 307-322.

Rae, A.L., C.L. McIntyre, R.J. Jones, and J.M. Manners. (1996). Antisense suppression of the lignin biosynthetic enzyme, caffeic acid *O*-methyl transferase, improves forage digestibility. SYM-21-07. In *Combined Conference Abstracts*, 29 September to 2 October, 1996, Canberra, AU: Australian Society of Plant Physiology.

Reed, J.D., C. Krueger, G. Rodriguez, and J. Hanson. (2000). Secondary plant compounds and forage evaluation. In *Forage Evaluation in Ruminant Nutrition*, ed. D.I. Givens, E. Owen, R.F.E. Axford and H.M. Omed. Wallingford, UK: CABI Publishing, pp. 433-448.

Reid, J.T., W.K. Kennedy, K.L. Turk, S.T. Slack, G.W. Trimberger, and R.P. Murphy. (1959). Effect of growth stage, chemical composition, and physical properties upon the nutritive value of forages. *Journal of Dairy Science* 42:567-571.

Reid, R.L. (1994). Milestones in forage research. In *Forage Quality, Evaluation and Utilization*, ed. G.C. Fahey, Jr., Madison, WI: American Society of Agronomy, Inc., pp. 1-58.

Rumbaugh, M.D., J.L. Caddel, and D.E. Rowe. (1988). Breeding and quantitative genetics In *Alfalfa and Alfalfa Improvement*, Agronomy Monograph 29, ed. A.A. Hansen et al., Madison, WI: ASA, CSSA, SSSA, pp. 777-808.

Seguin, P., C.C. Sheaffer, N.J. Ehlke, and R.L. Becker. (1999). Kura clover establishment methods. *Journal of Production Agriculture* 12:483-487.

Shahidi, F. (1995). Beneficial health effects and drawbacks of antinutrients and phytochemicals in foods: an overview. In *ACS Symposium Series 662: Antinutrients and Phytochemicals in Food*. American Chemical Society, Washington, pp. 1-9.

Sheaffer, C.C., D. Cash, N.J. Ehlke, J.C. Henning, J. Grimsbo Jewett, K.D. Johnson, M.A. Peterson, M.Smith, J.L. Hansen, and D.R. Viands. (1998). Entry × environment interactions for alfalfa forage quality. *Agronomy Journal* 90:774-780.

Sheaffer, C.C., N.P. Martin, J.F.S. Lamb, G.R. Cuomo, J. Grimsbo Jewett, and S.R. Quering. (2000). Leaf and stem properties of alfalfa entries. *Agronomy Journal* 92:733-739.

Shenk, J.S., M.O. Westerhaus, and M.R. Hoover. (1979). Analysis of forage by infrared reflectance. *Journal of Dairy Science* 62:807-812.

Skerman, P.J., D.G. Cameron, and F. Riveros. (1988). Leguminous browse. In *Tropical Forage Legumes*, 2nd Edition, ed. Skerman et al., Rome, Italy: FAO, pp. 488-587.

Spangenberg, G., Z.Y. Wang, R. Heath, V. Kaul, and R. Garrett. (1999). Biotechnology in pasture plant improvement: Methods and prospects. In *Proceedings of the XVIII International Grassland Congress*, Volume 3, ed. J.G. Buchanon-Smith, L.D. Bailey, and P. McCaughey. 8-17 June, 1997. Winnipeg and Saskatoon. Calgary, Canada: Association Management Centre. pp. 79-96.

Sulc, R.M., K.A. Albrecht, J.H. Cherney, M.H. Hall, S.C. Mueller, and S.B. Orloff. (1997). Field testing a rapid method for estimating alfalfa quality. *Agronomy Journal* 89:952-957.

Taylor, N.L. and K.H. Quesenberry. (1996). *Red Clover Science*. Dordrecht, The Netherlands: Kluwer Academic Publishers.

Terrill, T.H., W.R. Windham, C.S. Hoveland, and H.E. Amos. (1989). Forage preservation method influences on tannin concentration, intake, and digestibility of sericea lespedeza by sheep. *Agron. J.* 81:435.

Tesfaye, M., C.P. Vance, D.L. Allan, and D.A. Samac. (2000). Aluminum tolerance in transgenic alfalfa (*Medicago sativa* L.). In *Proceedings of the 37th North American Alfalfa Improvement Conference*. 16-19 July 2000, Madison, WI. pp. 306.

Tilley, J.M.A. and R.A. Terry. (1963). A two-stage technique for *in vitro* digestion of forage crops. *Journal of the British Grassland Society* 18:104-111.

Titchen, N.M. and L. Phillips. (1996). Environmental effects of legume based grassland systems. In *Legumes in Sustainable Farming Systems*, Occasional Symposium No. 30, 2-4 September, 1996, ed. D. Younie, British Grassland Society. pp. 257-261.

Van der Meer, H.G. and A.H.J. Van der Putten. (1995). Reduction of nutrient emissions from ruminant livestock farms. In *Grassland into the 21st Century*, Occasional Symposium No. 29. ed. G.E. Pollot, British Grassland Society. pp. 118-134.

Van Soest, P.J. (1995). What constitutes alfalfa quality: new considerations. In *Proceedings of the 25th National Alfalfa Symposium*, Syracuse, NY. Davis, CA: Certified Alfalfa Seed Council. pp. 1-14.

Van Wijk, A.J.P., J.G. Boonman, and W. Rumball. (1993). Achievements and perspectives in the breeding of forage grasses and legumes. In *Grasslands for Our World*, ed. M.J. Baker, Wellington, NZ: SIR Publishing. pp. 116-120.

Veronesi, F. and D. Rosellini. (2000). Biotech traits in alfalfa–European perspective. In *Proceedings of the 37th North American Alfalfa Improvement Conference*. 16-19 July 2000, Madison, WI. pp. 389-395.

Viands, D. (1995). What breeding objectives really will improve forage quality of alfalfa? In *Proceedings of the 25th National Alfalfa Symposium*, Syracuse, NY. Davis, CA: Certified Alfalfa Seed Council. pp. 24-28.

Vogel, K.P. and D.A. Sleper. (1994) Alteration of plants via genetics and plant breeding. In *Forage Quality, Evaluation and Utilization*, ed. G.C. Fahey, Jr., Madison, WI: American Society of Agronomy, Inc. pp. 891-921.

Volenec, J.J., J.H. Cherney, and K.D. Johnson. (1987). Yield components, plant morphology, and forage quality of alfalfa as influenced by plant population. *Crop Science* 27:321-326.

Waghorn, G.C., J.D. Reed, and L.R. Ndlovu. (1999). Condensed tannins and herbivore nutrition. In *Proceedings of the XVIII International Grassland Congress*, Volume 3, ed. J.G. Buchanon-Smith, L.D. Bailey, and P. McCaughey. 8-17 June, 1997.

Winnipeg and Saskatoon. Calgary, Canada: Association Management Centre. pp. 153-166.

Webb, K.J. (1996). Opportunities for biotechnology in forage legume breeding. In *Legumes in Sustainable Farming Systems*, Occasional Symposium No. 30, 2-4 September, 1996, ed. D. Younie, British Grassland Society. pp. 77-85.

Wheeler, J.L. and J.L. Corbett. (1989). Criteria for breeding forage of improved feeding value: Results of a Delphi survey. *Grass and Forage Science* 44:77-83.

Wilkins, R.J. (2000). Forages and their role in animal systems. In *Forage Evaluation in Ruminant Nutrition*, ed. D.I. Givens et al. Wallingford, UK: CABI Publishing, pp. 1-14.

Sulfur Metabolism and Protein Quality
of Soybean

Peter J. Sexton
Nam C. Paek
Seth L. Naeve
Richard M. Shibles

SUMMARY. Soybean is an important source of protein in livestock production, and is of growing importance for human consumption. As a sole dietary protein source, soybean seed protein is deficient in the amino acids methionine, cysteine, and threonine. Increasing the amount of methionine in the amino acid profile of soybean meal would enhance its value for producers and consumers. Methionine contains S, and so its production is necessarily linked to sulfur metabolism within the soybean plant. Sulfur is taken up from the soil in the form of sulfate. During vegetative growth, developing leaves appear to be the predominate site of sulfate reduction and incorporation of reduced S into amino acids. During reproductive growth, developing pods and seeds seem to be the predomi-

Peter J. Sexton is affiliated with University of Maine Cooperative Extension, P.O. Box 727, Presque Isle, ME 04769 USA.

Nam C. Paek is affiliated with School of Plant Science, Seoul National University, 103 Seodun-Dong Kwonsun-Ku, Suwon 441-744, Republic of Korea.

Seth L. Naeve is affiliated with Department of Agronomy and Plant Genetics, 411 Borlaug Hall, University of Minnesota, St. Paul, MN 55108 USA.

Richard M. Shibles is affiliated with Crops Research, 1563 Agronomy Hall, Iowa State University, Ames, IA 50011 USA.

Address correspondence to: Peter J. Sexton, University of Maine Cooperative Extension, P.O. Box 727, Presque Isle, ME 04769 USA (E-mail: psexton@umext. maine.edu).

[Haworth co-indexing entry note]: "Sulfur Metabolism and Protein Quality of Soybean." Sexton, Peter J. et al. Co-published simultaneously in *Journal of Crop Production* (Food Products Press, an imprint of The Haworth Press, Inc.) Vol. 5, No. 1/2 (#9/10), 2002, pp. 285-308; and: *Quality Improvement in Field Crops* (ed: A. S. Basra, and L. S. Randhawa) Food Products Press, an imprint of The Haworth Press, Inc., 2002, pp. 285-308. Single or multiple copies of this article are available for a fee from The Haworth Document Delivery Service [1-800-HAWORTH, 9:00 a.m. - 5:00 p.m. (EST). E-mail address: getinfo@haworthpressinc.com].

285

nate location of sulfate reduction. Sulfate reduction and methionine synthesis are complex and highly regulated processes. Synthesis of storage proteins within the developing seed is sensitive to the amount of methionine present such that provision of extra methionine blocks synthesis of poor quality storage proteins. Sulfur deficiency on the other hand dramatically enhances accrual of poor quality seed storage proteins. It appears that the plant manufacturers higher qualty storage proteins as long as methionine is present in adequate quanitities relative to the non-S-amino acids. Accumulation of poor-quality seed storage proteins thus appears to be a function, at least to some degree, of rate of methionine synthesis within the seed. Efforts to enhance soybean seed protein quality may require enhanced rates of methionine synthesis within the seed to be successful. *[Article copies available for a fee from The Haworth Document Delivery Service: 1-800-HAWORTH. E-mail address: <getinfo@haworthpressinc. com> Website: <http://www.HaworthPress.com> © 2002 by The Haworth Press, Inc. All rights reserved.]*

KEYWORDS. Sulfur metabolism, methionine, soybean, protein quality

INTRODUCTION

Soybean (*Glycine max* [L]. Merr.) is the world's most productive protein crop, its seeds supplying 63% of world protein meal consumed in 1998. Although direct human consumption and industrial uses represent growing markets, 98% of the meal is consumed in animal feedstuffs with poultry (46%) and swine (27%) production being the leading uses in the US (USB, 1999). Soybean meal is about 54% protein and is highly valued as a protein source for feeds because of low cost and consistency of protein content (Mounts et al., 1987).

As a sole dietary source soybean protein is deficient in the amino acids methionine, cysteine, and threonine. Methionine is the most limiting amino acid. Soybean meal contains only 56% of the methionine plus cysteine found in egg protein, the FAO reference protein (Wilson, 1987). The average S-amino content of soybean meal would need to be increased by 14, 91, and 119% to fully supply adult, infant, and growing swine needs with the correct complement of essential amino acids (de Lumen et al., 1997). To achieve a nutritionally balanced protein it is common practice in the feeds industry to supplement soybean protein with synthetic methionine, particularly for monogastric animals like swine and poultry. Having soybeans with greater S-amino content, obviating the need to supplement with synthetic methionine, would be economically beneficial. McVey et al. (1995) estimate that raising the S-amino content of soybeans by 1% would generate a benefit, split almost equally between producers and users, of US$ 2.4 billion annually.

On a dry weight basis soybean seeds range from 38 to 44% protein with about 70% of total protein being in the form of two seed storage proteins, glycinin (11 S) and β-conglycinin (7 S) (Murphy and Resurreccion, 1984). The 11 S and 7 S classes occur in roughly 60-40 ratio (Koshiyama, 1983). The 11 S protein is composed of six acidic and five basic subunits and can be divided into two protein groups by amino acid homology: Group I ($A_{1a}B_2$, $A_{1b}B_{1b}$, A_2B_{1a}) and Group II (A_3B_4, $A_4A_5B_3$) (Nielsen, 1985). Null mutants are reported for Group I (G_1, G_2, and G_3), for $A_4A_5B_3$ (G_4), and for A_3B_4 (G_5), each of which occurred by single, recessive mutation (Kitamura, 1995). The 7 S protein contains three subunit classes, designated as α' (76 kD), α (72 kD), and β (53 kD) (Derbyshire et al., 1976). Kitamura (1995) reported independently inherited null recessive mutations for the α- and α'-subunits, but a null mutant of the β-subunit has not been reported.

Sulfur amino acids comprise approximately 3 to 4.5% of the 11 S residues, which is similar to other high-quality dietary proteins (Fukushima, 1991; Nielsen et al., 1989). In contrast, the 7 S protein is composed of less than 1% S-amino acids (Harada et al., 1989; Sebastiani et al., 1990) principally because the β-subunit bears only one cysteine and no methionine residues (Coates et al., 1985). Therefore, the presence of the β-subunit, which can vary from 10 to 30% of storage proteins depending upon S and N nutrition (Paek et al., 1997, 2000; Sexton et al., 1998a) and other environmental factors, is principally responsible for the dilution, with non-S-amino acids, of the otherwise high quality soybean protein. Genetic regulation of storage protein subunits has been studied extensively at the molecular level (Fujiwara et al., 1992; Fujiwara and Beachy, 1994; Hirai et al., 1994, 1995; Ladin et al., 1987; Meinke et al., 1981; Naito et al., 1994a,b).

In the following sections we discuss plant sulfur metabolism, whole-plant sulfur dynamics of the soybean crop, and molecular approaches to improve soybean protein quality. We conclude with a set of working hypotheses which seem to us appropriate for guiding further research in improving the S-amino acid content of soybean seeds.

FORMS OF S WITHIN THE PLANT

There are a variety of S containing compounds typically found in plants including lipoic acid, coenzyme A, biotin, thiamin, glucosinolates, and sulfolipids (Goodwin and Mercer, 1983). However, these compounds represent a minor fraction of the organic S within the plant, with typically 80% of the organic S being found incorporated in proteins as either methionine or cysteine, and most of the remaining organic S being found in soluble peptides and free amino acids (Allaway and Thompson, 1967; Anderson, 1990). Glutathione, or for some legumes such as soybeans, homo-glutathione, is the most important S containing peptide (Klapheck, 1988; Macnicol and Bergman, 1984). Although beyond the scope of this review, the importance of glutathione in redox control and protection against free radicals should be pointed out (Leustek and Saito, 1999; Foyer et al., 1994). Because the subject treated here is soybean protein quality, our discussion will focus on methionine and cysteine, starting from sulfate uptake and ending at seed storage proteins.

SULFATE UPTAKE FROM SOIL AND MOVEMENT WITHIN THE PLANT

Sulfur is taken up from the soil as sulfate. The sulfate anion must traverse several membrane systems as it moves into and within the plant. Uptake from the soil solution appears to be mediated by proton symporters with a ratio of 3 H^+:1 SO_4^{2-} crossing the plasmalemma. Thus there appears to be a cost of 3 ATP per sulfate imported (Anderson, 1990). Sulfate symporters have been characterized at the molecular level and appear to span the plasmalemma 12 times (Smith et al., 1997; Takahashi et al., 1997). There are several sulfate transporters that differ in their affinity for sulfate with high affinity transporters being expressed in roots and lower affinity transporters being expressed in leaves as well as roots (Smith et al., 1997). Synthesis of the high affinity transporter in roots is sensitive to S nutritional status. It increases in response to S starvation (Vidmar et al., 1999) and decreases with provision of S. Once in the cytoplasm, sulfate may be transported across the tonoplast membrane into the vacuole, incorporated into organic sulfur compounds, or it may undergo long-distance transport to other parts of the plant. Reduction of sulfate predominately occurs in leaf chloroplasts (Renosto et al., 1993; Lunn et al., 1990).

During soybean vegetative growth, developing leaves are the primary sink for sulfate (Smith and Lang, 1988; Sunarpi and Anderson, 1996). Older leaves appear to reload sulfate they acquire from the transpiration stream into the phloem for transport to developing leaves. Once developing leaves reach about 70% expansion, they cease to be significant sinks for sulfate. Work with pea (*Pisum sativum*) (von Arb and Brunold, 1986) and dry bean (*Phaseolus vulgaris*) (Schmutz and Brunold, 1982) indicates that the activity of S reduction enzymes greatly decreases once the leaf is fully expanded, even though enzymes in the C and N reduction cycles remain active. It appears that once the developing leaf meets its own needs for reduced S, it decreases its rate of sulfate reduction and passes incoming sulfate on to developing leaves.

During reproductive growth the developing pods are the main sink for sulfate, while seeds appear to accumulate S in organic forms, implying that either sulfate is reduced in the pod wall and transported to the seed, or sulfate is quickly reduced when it does reach the seed so that inorganic S does not accumulate there (Sunarpi and Anderson, 1997). Work by Sexton and Shibles (1999) suggests the latter hypothesis to be correct, as the activity of ATP-sulfurylase (first enzyme in the S reduction pathway) is predominately located in seed tissue during reproductive growth.

From the above work, it appears that sulfate reduction predominately occurs first in developing leaves and then later in the lifecycle in the seeds of the soybean plant. These tissues are rich in protein. Why the soybean plant is designed to reduce sulfate "on-site" where sulfur amino acids are needed for protein synthesis, rather than to reduce sulfate in another organ and transport it to the developing organ, is a point for further research and reflection.

SULFATE REDUCTION AND CYSTEINE SYNTHESIS

The reduction of sulfate (SO_4^{-2}) to sulfide (S^{-2}) for incorporation into the plants' amino acid pool is a process that involves several steps and is quite energy intensive, requiring more energy (approximately 730 kJ mol^{-1}) than does the reduction of nitrate to ammonia (347 kJ mol^{-1}) (Leustek and Saito, 1999). Sulfate is a relatively stable molecule and requires activation by ATP to make adenosine phosphosulfate (APS) as the first step in the pathway (Figure 1) (Brunold, 1990; Hell, 1997). After this point, the precise mechanism for sulfur reduction is

FIGURE 1. Putative pathway of sulfate reduction to cysteine. Based on activity of ATP-sulfurylase, sulfate reduction appears to occur predominately in leaf chloroplasts during vegetative growth and in seed tissue during reproductive growth. Enzymes names are as follows: (1) ATP-sulfurylase; (2) APS sulfo-transferase; (3) sulfite reductase; and (4) OAS thiol-lyase.

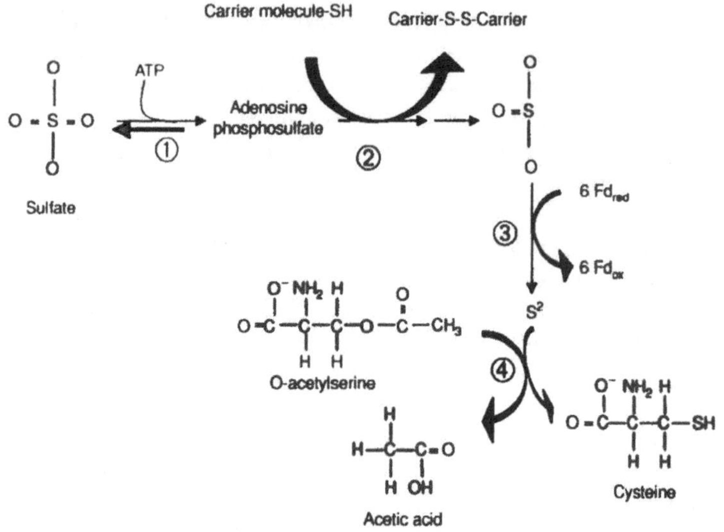

still a matter of debate. It appears that the sulfate group is passed from APS to a thiol compound, such as glutathione, to make a thiol-sulfonate. This reaction is catalyzed by APS-sulfotransferase. The resulting sulfite (SO_3^{2-}) is then either released and reduced in a free state by sulfite reductase to sulfide (S^{2-}), or remains bound to a carrier molecule where it is reduced. After the reduction step, the enzyme OAS-(thiol)lyase cat-alyzes the addition of sulfide (S^{2-}) to O-acetylserine (OAS). Cysteine and acetate are the end products of this reaction.

In addition to being able to act independently, the enzymes that form OAS and cysteine (serine acetyltransferase and OAS-(thiol)lyase, re-spectively) may form a complex known as cysteine synthase (Leustek and Saito, 1999). It appears that OAS-[thiol]lyase acts as a regulatory subunit that increases the activity of serine acetyltransferase when the two are complexed. The stability of this complex increases with the concentration of sulfide, and decreases as OAS concentrations increase.

The reduction of sulfate to sulfide and the incorporation of sulfide into cysteine is generally thought to be localized within plastids, especially in chloroplasts of green leaves (Brunold, 1990; Lunn et al., 1990; Renosto et al., 1993). The electrons with the required redox potential are provided by reduced ferredoxin in chloroplasts, and by NADPH in non-photosynthetic tissue (Hell, 1997). The cysteine synthase enzymes (serine acetyltransferase and OAS-(thiol)lyase) are also found in the cytoplasm and in mitochondria as well as in plastids (Lunn et al., 1990).

METHIONINE SYNTHESIS

The carbon backbone of methionine is derived from aspartate along a branched pathway which also produces lysine, threonine and isoleucine (Figure 2) (Azevedo et al., 1997; Ravanel et al., 1998). Through a series

FIGURE 2. Pathway of methionine synthesis, based primarily on Azevedo et al. (1997). This pathway is known to be localized in plastids up the point of homocysteine. Conversion of homocysteine to methionine is known to occur in the cytosol; it may also occur in plastids as well (Ravanel et al., 1998). Enzyme names are as follows: (1) aspartate kinase; (2) aspartate-semialdehyde dehydrogenase; (3) homoserine dehydrogenase; (4) homoserine kinase; (5) threonine synthase; (6) cystathionine synthase; (7) cystathionine β-lyase; (8) methionine synthase; and (9) S-adenosylmethionine synthetase. The multiple arrows from AdoMet to methionine are because AdoMet serves as a methyl group donor for a number of reactions, but its S is recycled to methionine.

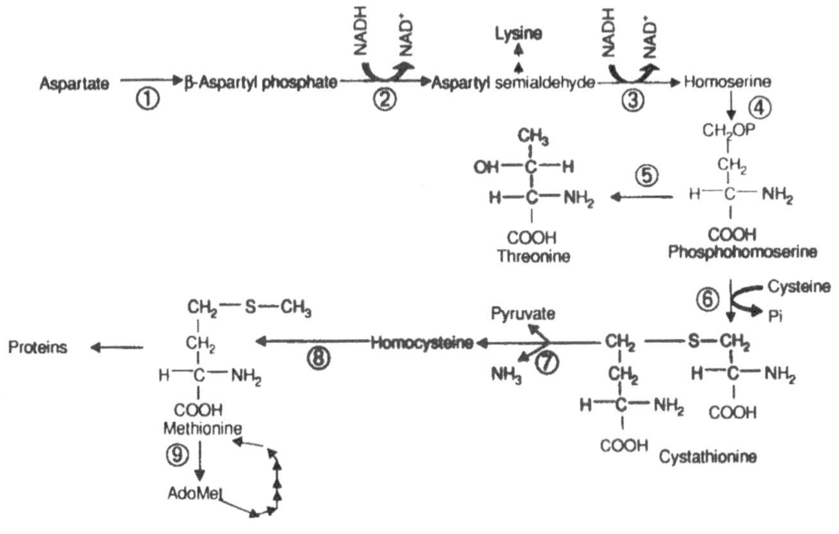

of steps, aspartate is converted into O-phosphohomoserine. This molecule is at a branchpoint in the pathway and may either be converted directly into threonine by threonine synthase, or be combined with cysteine to make cystathionine via cystathionine synthase. Cystathionine is then broken into homocysteine, pyruvate, and ammonia by the enzyme cystathionine lyase. The last step in methionine synthesis consists in the transfer of a methyl group from N-methyl-tetrahydrofolate to homocysteine by the enzyme methionine synthase which leaves methionine as the final product. From this point methionine may be incorporated into protein, or react with ATP to make S-adenosyl methionine (AdoMet) (Figure 2).

Acting as the major methyl donor in plants, and as a precursor for polyamine and ethylene synthesis, AdoMet plays an important role in cell metabolism and its concentration is highly regulated (Azevedo et al., 1997). While the flux of methionine going into AdoMet appears to be greater than that going into protein, AdoMet does not represent a major sink for methionine within the plant cell (Giovanelli et al., 1983). Whether used as a methyl donor or a precursor for synthesis of other molecules, the reduced S component of AdoMet is conserved and recycled back into methionine (Figure 2). Transgenic plants with suppressed AdoMet synthesis have shown increased levels of free methionine; however, this created a stunted phenotype with leather-like leaves (Boerjan, et al., 1994).

All biochemical steps for conversion of aspartate to homocysteine appear to be compartmentalized within plastids (Ravanel et al., 1998). However, it appears that methionine synthase and AdoMet synthetase occur in the cytosol. It is unclear if localization of these enzymes is exclusively cytosolic or if they occur in plastids as well. The presence of these enzymes in the cytosol may serve to allow recycling of reduced S from AdoMet spent in trans-methylation reactions (Figure 2).

Regulation of cysteine and methionine synthesis. The mechanisms by which rates of cysteine and methionine synthesis are regulated are not entirely clear. Nevertheless, the following salient points may be made:

- Steady state levels of mRNA for the SO_4^{2-} transporter, ATP sulfurylase, and APS sulfotransferase enzymes increase in response to S starvation and decrease in response to application of cystine or glutathione to plant roots (Lappartient and Touraine, 1996; Leustek and Saito, 1999);
- Provision of supplemental OAS has been reported to increase mRNA levels for a sulfate transporter in roots and to increase con-

centration of reduced S compounds within the plant (Smith et al., 1997);

- Methionine synthesis is suppressed with provision of supplemental threonine and lysine, which apparently decrease activity of the aspartate pathway (Ravanel et al., 1998);
- The activity of threonine synthase is positively related to AdoMet concentration (thus tending to favor threonine synthesis rather than methionine at the branchpoint when AdoMet concentration is great) (Azevedo et al., 1997);
- In yeast and bacteria, rate of methionine synthesis is negatively related to AdoMet concentration (Ravanel et al., 1998);
- Cystathionine synthase (at the starting point of methionine synthesis from cysteine) is down regulated with provision of supplemental methionine, and its activity increases when methionine is lacking (Azevedo et al., 1997).

Based on observations that rate of S assimilation is decreased in response to N deficiency, and vice versa, it has been hypothesized that reduction of nitrate and sulfate might be coordinated through products of one pathway influencing the rate of the other (Anderson, 1990). However, a simpler hypothesis might be that since protein synthesis is the main sink for products of both N and S reduction, a limitation in either S or N would lead to a decrease in rate of protein synthesis, which would cause the pool of non-limiting amino acids to increase, and eventually provide some negative feedback on reduction of the non-limiting element.

WHOLE PLANT SULFUR AND NITROGEN

When grown under sulfur sufficient conditions, the soybean plant acquires S and N in a rather parallel fashion. As mentioned above, N and S assimilation by plants appears to be coordinated (Clarkson et al., 1989; Anderson, 1990; Bell et al., 1995). On a mass basis, soybean acquires 15 to 20 times as much N as S. Organs within a soybean in its vegetative development stage contain similar N/S ratios, though the concentrations of these vary due to protein content of these organs. Agrawal and Mishra (1994), Gaines and Phatak (1982), and Sweeney and Granade (1993) all found sulfur fertilization of previously sulfur deficient plants to increase leaf and whole-plant S concentration, and to decrease N/S ratios; however *plants already sufficient in sulfur did not respond to sul-*

fur fertilization by increasing their concentration of protein-S. Gaines and Phatak (1982) fractionated N and S compounds and determined that protein-S levels responded first to additional S when S was scarce. Nonprotein-S accumulated in the plant only as protein levels plateaued. Although the total-N to total-S ratio fell nearly in half through S fertilization, the protein-N to protein-S ratio did not change. Across S fertilization treatments (from severe S-deficiency to sufficiency), Sexton et al. (1998b) found S accrual to climb through higher levels of S fertilization than did N or dry matter accrual. They found the S accrual rate in shoot tissue to increase 84% across treatments of 37 to 62 mg available S plant^{-1}, while that of N and dry matter increased only 24 and 3%, respectively. It appeared that S was limiting protein production at only very low fertilization levels, while at higher S levels, the plants began to accumulate SO_4^{-2}. Stems and roots were the major organs for SO_4^{-2} storage in their studies (Sexton et al., 1998b).

Looking across the life-cycle of the soybean plant, S uptake parallels that of N with accumulation continuing well into the seed filling period (Matheny and Hunt, 1981; Hanway et al., 1984; Sexton et al., 1998b). Initially leaves and stems accrue S, but as the plants go into reproductive growth, S accrual by these tissues stops. By the time the seed is fully expanded (R6 growth stage according to Fehr and Caviness, 1977) the S content of leaves, and to a lesser extent pods, greatly diminishes, apparently as a result of mobilization of vegetative S to seed tissue. Both Matheny and Hunt (1981) and Sexton et al. (1998b) conclude that mobilization of S from vegetative to reproductive tissue is less efficient than that of N, contributing to a greater N:S ratio in seed than is typically observed in leaves.

SULFUR REMOBILIZATION DURING REPRODUCTIVE GROWTH

Seed protein production puts a large demand on the soybean's vegetative tissues for mobilization of stored N (Sinclair and DeWitt, 1975; Shibles and Sundberg, 1998). Plant nutrient stresses in crop species often are most evident late in the life cycle (Marschner, 1995). The combined effects of a reduced uptake and increased usage during seed growth cause plants to rely heavily on nutrients stored in vegetative tissues. The rapid mobilization of highly mobile nutrients, such as N and P, from leaves and subsequent loss of photosynthetic activity cause vegetative tissues to appear to be "self-destructing" through nutrient removal and mobilization to seed (Nooden, 1988). Soybean seed-N is composed of approximately one-half mobilized N under normal field growth conditions (Hanway and Weber, 1971; Loberg et al, 1984; Imsande and Edwards, 1988). This corresponds to a mobilization of 66 to 79% of its vegetative-N to seed (Vasilas et al., 1995).

Since methionine and cysteine make up 90% of the S in most plants (Allaway and Thompson, 1966) and most N is in the amino form (Henry et al., 1992; Marschner, 1995), one might expect vegetative-S storage and mobilization to follow that of N. As N is scavenged from protein in vegetative tissue (Feller et al., 1977; Peoples and Dalling, 1988), S-amino acids would likely be released and available for mobilization. Sulfur's relative mobility in plants has been contested, however. Based upon studies with applied ^{35}S, S has historically been considered an immobile nutrient (Marschner, 1995). Also, chlorosis due to S-stress has been noted to begin with young leaves, indicating a lack of S mobility (Bouma, 1975; Hell, 1997). Nitrogen stress may also complicate the argument of S mobility (Loneragan et al., 1976). Nitrogen stress has been demonstrated to induce a mobilization of nitrogen through the action of increased protein hydrolytic activity and amino acid export from developed leaves of cereal grasses (Mei and Thimman, 1984; Guitmann et al., 1991). This plant response is similar to that shown by senescing leaves (Feller, 1977). Protease activity in leaves by either mechanism allows the release of amino acids including methionine and cysteine. Sunarpi and Anderson (1997a, d) demonstrated that N-stress enhances the mobilization of S from growing leaves, whereas super-optimal N inhibits S export from mature leaves. They believe that soybean leaves mobilize S in two distinct stages. The first involves the loss of free low molecular weight S compounds, principally SO_4^{-2} and homo-glutathione (hGSH), from an ethanol-soluble S fraction. These soluble-S compounds are mobilized from plants irrespective of N nutrition, but the rate is increased under N stress. The second stage of S mobilization involves loss of S from the ethanol-insoluble pool. This loss of S accompanies a loss in insoluble-N, and does not occur in leaves of plants grown under high N conditions. This implies that S mobilization is occurring by a proteolytic event that is triggered by a N-stress and involves the export of organic N and S.

There appears to be a preferential loss of Rubisco in soybean under S deficiency, with Rubisco fraction falling from 50% under adequate S to 10% of the soluble protein fraction under severe S stress (Sexton et al., 1997). Ferreira and Teixeira (1992) found no evidence of Rubisco degradation when *Lemna* (*Lemna minor* L.) was subjected to incubation in either darkness or lack of N. But, when *Lemna* was incubated in the absence of S, there was rapid and preferential degradation of Rubisco that went nearly to completion. Preferential degradation of proteins during senescence in non-stressed plants has been noted previously (Brady, 1988). Rubisco's high methionine + cysteine content (Kawashima and Wildman, 1970) would create a 50 mM concentration of free S-amino acids in *Lemna* chloroplasts, indicating Rubisco's potential value to the plant for S as well as N mobilization (Ferreira and Teixeira, 1992).

Alternatively, Sunarpi and Anderson (1997b) found no evidence of increased S mobilization from older to younger soybean leaves due to S-stress. From the results of a recent study (Naeve and Shibles, unpublished), a concurrent N- and S-stress was required to hasten the senescence process, or promote proteolysis through some unrelated mechanism. Sulfur stress alone or provision of surplus sulfate seemed to have no effect on S distribution or mobilization patterns in soybean during seed filling in this study.

Sulfate stored in vegetative tissue may be of only limited value for mobilization to developing seeds. Mature leaves of *Macroptilium atropurpureum* have been shown to retain stored SO_4^{-2} (Bell et al., 1995). Even under S stress, mature leaves seem to retain SO_4^{-2} (Janzen and Bettany, 1984; Adiputra and Anderson, 1995; Sunarpi and Anderson, 1997b). Bell et al. (1994) found SO_4^{-2} to be exchanged very slowly across the tonoplast membrane. Vacuolar SO_4^{-2} turns over slowly with cytosolic S (Herschbach and Rennenberg, 1996), and rate constants for SO_4^{-2} exchange across leaf vacuolar tonoplasts are at least two orders of magnitude less than that found in whole-leaf cell populations (Bell et al., 1994). Loss of SO_4^{-2} from leaves may be limited by sulfate transfer out of vacuoles.

Soybean leaf-S concentration declines during seed filling (Sweeney and Granade, 1993; Fantanive et al., 1996; Sexton et al., 1998b), but so does the N/S ratio (Sweeney and Granade, 1993; Imsande and Schmidt, 1998). Likewise, the harvest index for S may be 15-20% lower than that for N (Sexton et al., 1998b). The nitrogen to sulfur ratio in leaves drops from 17.2 to 11.5 indicating that about 50% more S relative to N remains in abscised leaves than in healthy ones. Soybean vegetative tissues may accumulate ample amounts of sulfur for the production of high quality and high protein seed, while not efficiently mobilizing it from vegetative to reproductive tissues (Sunarpi and Anderson, 1997c; Sexton et al., 1998b).

Using ^{35}S pulse-chase techniques, Sunarpi and Anderson (1997c) examined S mobilization in reproductive soybean. They reported that soybean leaves did not act as large reservoirs for S. Although total leaf- and pod-S content dropped by 49 and 67%, respectively, less than 13% of the seeds' S needs were met from S remobilized from vegetative tissue. The plants acquired 87% of the seeds' S needs during seed filling.

In another recent pulse-chase study (Naeve and Shibles, unpublished data) leaves of soybean grown in a sulfur sufficient environment supplied the seed with ca. 20% of its total S requirement through mobilization. Pods contributed ca. 10% of seed-S content. Sunarpi and Anderson (1997c) also found pod SO_4^{-2} levels to drop rapidly at the beginning of seed enlargement. They implicated pods as being important in S storage. Altogether about 38% of the developing seeds' S needs were met by mobilized sulfur in the study by Naeve and Shibles (unpublished data). Additionally, S taken up later in seed development was more likely to be mobilized to the seed, than was S acquired early in development. Plants pulsed with ^{35}S labeled SO_4^{-2} after R5.0 mobilized a larger portion of their total label to seeds than did those pulsed prior to rapid seed growth. In a study where timing of sulfur deficiency was varied to occur either during vegetative growth or during reproductive growth, Sexton et al. (1998c) reported that seed protein quality (a reflection of S availability) was insensitive to S deficiency occurring during vegetative growth, but was very sensitive to S deficiency occurring during reproductive growth. This also suggests that S acquired during vegetative growth plays a minor role in meeting the S needs of developing seed tissue. Thus, from the information reviewed thus far, it appears that most of the protein S within the seed is derived from sulfate taken up during seed filling, rather than from S mobilized out of vegetative tissue, and that this sulfate is probably reduced within the seed itself.

NUTRITIONAL EFFECTS ON PROTEIN QUALITY

The soybean plant possesses considerable plasticity in matching synthesis of storage proteins and protein quality to its nutritional environment. Several studies have shown enhanced protein quality with provision of reduced S as methionine during seed filling. Grabau et al. (1986) provided stem infusions of methionine during seed filling and found soybean to be able to utilize the exogenous source of methionine to produce seed with 23 and 31% increases in methionine and cysteine concentrations, respectively. Soybean cotyledons cultured in a methionine supplemented medium contain no β-subunit of β-conglycinin, a storage protein which is practically devoid of S amino acids and is associated with poor protein quality (Thompson et al., 1984; Holowach et al., 1984a,b; Holowach et al., 1986; Thompson and Madison, 1990; Horta and Sodek, 1997). The same result occurs *in planta*, soybean supplemented with methionine in hydroponic culture shows enhanced accumulation of 11 S proteins and no accumulation of the β-subunit (Paek et al., 2000). On the other hand, soybean plants grown under S-deficient conditions accumulate large amounts of the β-subunit while the better quality 11 S storage proteins are present in lesser amounts relative to controls (Gayler and Sykes, 1985; Paek et al., 2000; Sexton et al., 1998a). To provide a contrast, under severe S deficiency the β-subunit of β-conglycinin may comprise up to 40% of total seed storage proteins, whereas with the provision of supplemental methionine this subunit is not expressed.

While S, and in particular methionine, availability influence protein quality, so does the availability of N effect quality and quantity of soybean seed protein. Provision of reduced N as urea results in increased total protein within the seed, but because synthesis of the β-subunit is disproportionately favored, the 11 S:7 S ratio declines (Paek et al., 1997; Paek et al., 2000). Conversely, N deficient plants have been reported not to accumulate the β-subunit (Ohtake et al., 1994). Thus it appears that the β-subunit may be a kind of barometer of the amount of reduced N versus S in the developing seed. It seems that when N is available but not S, then β-subunit accumulation is promoted as a way to continue protein deposition; when adequate reduced S is present relative to N, the seed produces better quality proteins and the β-subunit is not accumulated. Because methionine is a strong inhibitor of the sulfur-poor β-subunit of the β-conglycinin protein (Naito et al., 1988; Fujiwara, et al., 1992; Hirai et al., 1995), the accumulation of this storage protein late in seed development (Gayler and Sykes, 1985) may indicate a shortage of reduced sulfur compounds available to the seed during the latter part of seed filling.

MOLECULAR STUDIES

Sulfur-containing amino acids, especially methionine, in dietary soybean protein is the primary limiting essential amino acid. It has been suggested that the nutritional quality of soybean storage protein can be improved by:

- null mutation of the methionine-deficient β-conglycinin subunits and, consequently, increase of the concentration of methionine-sufficient glycinin subunits;
- engineering for increased free methionine level in seeds;
- genetic modification of soybean storage protein subunits with increased numbers of methionine residues;
- the introduction of methionine-rich foreign storage protein genes from other legumes or related species.

In corresponding order, each path has to meet the following constraint in order to successfully increase quality without compromising quantity of seed protein:

- the sulfur-poor β-conglycinin is not accumulated without altering the concentration of total protein;
- the regulation of sulfur assimilation can be controlled to increase the production of methionine and increased methionine can be transferred and stored in developing seeds without affecting the normal growth of the plant;
- the β-conglycinin and glycinin subunits engineered with more methionine residues are transferred and successfully stored into protein bodies, and rate of methionine synthesis increases to satisfy increased demand;
- transferred storage protein genes are expressed, do not contribute to anti-quality factors, and rate of methionine synthesis increases to satisfy increased demand.

Kitamura (1995) has reported null mutants that do not accumulate two subunits, α- and α'-subunits of β-conglycinin. No null mutant of β-subunit in β-conglycinin has been reported yet. It is doubtful, however, that such a mutant would be of great practical benefit. Our experimental results (Paek et al, 1997, 2000) showed that under methionine-rich conditions during seed development the sulfur-poor β-subunit of β-conglycinin was not accumulated at all. It appears that this subunit acts as a protein reservoir for continued protein accumulation in the developing soybean seed when methionine supply becomes limiting. Simple null mutation of β-conglycinin seems not to be a useful approach because the synthesis of methionine-rich glycinin subunits are controlled by *in vitro* threshold level of methionine concentration. Because the β-subunit is only synthesized when the methionine is limiting, blocking its expression may only improve quality at the expense of decreased concentration of total protein.

Increasing the concentration of free methionine within the seed may seem at first to be a feasible pathway to increasing methionine concentration. However, the level of free methionine in soybean seed is low and appears to be well regulated (Macnicol and Randall, 1987; Zarkadas et al., 1994, cited in Clarke and Wiseman, 2000). Also work in our lab (Paek et al., unpublished data) has shown that excess free methionine may have deleterious effects on plant growth and development, including: very thick leaves, leaf rigosity, and decreased pod set similar to the phenotype of partial male sterile lines. Transgenic tobacco plants with suppressed AdoMet concentration which led to increased free methionine also showed an abnormal phenotype (Boerjan et al., 1994). It seems that in order to maintain proper cell metabolism, increasing methionine concentration calls for providing a place, that is seed storage proteins, for its sequestration.

There are some null mutants that do not produce one of the subunits of glycinin, such as *gy4* ($A_5A_4B_3$) or *gy5* (A_3A_4) (Kitamura, 1995). It has been thought that genetically manipulated methionine-rich *Gy4* or *Gy5* genes could be introduced into the respective mutants and, consequently, the relative methionine concentration might be increased. However, the amino acid sequences of glycinin subunits are highly conserved and small changes of glycinin amino acid sequence seem not to be tolerable for transfer from the endoplasmic reticulum to protein bodies. In particular, the amino acid sequence of glycinin around the Asn-Gly bonds is important for recognition and post-translational cleavage by asparaginyl endopeptidase (Jung et al., 1998). Even minor changes may affect the specificity of the asparaginyl endopeptidase responsible to cleave the proglycinins to the acidic and the basic subunits. Also cysteine residues in the amino acid sequence are very important to make disulfide bonds between the two cleaved subunits. These two limitations appear to be very difficult to overcome for successfully producing genetically modified methionine-rich glycinin in seeds. So far, there have been no successful introductions of the methionine-enhanced glycinin in soybean.

The introduction of methionine-rich seed storage proteins from other grain legumes and related species has been suggested as the most promising approach (Müntz et al., 1998). Methionine-rich 2 S albumin from Brazil nut was introduced successfully and accumulated in soybean seed. This gene was put under control of a seed specific promoter and transferred by an *Agrobacterium* transformation system. The resulting transgenic soybean seed showed increased concentration of methionine and it appeared that the methionine problem might have been resolved. However, the Brazil nut is a known allergic food. The purified Brazil nut 2 S albumin fraction in transgenic soybean seed protein was tested for its allergenicity (Nordlee et al., 1996). On skin-prick testing, it had positive reactions similar to Brazil nut extracts. This means that the 2 S albumin accumulated in transgenic soybean is probably a major Brazil nut allergen. However, this demonstrated the possibility that methionine concentration may be increased by introduction of a foreign methionine-rich storage protein gene.

As an alternative, the Gm2S-1 (8 kDa) sulfur-rich protein was identified from the albumin fraction of soybean seed (de Lumen et al., 1999). A high methionine sunflower protein, a 2 S albumin, has been expressed in lupine and might provide another opportunity (Tabe et al., 1997). These might be candidates for over expression to improve the nutritional value of soybean and other grain legumes. As pointed out by Clarke and Wiseman (2000), the digestibility of introduced proteins must be considered.

The implicit assumption behind efforts to introduce high methionine storage proteins is that the additional methionine requirement created by these proteins will stimulate increased production of methionine within the plant cell. Successful efforts to increase methionine concentration in narbon bean (Muntz et al., 1997) and lupine (Tabe et al., 1997) did not increase S accumulation by seed tissue, but they changed S partitioning within the seed. In the case of the narbon bean, the S for increased methionine came at the expense of a dipeptide containing cysteine which was abundant in wild type seed. In the case of the lupine, the S for methionine came at the expense of a decreased sulfate concentration within the seed. Where the S for extra methionine synthesis would come from in soybean is a matter of doubt. Soybeans do not appear to contain substantial amounts of sulfate in seed tissue (Sexton et al., 1998b; Sunarpi and Anderson, 1997). Less than expected increases in methionine concentration with insertion of a high-methionine Brazil nut albumin in soybean was associated with decreased expression of endogenous proteins high in methionine (Townsend and Thomas, 1994). Apparently there were metabolic limitations on rate of methionine synthesis which lead to competition for free methionine between endogenous high-methionine proteins and the introduced protein. There may be a need to develop genotypes with greater rates of methionine synthesis within the seed as well as greater demand for methionine by seed storage proteins.

CONCLUSION:
TOWARDS INCREASED METHIONINE CONCENTRATION

We propose the following set of working hypotheses for productively increasing the methionine concentration of soybean seed:

- Expression of the β-subunit of β-conglycinin is a function of supply of reduced N versus reduced S. Blocking expression of this subunit will only increase quality at the expense of decreased protein concentration. The expression of this subunit is naturally suppressed in the presence of methionine.
- Substantially increasing free methionine concentration will be difficult since its regulation is complex and even if successful, it will lead to an unproductive phenotype. Therefore a sink, or a place to sequester methionine (met-rich storage protein) is needed.
- Based on observations that insertion of methionine-rich protein may lead to suppression of native methionine-rich proteins and increased synthesis of the poor quality β-subunit of β-conglycinin, we hypothesize that the metabolic capacity of the plant for methionine synthesis will need to be enhanced to achieve the full benefit of methionine-rich storage proteins added to the soybean genome.
- Based on observations of S transfer between organs during reproductive growth, and localization of ATP-sulfurylase activity, we postulate that the majority of the methionine within the seed is synthesized there rather than being transported to the seed from leaves or other tissues. Therefore, efforts to increase rate of methionine synthesis should be localized within the seed, especially during the latter part of seed filling, as accumulation of the β-subunit during late seed-filling (Gayler and Sykes, 1985) suggests that methionine synthesis may not be keeping pace with protein deposition at that time.

REFERENCES

Adiputra, I.G.K. and J.W. Anderson. 1992. Distribution and redistribution of sulphur taken up from nutrient solution during vegetative growth in barley. Physiol. Plant. 85:453-460.

Agrawal, H.P. and A.K. Mishra. 1994. Sulphur nutrition of soybean. Comm. Soil Sci. Plant Anal. 25:1303-1312.

Allaway, W.H. and J.F. Thompson. 1966. Sulfur in the nutrition of plants and animals. Soil Sci. 101:240-247.

Anderson, J.W. 1990. Sulfur metabolism in plants. Pp. 327-381. In: B.J. Miflin and P.J. Lea (ed.) The Biochemistry of Plants. Vol. 16, Intermediary nitrogen metabolism. Academic Press, San Diego, CA.

Azevedo, R.A., P. Arruda, W.L. Turner, and P.J. Leas. 1997. The biosynthesis and metabolism of the aspartate derived amino acids in higher plants. Phytochem. 46:395-419.

Bell, C.I., D.T. Clarkson, and W.C. Cram. 1995. Partitioning and redistribution of sulphur during S-stress in *Macroptilium atropurpureum* cv. Siratro. J. Exp. Bot. 46:73-81.

Bell, C.I., W.C. Cram, and D.T. Clarkson. 1994. Compartmental analysis of $^{35}SO_4{}^{2-}$ exchange kinetics in roots and leaves of a tropical legume *Macroptilium atropurpureum* cv. Siratro. J. Exp. Bot. 45:879-886.

Boerjan, W., G. Bauw, M. van Montagu, and D. Inze. 1994. Distinct phenotypes generated by overexpression and suppression of S-adenosyl-L-methionine synthetase reveal developmental patterns of gene silencing in tobacco. Plant Cell 6:1401-1414.

Bouma, D. 1975. The uptake and translocation of sulphur in plants. Pp. 79-86. In: K. D. McLachlan (ed.) Sulphur in Australasian agriculture. Sydney University Press.

Brady, C.J. 1988. Nucleic acid and protein synthesis. Pp. 147-179. In: L.D. Nooden and A.C. Lepold (eds.) Senescence and aging in plants. Academic Press, San Diego.

Brunold, C. 1983. Changes in ATP sulfurylase and adenosine 5'-phosphosulfate sulfotransferase activity during autumnal senescence of beech leaves. Physiol. Plant. 59:319-323.

Brunold, C. 1990. Reduction of sulfate to sulfide. Pp. 53-65. In: Rennenberg et al. (eds.) Sulfur nutrition and sulfur assimilation in higher plants. SPB Academic Publ., Hague, The Netherlands.

Clarke, E.J. and J. Wiseman. 2000. Developments in plant breeding for improved nutritional quality of soya beans I. Protein and amino acid content. J. Agri. Sci. 134:111-124.

Clarkson, D.T., L.R. Saker, and J.V. Purves. 1989. Depression of nitrate and ammonium transport in barley plants with diminished sulphate status: Evidence of co-regulation of nitrogen and sulphate intake. J. Exp. Bot. 40:953-963.

Coates, J.B., J.B. Mederiros, V.H. Thanh, and N.C. Neilsen. 1985. Characterization of the subunits of β-conglycinin. Arch. Biochem. Biophys. 243:184-194.

de Lumen, B.O., A.F. Galvez, M.J. Revilleza, and D.C. Krenz. 1999. Molecular strategies to improve the nutritional quality of legume proteins. Adv. Exp. Med. Biol. 464:117-126.

Derbyshire E., D.B. Wright, and D. Boulter. 1976. Legumin and vicilin, storage proteins of legume seeds. Phytochem. 15: 3-24.

Fantanive, A.V., A.M. de la Horra, and M. Moretti. 1996. Foliar analysis of sulfur in different soybean cultivar stages and its relation to yield. Comm. Soil Sci. 27: 179-186.

Fehr, W.R. and C.E. Caviness. 1977. Stages of soybean development. Iowa State Univ. Spec. Rpt. 80, Coop. Ext. Service. Iowa State University, Ames, IA.

Feller, K.U., T.T. Soong, and R.H. Hageman. 1977. Leaf proteolytic activities and senescence during development of field grown corn (*Zea mays* L.). Plant Physiol. 59:290-294.

Ferreira, R.M.B. and A.R.N. Teixeira. 1992. Sulfur starvation in *Lemna* leads to degradation of ribulose-bisphosphate carboxylase without plant death. J. Biol. Chem. 267:7253-7257.

Foyer, C.H., P. Descourvieres, and K.J. Kunert. 1994. Protection against oxygen radicals: An important defense mechanism studied in transgenic plants. Plant, Cell and Environment 17:507-523.

Fujiwara T. and R.N. Beachy. 1994. Tissue-specific and temporal regulation of a β-conglycinin gene: Roles of the RY repeat and other cis-acting elements. Plant Mol. Biol. 24: 261-272.

Fujiwara, T., M.Y. Hirai, M. Chino, Y. Komeda, and S. Naito. 1992. Effects of sulfur nutrition on expression of the soybean seed storage protein genes in transgenic petunia. Plant Physiol. 99:263-268.

Fukushima, D. 1991. Recent progress of soybean protein foods: Chemistry, technology, and nutrition. Food Rev. Int. 7:323-351.

Gaines, T.P. and S.C. Pathak. 1982. Sulfur fertilization effects on the constancy of the protein N:S (nitrogen:sulfur) ratio in low and high sulfur accumulating crops. Agron. J. 74:415-418.

Gayler, K.R. and G.E. Sykes. 1985. Effects of nutritional stress on the storage proteins of soybeans. Plant Physiol. 78:582-585.

Giovanelli, J., S.H. Mudd, and A.H. Datko. 1983. *In vivo* metabolism of 5'-methyl-thioadenosine in *Lemna*: Conversion to methionine. Plant Physiol. 71:319-326.

Goodwin, T.W. and E.I. Mercer. 1983. Introduction to plant biochemistry. 2nd ed. Pergamon Press, Oxford, England.

Grabau, L.J., D.G. Blevins, and H.C. Minor. 1986. Stem infusions enhanced methionine content of soybean storage protein. Plant Physiol. 82:1013-1018.

Guitman, M.R., P.A. Arnozis, and A.J. Barnex. 1991. Effect of source sink relations and nitrogen nutrition senescence and N mobilization in the flag leaf of wheat. Physiol. Plant. 82:278-284.

Hanway, J.J., E.J. Dunphy, G.L. Loberg, and R.M. Shibles. 1984. Dry weights and chemical composition of soybean plant parts throughout the growing season. J. Plant Nut. 7:1453-1475.

Hanway, J.J. and C.R. Weber. 1971. Dry matter accumulation in eight soybean (*Glycine max* (L.) Merrill) varieties. Agron. J. 63:227-230.

Harada, J.J., S.J. Barker, and R.B. Goldberg. 1989. Soybean beta-conglycinin genes are clustered in several DNA regions and are regulated by transcriptional and posttranscriptional processes. Plant Cell 1:415-425.

Hell, R. 1997. Molecular physiology of plant sulfur metabolism. Planta. 202:138-148.

Henry, L.T., C.D. Raper, and J.W. Rideout. 1992. Onset of and recovery from nitrogen stress during reproductive growth of soybean. Int. J. Plant Sci. 153:178-185.

Herschbach, C. and H. Rennenberg. 1996. Storage and remobilisation of sulphur in beech trees (*Fagus sylvatica*). Physiol. Plant. 98:125-132.

Hirai, M.Y., T. Fujiwara, M. Chino, and S. Naito. 1995. Effects of sulfate concentrations on the expression of a soybean seed storage protein gene and its reversibility in transgenic *Arabidopsis thaliana*. Plant Cell Physiol. 36:1331-1339.

Hirai, M.Y., T. Fujiwara, K. Goto, Y. Komeda, M. Chino, and S. Naito. 1994. Differential regulation of soybean seed storage protein gene promoter-GUS fusions by exogenously applied methionine in transgenic *Arabidopsis thaliana*. Plant Cell Physiol. 35:927-934.

Holowach, L.P., J.F. Thompson, and J.T. Madison. 1984a. Effects of exogenous methionine on storage protein composition of soybean cotyledons cultured *in vitro*. Plant Physiol. 74:576-583.

Holowach, L.P., J.F. Thompson, and J.T. Madison. 1984b. Storage protein composition of soybean cotyledons grown *in vitro* in media of various sulfate concentrations in the presence and absence of exogenous L-methionine. Plant Physiol. 74:584-589.

Holowach, L.P., J.F. Thompson, and J.T. Madison. 1986. Studies on the mechanism of regulation of the mRNA level for a soybean storage protein subunit by exogenous L-methionine. Plant Physiol. 80:561-567.

Horta, A. C. G. and L. Sodek. 1997. Free amino acid and storage protein composition of soybean fruit explants and isolated cotyledons cultured with and without methionine. Ann. Bot. 79: 547-552.

Imsande, J. and D.G. Edwards. 1988. Decreased rates of nitrate uptake during pod fill by cowpea, green gram, and soybean. Agron. J. 80:789-783.

Imsande, J. and J.M. Schmidt. 1998. Effect of N source during soybean pod filling on nitrogen and sulfur assimilation and remobilization. Plant Soil 202:41-47.

Janzen, H.H. and J.R. Bettany. 1984. Sulfur nutrition of rapeseed. II. Effect of time of sulfur application. J. Soil Sci. Soc. Am. 48:107-112.

Jung, R., M.P. Scott, Y.-W. Nam, T.W. Beaman, R. Bassüner, I. Saalbach, K. Müntz, and N.C. Nielsen. 1998. The role of proteolysis in the processing and assembly of 11S seed globulins. Plant Cell 10:343-357.

Kawashima, N. and S.G. Wildman. 1970. Fraction I protein. Annu. Rev. Plant Physiol. 21:416-418.

Kim, H., M.Y. Hirai, H. Hayashi, M. Chino, S. Naito, and T. Fujiwara. 1999. Role of O-acetyl-L-serine in the coordinated regulation of the expression of a soybean seed storage-protein gene by sulfur and nitrogen nutrition. Planta 209:282-289.

Kitamura, K. 1995. Genetic improvement of nutritional and food processing quality in soybean. JARQ 29:1-8.

Klapheck, S. 1988. Homoglutathione: Isolation, quantification and occurrence in legumes. Physiol. Plant. 74:727-732.

Koshiyama, I. 1983. Storage proteins of soybean. Pp. 427-450. In: W. Gottschalk and H.P. Muller (eds.), Proteins: Biochemistry, genetic and nutritive value. M. Nijhoff, W. Junk, Publ. Hague, The Netherlands.

Ladin, B.F., M.L. Tierney, D.W. Meinke, P. Hosangadi, M. Veith, and R.N. Beachy. 1987. Developmental regulation of beta-conglycinin in soybean axes and cotyledons. Plant Physiol. 84:35-41.

Lappartient, A. and B. Touraine. 1996. Demand-driven control of root ATP sulfurylase activity and SO_4^{2-} uptake in intact canola. Plant Physiol. 111: 147-157.

Leustek, T. and K. Saito. 1999. Sulfate transport and assimilation in plants. Plant Physiol. 120: 637-643.

Loberg, G.L., R. Shibles, D.E. Green, and J.J. Hanway. 1984. Nutrient mobilization and yield of soybean genotypes. J. Plant Nut. 7:1311-1327.

Loneragan, J.F., K. Snowball, and A.D. Robson. 1976. Remobilization of nutrients and its significance in plant nutrition. In: Transport and Transfer Processes in Plants, Proceedings of a Symposium. pp. 463-469.

Lunn, J.E., M. Droux, J. Martin, and R. Douce. 1990. Localization of ATP sulfurylase and O-Acetylserine(thiol)lyase in spinach leaves. Plant Physiol. 94:1345-1352.

Macnicol, P.K. and L. Bergmann. 1984. A role for homoglutathione in organic sulfur transport to the developing mung bean seed. Plant Sci. Lett. 36:219-223.

Macnicol, P.K. and P.J. Randall. 1987. Changes in the levels of major sulfur metabolites and free amino acids in pea cotyledons recovering from sulfur deficiency. Plant Physiol. 83:354-359.

Marschner, H. 1995. Mineral nutrition of higher plants. 2nd ed. Academic Press, London, San Diego.

Matheny, T.A. and P.G. Hunt. 1981. Effects of irrigation and sulphur application on soybean grown on a Norfolk loamy sand. Commun. Soil Sci. Plant Anal. 12: 147-159.

McVey, M.J., G.R. Pautsch, and C.P. Baumel. 1995. Estimated domestic producer and end user benefits from genetically modifying U.S. soybeans. J. Prod. Agric. 8: 209-214.

Mei, H. and K.U. Thimann. 1984. The relation between nitrogen deficiency and leaf senescence. Physiol. Plant. 62:157-161.

Meinke, D.W., J. Chen, and R.N. Beachy. 1981. Expression of storage-protein genes during soybean seed development. Planta 153:130-139.

Muntz, K., V. Christov, R. Jung, G. Saalbach, I. Saalbach, D. Waddell, T. Pickardt, and O. Schieder. 1997. Genetic engineering of high methionine proteins in grain legumes. Pp. 71-86. In: Cram, W.J., L.J. DeKok, I. Stulen, C. Brunold (eds) Sulphur metabolism in higher plants. Backhuys Publishers, Leiden, The Netherlands.

Müntz, K., V. Christov, G. Saalbach, I. Saalbach, D. Waddell, T. Pickardt, O. Schieder, and T. Wustenhagen. (1998). Genetic engineering for high methionine grain legumes. Nahrung 42:125-127.

Naito, S., P.H. Dube, and R.N. Beachy. 1988. Differential expression of a conglycinin α' and β subunit genes in transgenic plants. Plant Mol. Biol. 11:109-123.

Naito, S., M.Y. Hirai, M. Chino, and Y. Komeda. 1994a. Expression of a soybean (*Glycine max* [L.] Merr.) seed storage protein gene in transgenic *Arabidopsis thaliana* and its response to nutritional stress and to abscisic acid mutations. Plant Physiol. 104:497-503.

Naito, S., K. Ibana-Higano, K. Kumagai, T. Kanno, E. Nambara, T. Fujiwara, M. Chino, and Y. Komeda. 1994b. Maternal effects of *mto1* mutation, that causes overaccumulation of soluble methionine, on the expression of a soybean β-conglycinin gene promoter-GUS fusion in transgenic *Arabidopsis thaliana*. Plant Cell Physiol. 35:1057-1063.

Nielsen, N.C. 1985. The structure and complexity of the 11S polypeptides in soybeans. J. Am. Oil Chem. Soc. 62:1680-1686.

Nielsen, N.C., C.D. Dickinson, T. Cho, V.H. Thanh, B.J. Scallon, R.L. Fischer, T.L. Sims, G.N. Drews, and R.B. Goldberg. 1989. Characterization of the glycinin family in soybean. Plant Cell 1:313-328.

Nooden, L.D. 1988. Whole plant senescence. Pp. 392-441. In: L.D. Nooden and A.C. Lepold (eds.) Senescence and aging in plants. Academic Press, San Diego, CA.

Nordlee, J.A., S.L. Taylor, J.A. Townsend, L.A. Thomas, and R.K. Bush. 1996. Identification of a Brazil-nut allergen in transgenic soybeans. N. Engl. J. Med. 14: 688-692.

Ohtake, N., M. Suzuki, Y. Takahashi, T. Fujiwara, M. Chino, T. Ikarashi, and T. Ohyama. 1996. Differential expression of β-conglycinin genes in nodulated and non-nodulated isolines of soybean. Physiol. Plant. 96:101-110.

Paek, N.C., J. Imsande, R.C. Shoemaker, and R.M. Shibles. 1997. Nutritional control of soybean seed storage protein. Crop Sci. 37: 498-503.

Paek, N.C., P.J. Sexton, S.L. Naeve, and R.M. Shibles. 2000. Differential accumulation of soybean seed storage protein subunits in response to sulfur and nitrogen nutritional sources. Plant Prod. Sci. 3:268-174.

Peoples, M.B. and M.J. Dalling. 1988. The interplay between proteolysis and amino acid metabolism during senescence and nitrogen reallocation. Pp. 181-217. In: L.D. Nooden and A.C. Lepold (eds.) Senescence and aging in plants. Academic Press, San Diego.

Ravanel, S., B. Gakiere, D. Job, and R. Douce. 1998. The specific features of methionine biosynthesis and metabolism in plants. Plant Biol. 95:7805-7812.

Renosto, F., H.C. Patel, R. Martin, C. Thomassian, G. Zimmerman, and I.H. Segel. 1993. ATP sulfurylase from higher plants: Kinetic and structural characterization of the chloroplast and cytosol enzymes from spinach leaf. Arch. Biochem. Biophys. 307:272-285.

Schmutz, D. and C. Brunold. 1982. Regulation of sulfate assimilation plants. Plant Physiol. 70: 524-527.

Sebastiani, F.L., L.B. Farrell, M.A. Schuler, and R.N. Beachy. 1990. Complete sequence of a cDNA of a subunit of β-conglycinin. Plant Mol. Biol. 15: 197-201.

Sexton, P.J. and R.M. Shibles. 1999. Activity of ATP-sulfurylase in reproductive soybean. Crop Sci. 39:131-135.

Sexton, P.J., W.D. Batchelor, and R.M. Shibles. 1997. Sulfur availability, rubisco content, and photosynthetic rate of soybean. Crop Sci. 37:1801-1806.

Sexton, P.J., S.L. Naeve, and N.C. Paek, and R.M. Shibles. 1998a. Sulfur availability, cotyledon nitrogen:sulfur ratio, and relative abundance of seed storage proteins of soybean. Crop Sci. 38:983-986.

Sexton, P.J., N.C. Paek, and R.M. Shibles. 1998b. Soybean sulfur and nitrogen balance under varying levels of available sulfur. Crop Sci. 38:975-982.

Sexton, P.J., N.C. Paek, and R.M. Shibles. 1998c. Effects of nitrogen source and timing of sulfur deficiency on seed yield and expression of 11S and 7S seed storage proteins. Field Crops Res. 59:1-8.

Shibles, R. and D.N. Sundberg. 1998. Relation of leaf nitrogen content and other traits with seed yield of soybean. Plant Prod. Sci. 1:3-7.

Sinclair, T.R., and C.T. DeWitt. 1975. Photosynthate and nitrogen requirements for seed production by various crops. Science. 189:565-567.

Smith, F.W., M.J. Hawkesford, P.M. Ealing, D.T. Clarkson, P.J. Vanden, A.R. Belcher, and A.G. Warrilow. 1997. Regulation of expression of a cDNA from barley roots encoding a high affinity sulphate transporter. Plant J. 12: 875-884.

Smith, I.K. and A.L. Lang. 1988. Translocation of sulfate in soybean (*Glycine max* L. Merr). Plant Physiol. 86:798-802.

Sunarpi and J.W. Anderson. 1996. Distribution and redistribution of sulfur supplied as [^{35}S] sulfate to roots during vegetative growth of soybean. Plant Physiol. 110: 1151-1157.

Sunarpi, and John W. Anderson. 1997a. Effect of nitrogen on the export of sulfur from leaves in soybean. Plant Soil 188:177-187.

Sunarpi, and J.W. Anderson. 1997b. Inhibition of sulphur redistribution into new leaves of vegetative soybean by excision of the maturing leaf. Physiol. Plant. 99:538-545.

Sunarpi, and John W. Anderson. 1997c. Allocation of S in generative growth of soybean. Plant Physiol. 114:687-693.

Sunarpi, and John W. Anderson. 1997d. Effect of nitrogen nutrition on remobilization of protein sulfur in the leaves of vegetative soybean and associated changes in soluble sulfur metabolites. Plant Physiol. 115:1671-1680.

Sweeney, D.W. and G.V. Granade. 1993. Yield, nutrient, and soil sulfur response to ammonium sulfate fertilization of soybean cultivars. J. Plant Nut. 16:1083-1098.

Tabe, L., L. Molvig, R. Khan, H. Schroeder, S. Gollasch, T. Wardley-Richardson, A. Moore, S. Craig, D. Spencer, B. Eggum, and T.J.V. Higgins. 1997. Modifying the sulphur amino acid content of transgenic legumes. Pp. 87-93. In: Cram, W.J., L.J. DeKok, I. Stulen, and C. Brunold (eds.) Sulphur metabolism in higher plants. Backhuys Publishers, Leiden, The Netherlands.

Takahashi, H., M. Yamazaki, N. Sasakura, A. Watanabe, T. Leustek, J. de Aleida-Engler, G. Engler, M. Van Montagu, and K. Saito. 1997. Regulation of cysteine biosynthesis in higher plants: a sulfate transporter induced in sulfate-starved roots plays a central role in *Arabadopsis thaliana*. Proc. Natl. Acad. Sci. USA. 94: 11102-11107.

Thompson J.F., J.T. Madison, L.P. Holowach, and G.L. Creason. 1984. The effect of methionine on soybean storage protein gene expression. Curr. Top. Plant Biochem. Physiol. 3:1-8.

Thompson, J.F. and J.T. Madison. 1990. The effect of sulfate and methionine on legume proteins. Pp. 145-158. In: H. Rennenberg et al., (eds.) Sulfur nutrition and sulfur assimilation in higher plants, SPB Academic Publ., Hague, The Netherlands.

Townsend, J.A. and L.A. Thomas. 1994. Factors which influence the *Agrobacterium*-mediated transformation of soyabean. Journal of Cell Biochemistry. Suppl. 18A, Abstr. X1-014.

Vasilas, B.L., R.L. Nelson, J.J. Fuhrmann, and T.A. Evans. 1995 Relationship of nitrogen utilization patterns with soybean yield and seed fill. Crop Sci. 35:809-813.

Vidmar, J.J., Schjoerring, J.K., Touraine, B. and A.D.M. Glass. 1999. Regulation of the *hvst1* gene encoding a high-affinity sulfate transporter from Hordeum vulgare. Plant Mol. Biol. 40:883-892.

von Arb, C. and C. Brunold. 1986. Enzymes of assimilatory sulfate reduction in leaves of *Pisum sativum*: Activity changes during ontogeny and *in vivo* regulation by H_2S and cyst(e)ine. Physiol. Plant. 67:81-86.

Wilson, R.F. 1987. Seed metabolism. Pp. 643-686. In: Soybeans: Improvement, Production, and Uses. 2nd ed. ASA, Madison, Wisconsin.

Breeding Oilseed Crops
for Improved Oil Quality

Leonardo Velasco
José M. Fernández-Martínez

SUMMARY. Vegetable oils are one of the most valuable commodities in world trade. They are subject to specific quality requirements, both for food and non-food uses, there being a continuous demand for new oil types. Thus, plant breeders have made great efforts over the past four decades to develop those quality features demanded by the industry, mainly related to the fatty acid composition of the seed oil. Initially, breeders had to focus on the natural variation occurring within each oilseed crop and closely related species. From the 1970s onwards, the induction of mutations by treatment of seeds with mutagenizing agents was revealed as an effective system for modifying the fatty acid profile. In fact, mutagenesis has proved to be one of the most successful approaches for creating novel oil types. Nowadays, breeding for improved seed oil quality is in a transitional stage, both from a conceptual and a methodological point of view. First, the concept of oil quality is changing. It is not only defined by its fatty acid composition but also by other parameters, the most important being the triacylglycerol profile and the tocopherol content and composition. Second, molecular techniques for gene identification and manipulation are opening up new possibilities, much more powerful and less random than the traditional ones, for the

Leonardo Velasco is Associate Scientist, and José M. Fernández-Martínez is Professor, Instituto de Agricultura Sostenible (CSIC), Alameda del Obispo s/n, Apartado 4084, E-14080 Córdoba, Spain.

Address correspondence to: Leonardo Velasco or José M. Fernández-Martínez at the above address (E-mail: ia2veval@uco.es or cs9femaj@uco.es).

[Haworth co-indexing entry note]: "Breeding Oilseed Crops for Improved Oil Quality." Velasco, Leonardo, and José M. Fernández-Martínez. Co-published simultaneously in *Journal of Crop Production* (Food Products Press, an imprint of The Haworth Press, Inc.) Vol. 5, No. 1/2 (#9/10), 2002, pp. 309-344; and: *Quality Improvement in Field Crops* (ed: A. S. Basra, and L. S. Randhawa) Food Products Press, an imprint of The Haworth Press, Inc., 2002, pp. 309-344. Single or multiple copies of this article are available for a fee from The Haworth Document Delivery Service [1-800-HAWORTH, 9:00 a.m. - 5:00 p.m. (EST). E-mail address: getinfo@haworthpressinc.com].

modification of quality traits. In this review, we aim to offer an overview of seed oil quality as well as its genetic improvement by traditional and biotechnological means. *[Article copies available for a fee from The Haworth Document Delivery Service: 1-800-HAWORTH. E-mail address: <getinfo@ haworthpressinc.com> Website: <http://www.HaworthPress.com> © 2002 by The Haworth Press, Inc. All rights reserved.]*

KEYWORDS. Antioxidants, biotechnology, fatty acids, genetic resources, mutagenesis, oilseed breeding, quality, seed oils, tocopherols, triacylglycerols

INTRODUCTION

The modification of the fatty acid profile of the seed oil has been one of the main tasks faced by oilseed breeders over the past forty years. Success in this field has been of paramount importance for the worldwide expansion of some oilseed crops. Thus, the elimination of erucic acid from rapeseed oil, which naturally contains about 40% of this harmful fatty acid, was the first step towards the development of canola (zero-erucic, low-glucosinolate rapeseed) as one of the major sources of vegetable oil in the world (Stefansson, Hougen, and Downey, 1961). Other landmarks in oilseed breeding for seed oil quality have been the development of high oleic, low linolenic acid canola (Rücker and Röbbelen, 1997), low linolenic acid linseed (Green, 1986a) and soybean (Fehr and Hammond, 1998), high oleic acid sunflower (Soldatov, 1976), high saturated sunflower (Osorio et al., 1995), and sunflower lines with modified tocopherol composition (Demurin, Škorić, and Karlovic, 1996).

Traditionally, the concept of oil quality has been almost exclusively associated with the fatty acid composition of the oil. Recently, however, this concept has evolved and oil chemists and nutritionists are emphasizing other components of vegetable oils that influence their physical and chemical properties. Similarly, the way in which plant breeders afford the genetic modification of quality parameters of seed oils has been changing in recent years, mainly due to the availability of new powerful tools for the identification and manipulation of genes. Within this context, this review deals with key aspects related to seed oil quality, its genetic improvement and the possible lines of evolution. Oilseed crops are treated as a group but no attempt has been made to review the breeding of all of them.

CHEMISTRY AND OCCURRENCE OF QUALITY COMPONENTS IN VEGETABLE OILS

Vegetable oils are predominantly (92-98%) triacylglycerols, the most important of the remaining components being polar lipids (phospholipids and galactolipids), mono- and diacylglicerols, free fatty acids, and polyisoprenoid lipids (Åppelqvist, 1989). The latter group comprises sterols and sterol derivatives, tocopherols, carotenoids and chlorophylls, some of these compounds having important vitamin and antioxidant properties. The presence of some of these constituents is markedly affected by external factors. Thus monoacylglycerol, diacylglycerol, and free fatty acids are usually present only in trace amounts in mature seeds, but in higher concentrations in immature, damaged or germinated seeds (McKillican, 1966). Similarly, some chlorophyll-bearing seeds such as canola contain these pigments at very low concentrations at full maturity, although cool environmental conditions or frost during seed ripening may increase the chlorophyll content considerably, thus downgrading oil quality (Uppström, 1995). From the point of view of seed oil quality improvement, the three main parameters that are usually considered as a target in breeding programmes are the fatty acid composition, the triacylglycerol composition, and the quantity and composition of the antioxidants present in the oil.

Fatty Acid and Triacylglycerol Composition

Triacylglycerols are glycerol molecules containing one fatty acid esterified to each of the three hydroxyl groups. The stereochemical positions of the three fatty acids in the glycerol molecule are designed *sn-1*, *sn-2*, and *sn-3* (Figure 1). Both the relative amount of fatty acids that are present in the oil and their distribution in triacylglycerol molecular species determine the physical, chemical, physiological and nutritional properties of vegetable oils (Padley, Gunstone, and Harwood, 1994).

Fatty acids differ in their number of carbon atoms and/or number and position in the carbon chain of double bonds and functional groups (hidroxy, epoxy, etc.). Depending on the presence or absence of double bonds, the fatty acids are divided into saturated, which do not contain double bonds, and unsaturated, which contain at least one double bond. As an example of the nomenclature commonly used to indicate the three parameters, linoleic acid is represented as 18:2 (n-6), which expresses that this fatty acid consists of a chain with 18 carbon atoms, two double

FIGURE 1. Structure of a triacylglycerol molecule. *sn*-1, *sn*-2 and *sn*-3 refer to the carbon numbers of the glycerol. R are fatty acids.

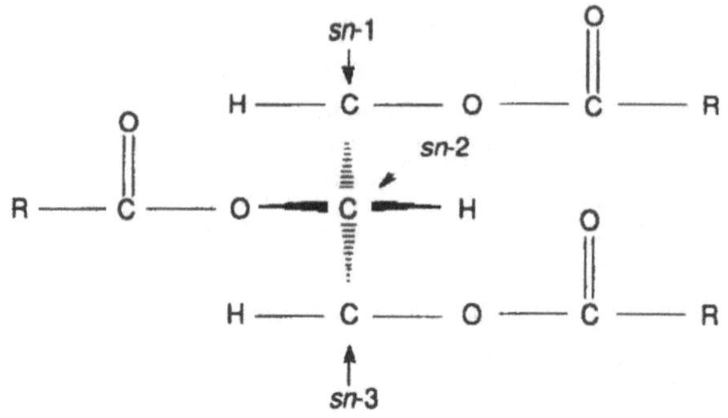

bonds, with the sixth carbon from the methyl end being the first unsaturated one. The distance between the methyl end of the carbon chain and the first double bond is of utmost importance for the nutritional and pharmaceutical properties of fatty acids (Åppelqvist, 1989). The unsaturated fatty acids can also have two possible configurations, cis or trans, depending on the relative position of the alkyl groups. This is of great relevance from a nutritional point of view, since trans fatty acids have a detrimental effect on human health (Willet and Ascherio, 1994). Most naturally occurring unsaturated fatty acids have the cis orientation, although several common industrial processes such as hydrogenation, which is applied for example for margarine production, induce cis-trans isomeration (Tatum and Chow, 1992).

All the fatty acids are linked in the biosynthetic pathway through modifications such as elongation and desaturation (Figure 2). This fact determines that the alteration of any of the biosynthetic steps influences the whole fatty acid profile. A description of the biosynthesis of the various fatty acids and their enzymatic relationships is out of the scope of this manuscript; a comprehensive review can be found in Harwood (1996).

The most common fatty acids in vegetable oils and their outstanding properties are given in Table 1. In general, saturated fatty acids have a hypercholesterolemic effect, whereas unsaturated fatty acids act lowering serum cholesterol. The exception to this rule is stearic acid, which exhibits a neutral effect (Mensink, Temme, and Hornstra, 1994). Linoleic, α-linolenic, and γ-linolenic acids are three of the essential fatty acids, i.e., they must be included in the diet because the human body is not

FIGURE 2. Schematic representation of the biosynthetic pathway of the principal fatty acids. Trivial names of the fatty acids are given in Table 1. a = elongation of saturated fatty acids; b = desaturation pathway to α-linolenic acid, typical of linseed; c = elongation of monounsaturated fatty acids, typical of the Brassicaceae family; d = desaturation pathway to γ-linolenic acid, typical of borage and evening primrose.

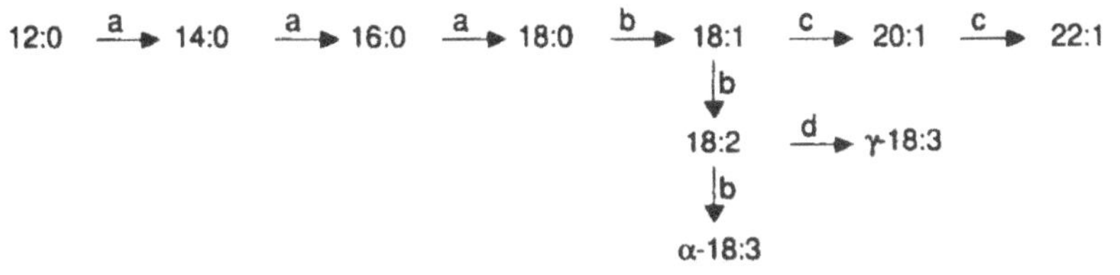

$$12:0 \xrightarrow{\ a\ } 14:0 \xrightarrow{\ a\ } 16:0 \xrightarrow{\ a\ } 18:0 \xrightarrow{\ b\ } 18:1 \xrightarrow{\ c\ } 20:1 \xrightarrow{\ c\ } 22:1$$

$$\downarrow b$$

$$18:2 \xrightarrow{\ d\ } \gamma\text{-}18:3$$

$$\downarrow b$$

$$\alpha\text{-}18:3$$

TABLE 1. The major fatty acids in vegetable oils and their outstanding properties

Trivial name	Symbol	Nutritional properties	Physical properties	Oxidative stability	Richest source
Lauric	12:0	Hipercholesterolemic	Solid	High	Palmkernel
Myristic	14:0	Hipercholesterolemic	Solid	High	Palmkernel
Palmitic	16:0	Hipercholesterolemic	Solid	High	Palm
Stearic	18:0	Neutral	Solid	High	Cocoa
Oleic	18:1 (n-9)	Hypocholesterolemic	Liquid	High	Olive
Linoleic	18:2 (n-6)	Hypocholesterolemic Essential fatty acid[a]	Liquid	Low	Safflower
α-Linolenic	18:3 (n-3)	Hypocholesterolemic Essential fatty acid	Liquid	Very low	Linseed
γ-Linolenic	18:3 (n-6)	Hypocholesterolemic Essential fatty acid Medical applications[b]	Liquid	Very low	Borage
Erucic	22:1 (n-9)	Toxic	Liquid	High	Crambe

[a]Fatty acids that are not synthesized by human and have to be obtained from the diet
[b]See review in Horrobin, 1992

able to manufacture them (Horrobin, 1992). These fatty acids are polyunsaturated, i.e., they contain more than one double bond in the carbon chain. Polyunsaturated fatty acids are more susceptible to autoxidation than monounsaturated or saturated fatty acids. The double bonds react rapidly with oxygen in the air in a process involving the production of free radicals, which are implicated in a number of diseases, tissue injuries and in the process of aging (Shahidi, 1996). Furthermore, the breakdown products of fatty acid autoxidation are the major source of off-flavors in oils, which reduce their shelf life (Tatum and Chow, 1992). Therefore, although polyunsaturated fatty acids such as α-linoleic are beneficial *per se*, their susceptibility to autoxidation make them undesirable at high levels in vegetable oils. For this reason linseed oil, which naturally contains over 45% of linolenic acid, is not used for human consumption but as a drying oil in the formulation of paints, varnishes, inks and industrial coatings (Green and Marshall, 1981). Oleic acid is nowadays considered as the preferred fatty acid for edible purposes, as it combines a hypocholesterolemic effect and a high oxidative stability (Mensink and Katan, 1989; Yodice, 1990).

Many other fatty acids are unsuitable for human or animal consumption if present at high concentrations in the oil, but they are of great interest for industrial uses. For example, erucic acid, present at high concentrations in oils from the Brassicaeae (e.g., rapeseed, mustards and crambe), petroselinic acid, present in the Apiaceae (e.g., coriander), vernolic acid (e.g., *Vernonia* spp.), or ricinoleic acid (castor bean).

Each oil has a characteristic pattern of triacylglycerols, which depends on the available fatty acids and the specificity of the biosynthetic enzymes (Fernández-Moya, Martínez-Force, and Garcés, 2000). The fatty acids are not distributed randomly between the different *sn*-carbon atoms of the triacylglycerol molecule. As a general rule, saturated fatty acids are confined to positions *sn*-1 and *sn*-3, whereas polyunsaturated fatty acids are located mainly at the *sn*-2 position (Stymne and Stobart, 1987). One exception is palm oil, which contains higher levels of saturated fatty acids at the *sn*-2 position. This has been suggested to have negative biological effects and to be involved in the atherogenic process (Renaud, Ruf, and Petithory, 1995). In rapeseed oil, erucic acid is excluded from the *sn*-2 position, which results in a theoretical breeding limit for increasing the concentration of this fatty acid of 66% of the total fatty acids (Taylor et al., 1994).

Antioxidants

Vegetable oils contain a wide spectrum of substances with antioxidant activity that are responsible for protecting unsaturated fatty acids from oxidation processes. The most important antioxidants in vegetable oils are chromanols, carotenoids, and phenolic compounds.

Chromanols

Structurally, chromanols consist of a chroman head with two rings, one phenolic and one heterocyclic, the latter substituted with a phytyl tail. The most important chromanols are tocopherols, with saturated phytyl tails, tocotrienols, having unsaturated phytyl tails, and plasto-chromanol-8, with a saturated phytyl tail longer than that of tocoph-erols. Both tocopherols and tocotrienols occur in four forms, named α, β, γ, and δ, differing in the number of methyl substituents and the pat-tern of substitution in the phenolic ring (Figure 3).

The chromanols exhibit antioxidant activity both *in vivo* and *in vitro*. *In vivo* they exert vitamin E activity, protecting cellular membrane lipids against oxidative damage (Muggli, 1994). *In vitro* they inhibit lipid oxidation in oils, fats, and foods containing them (Kamal-Eldin and Appelqvist, 1996). α-Tocopherol is the chromanol derivative ex-hibiting the greatest vitamin E effect (relative activity 100), followed by β-tocopherol (50), α-tocotrienol (30), and γ-tocopherol (10) (Padley, Gunstone, and Harwood, 1994). Conversely, the best *in vitro* antioxi-dant activity ranks in the order γ-tocopherol (relative antioxidant activ-ity 100), δ-tocopherol (68), β-tocopherol (64), and α-tocopherol (35) (Pongracz, Weiser, and Matzinger, 1995). The *in vitro* antioxidant ac-tivities of tocotrienols are unknown, whereas it has recently been shown that plastochromanol-8 is a more powerful antioxidant than α-tocopherol (Olejnik, Gogolewski and Nogala-Kalucka, 1997).

FIGURE 3. Chemical formulae of (A) tocopherols and (B) tocotrienols. Me = methyl groups.

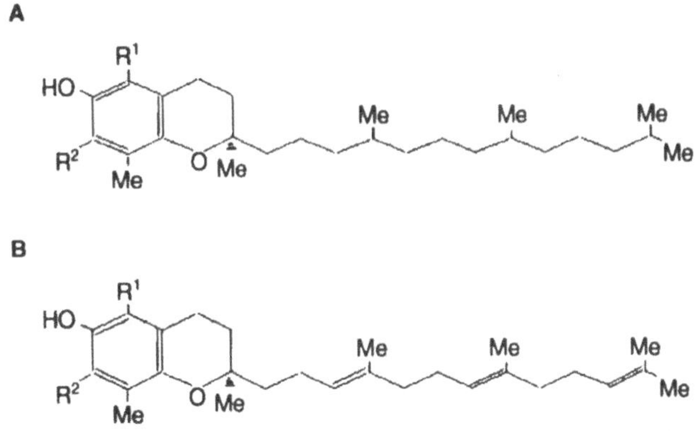

A

B

R¹ = Me; R² = Me: α-tocopherol/tocotrienol
R¹ = Me; R² = H: β-tocopherol/tocotrienol
R¹ = H; R² = Me: γ-tocopherol/tocotrienol
R¹ = H; R² = H: δ-tocopherol/tocotrienol

The tocopherol derivatives are the predominant chromanol form in vegetable oils. The most relevant exceptions are palm oil, which in addition to tocopherols contains large amounts of tocotrienols (Padley, Gunstone, and Harwood, 1994), and linseed oil, which contains both tocopherols derivatives and plastochromanol-8 (Velasco and Goffman, 2000).

Carotenoids

Carotenoids are a family of C_{40} polyunsaturated hydrocarbons (carotenes) and their oxygenated derivatives (xanthophylls). Similarly to chromanols, they play an important role *in vivo* as source of provitamin A as well as *in vitro*, protecting oils from oxidation (Henry, Catignani, and Schwartz, 1998). β-Carotene is the most nutritionally active carotene as provitamin A (Ong and Choo, 1996). The *in vitro* activity of carotenoids seems to be related to the inhibition of photoxidation, acting as a filter for light of short wavelengths (Warner and Frankel, 1987). Palm oil is the richest oil source of carotenoids with a concentration of 500-700 ppm (Ong and Choo, 1996). The concentration of carotenoids in seed oils is considerably lower (Uppström, 1995).

Phenolic Compounds

Phenolic compounds also have an important antioxidant effect. In oilseeds, the phenolics remain in the meal after oil extraction, producing astringency and a bitter taste. Further, they form complexes with proteins, lowering their nutritional value. Therefore phenolic compounds are undesirable in oilseeds (Naczk et al., 1998). Their major value is in virgin olive oil, which is obtained from the fruit mesocarp by mechanical pressing. This oil contains a considerable amount of phenolics, many of which have not yet been completely identified (Shukla, Wanasundara, and Shahidi, 1996). It has been shown that the antioxidant activity of some phenolic compounds in olive oil, e.g., *o*-diphenolic compounds, is much grater than that of α-tocopherol (Blekas, Tsimidou, and Boskou, 1995).

Synergistic Effects

Oil's oxidative stability is enhanced by synergism between different kinds of compounds. For example, synergistic effects between α-tocopherol and β-carotene (Terao et al., 1980; Heinonen et al., 1997), γ-tocopherol and oleic acid (Demurin, Škorić, and Karlovic, 1996), and between tocopherols and certain phospholipids (Lambelet, Saucy, and Löliger, 1984) have been demonstrated.

THE CONCEPT OF OIL QUALITY

Oil quality is a relative concept that depends on the end-use of the oil. Vegetable oils are intended for food and non-food applications. The former include salad and cooking oils as well as oils for the food industry (margarines, shortenings, etc.). The latter comprises countless industrial sectors such as lubricants, surfactants, surface coatings, cosmetics, plastics, etc. In general, those oil characteristics that are undesirable for a particular application are required for others. Therefore breeding for improved oil quality is in some ways a continuous exercise of divergent selection. Some examples will be useful to illustrate this.

Erucic Acid in Rapeseed and Mustards

Erucic acid is the main seed oil fatty acid in the genus *Brassica* (rapeseed and mustards) as well as in many other genera of the Brassicaceae (Velasco, Goffman, and Becker, 1998). This fatty acid has been shown to be cardiotoxic in animals and therefore also potentially toxic in humans (Beare-Rogers and Nera, 1972). Simultaneously, erucic acid is a very valuable feedstock for the oleochemical industry (Lühs and Friedt, 1994). Therefore efforts have been made to eliminate erucic acid from *Brassica* seed oils for human consumption as well as to develop lines with a maximum erucic acid content for industrial applications. In the former objective, breeders have been successful and zero-erucic lines have been developed for rapeseed (*B. napus*; Stefansson, Hougen, and Downey, 1961), turnip rape (*B. rapa*; Downey, 1964), Indian mustard (*B. juncea*; Kirk and Oram, 1981), Ethiopian mustard (*B. carinata*; Alonso, Fernández-Serrano, and Fernández-Escobar, 1991), and white mustard (*Sinapis alba* L.; Raney, Rakow, and Olson, 1995a). For the latter objective, however, the theoretical limit of 66% (see above) has not yet been surpassed, even when several breeding strategies such as interspecific crossing (Lühs and Friedt, 1995), mutagenesis (Velasco, Fernández-Martínez and De Haro, 1998), and genetic engineering (Lassner et al., 1995) have been attempted.

Linolenic Acid in Linseed

The genus *Linum* comprises species with either high or low linolenic acid content (Velasco and Goffman, 2000). Linseed (*L. usitatissimum*) is one of the species with a high concentration of linolenic acid in its seed oil. As stated above, linolenic acid is an essential fatty acid that possesses many beneficial healthy effects although, due to its susceptibility to autoxidation, an oil rich in linolenic acid is in practice undesirable for human consumption. Its main utility is for non-food applications, especially in the industry of paints, varnishes, inks and coatings (Green and Marshal, 1981). Therefore selection has been focused on lowering linolenic acid content in order to open up food applications to linseed oil, and on increasing linolenic acid content to promote its industrial applications. A drastic reduction of the linolenic content, from about 45-55% to levels below 2% has been achieved through mutagenesis (Green, 1986a; Rowland, 1991). The increase in linolenic acid content, up to 65% of the total fatty acids, has been achieved by conventional breeding (Green and Marshal, 1981). A further increase seems to be feasible by using the variability found in wild species (Bickert, Lühs, and Friedt, 1994) and *in vitro* breeding (Friedt, Bickert, and Schaub, 1995).

Saturated Fatty Acids in Sunflower

A sunflower oil rich in saturated fatty acids is desirable for the industry of margarines and related products, because its semi-solid consistency makes it unnecessary detrimental physical transformations such as hydrogenation or transterification (Álvarez-Ortega et al., 1997). These processes produce *trans* and positional isomers related to heart disease (Willett and Ascherio, 1994). On the other hand, saturated fatty acids with a chain length of ≤ 16 carbons are undesirable in edible oils due to their negative effect on serum total cholesterol concentrations (Wardlaw and Snook, 1990). Both breeding objectives have been tackled by using mutagenesis. Sunflower mutants with increased levels of saturated fatty acids were obtained by Ivanov et al. (1988), Osorio et al. (1995), and Fernández-Martínez et al. (1997). Similarly, sunflower mutants with reduced saturated fatty acid content were isolated by Vick and Miller (1996).

Tocopherol Profile

Among the four tocopherol derivatives, α-tocopherol exhibits the greatest vitamin E activity whereas γ-tocopherol is the best antioxidant *in vitro*. Therefore the development of cultivars with contrasting tocopherol profiles has been suggested as a means of adjusting the quality of the oil to specific end uses (Pongracz, Weiser, and Matzinger, 1995). This objective has been accomplished in sunflower, where Demurin, Škorić, and Karlovic (1996) developed both lines with 50% of β-tocopherol and lines with 95% of γ-tocopherol, compared with standard sunflower lines with 95% of γ-tocopherol. Modified lines were obtained by intraspecific germplasm evaluation. In rapeseed, Goffman, Velasco, and Becker (1998) explored interspecific variation as a way to develop high α- and high γ-tocopherol lines.

GENETIC CONTROL

Most of the above mentioned traits defining seed oil quality have been found to be governed by a reduced number of genes. This fact implies that the practical management of single quality traits in breeding programmes is relatively easy if compared with polygenic traits. Nevertheless, breeding programmes do not focus on single traits, but many of them have to be simultaneously taken into account. Therefore, it is of great importance to obtain precise information on how individual traits are inherited, how different traits influence to each other, and how their expression is influenced by the environment.

Inheritance Studies

In general, the fatty acid composition of the seed oil is determined by the genotype of the developing embryo. This is extremely important in breeding for a modified fatty acid profile, as selection can be carried out at a single-seed level. There are, however, exceptions to this rule in which the maternal genotype has a significant influence on the phenotypic expression of the fatty acid composition in the embryo. One well-characterized example for this is the low linolenic acid content in the soybean mutant A5 (Graef et al., 1988; Rennie and Tanner, 1991; Fehr et al., 1992). In these cases, selection using the half-seed technique (Downey and Harvey, 1963) is less effective.

In most cases the concentration of individual fatty acids in the seed oil is controlled by one to three major genes, with several alleles for each locus in most cases (Velasco, Pérez-Vich, and Fernández-Martínez, 1999a). Different mechanisms of genetic control have been described.

In rapeseed and mustards, erucic acid is governed by multiple alleles at two genes acting in an additive manner (Jönsson, 1977). Multiple alleles at one to three loci is also the most common mode of inheritance of altered fatty acid traits in soybean, though in this case having no additive effects but acting independently. This has been reported for reduced palmitic acid content (Fehr et al., 1991a), high palmitic acid content (Schnebly et al., 1994), high stearic acid content (Bubeck, Fehr, and Hammond, 1989), high oleic acid content (Rahman, Takagi, and Kinoshita, 1996) and low linolenic acid content (Rahman et al., 1998).

Sunflower is probably the oilseed crop in which the most complex genetic systems for fatty acid inheritance have been identified. The inheritance of high oleic acid content has been a subject of study and controversy since the high oleic acid mutant was obtained by Soldatov (1976). There is general agreement on the presence of a principal gene named *Ol1* and several other genes, but their number and mode of action remain unclear. We have recently proposed a genetic model including a total of five genes (Velasco, Pérez-Vich, and Fernández-Martínez, 2000), though studies are still under way. The inheritance of high palmitic acid content in sunflower was also found to be rather complex with a principal gene named *P1* and two other genes *P2* and *P3* showing interchangeable effects (Pérez-Vich et al., 1999). High stearic acid content in sunflower was found to be controlled by two unlinked loci, with one of them having a greater phenotypic effect (Pérez-Vich, Garcés, and Fernández-Martínez, 1999).

More simple is the inheritance of modified fatty acid contents in other oil crops such as safflower and flax. In the former, studies have shown that the levels of oleic and linoleic acid are controlled by three alleles at a single locus, whereas increased stearic acid content is determined by two alleles at another single locus (Ladd and Knowles, 1971). In flax, reduced linolenic acid content is produced by alleles at two unlinked loci (Green, 1986b), while high palmitic acid content is determined by one single locus (Ntiamoah, Rowland, and Taylor, 1995).

The inheritance of other seed oil quality traits apart from individual fatty acids has been scarcely studied. In sunflower, Demurin (1993) identified two non-allelic unlinked genes controlling altered tocopherol composition. The *Tph-1* gene produced increased β-tocopherol content, while the *Tph-2* gene produced increased γ-tocopherol content. The simple inheritance of altered tocopherol composition in sunflower has enabled the transference of the traits to inbred lines for production of hybrids with improved oil stability (Jocic et al., 2000).

Relationship Between Quality Traits

The recombination of genes controlling different oil seed quality traits is necessary in many cases to go a step further in the improvement of oil quality. In some cases the genes are unlinked, so that there is no major restriction for their recombination. For example, it has been possible to recombine different altered fatty acid profiles such as low linolenic with low or high palmitic acid content in soybean (Nickell, Wilcox, and Cavins, 1991) and linseed (Ntiamoah, Rowland, and Taylor, 1995). Similarly, genes controlling fatty acid levels seem to be independently inherited from those controlling the tocopherol profile, being it therefore feasible to make a number of combinations between different tocopherol and fatty acid profiles (Demurin, Škorić, and Karlovic, 1996). Conversely, it has not been possible to obtain a complete recombination of high stearic with either high palmitic or high oleic acid content in sunflower, in the former case because of an epistatic interaction (Pérez-Vich, Garcés, and Fernández-Martínez, 2000a) and in the latter as a result of a genetic linkage (Pérez-Vich, Garcés, and Fernández-Martínez, 2000b).

Environmental Effects

As stated above, seed oil quality traits are qualitative rather than quantitative, i.e., they are governed by a reduced number of genes. Their phenotypic expression is therefore less affected by the environment than in the case of quantitative traits such as oil or protein contents. Temperature is the main factor affecting oil quality traits. In general, lower temperatures cause an increase in the unsaturation level, affecting mainly to the oleic to linoleic acid ratio (Canvin, 1965). However, not all the genotypes of the same species are identically affected by the temperature, existing a genotype × environment interaction. For example, the oleic to linoleic acid ratio is much more stable across environments in mutants with very high oleic acid content than in standard types, as demonstrated in sunflower (Garcés and Mancha, 1989) and safflower (Knowles, 1989). A genotype × environment interaction affecting the fatty acid profile was also reported by Rennie and Tanner (1989) after evaluating several soybean genotypes grown under contrasting temperature conditions. Other factors such as light (Brockman, Norman, and Hildebrand, 1990) and soil water availability (Bouchereau et al., 1996) may also affect the composition of the seed oil.

SCREENING PROCEDURES

An indispensable prerequisite for improving seed oil quality traits is the availability of adequate screening procedures to measure them. This is critical at the first stages of breeding programs, where the breeder needs to analyze large numbers of very small samples. Therefore analytical methods have to be fast, cheap and suitable for small samples, but at the same time they should be reliable enough not to decrease the selection efficiency. Traditionally, plant breeders have developed their own screening techniques or have adapted standard procedures to their particular purposes. Here, three basic groups of screening techniques are being considered, two of them dealing with the analysis of the phenotype and a third group founded on genotypic identification.

Single-Trait Approach

Many fast screening procedures have been developed to analyze one single trait related to seed oil quality. Some of them have been specifically developed for screening purposes. The most successful one, developed to screen for linolenic acid levels, is the thiobarbituric acid procedure (McGregor, 1974). This procedure has been used in large-scale screening of half seeds within mutagenesis programs in linseed (Green and Marshall, 1984; Rowland and Bhatty, 1990; Rowland, 1991) and rapeseed (Auld et al., 1992). It is of interest to mention that all the currently existing linseed germplasm with less than 2% of linolenic acid derive from the above mentioned mutagenesis programs in which the thiobarbituric acid procedure was used for screening. Other screening procedures consisted in the adaptation of standard analytical procedures. Craig and Murty (1958) developed the methodology for analyzing the fatty acid composition of the seed oil by gas-liquid chromatography, which was the key for the immediate development of zero-erucic acid rapeseed (Stefansson, Hougen, and Downey, 1961). In a further step, Downey and Harvey (1963) adapted this technique for analyzing fatty acids in single seeds, leading to the development of zero-erucic turnip rape (Downey, 1964). Other techniques have also been adapted for plant breeding purposes in the field of oil quality. For example, Möllers et al. (1997) and Goffman, Velasco, and Thies (1999) developed chromatographic methods suitable for fast screening of trierucin and tocopherol contents, respectively, in single half seeds of rapeseed.

Multi-Trait Approach

The possibility of a nondestructive screening for many different quality traits in small samples had been the breeders' dream for many years. It became a reality in the 60s, when Norris and Hart (1965) developed near infrared spectroscopy (NIRS) for the analysis of seeds. The first application of NIRS for the analysis of seed oil quality parameters was the analysis of erucic acid content in intact rape seeds (Koester and Paul, 1989). To date, methods have been developed for the analysis of the fatty acid composition of the oil in seeds of Ethiopian mustard (Velasco, Fernández-Martínez, and De Haro, 1997), Indian mustard (Font et al., 1998), rapeseed (Velasco and Becker, 1998), sunflower (Pérez-Vich, Velasco, and Fernández-Martínez, 1998) and soybean (Pazdernik, Killam, and Orf, 1997). Fatty acid analysis by NIRS has the main advantage of being simultaneous to the analysis of other seed quality traits. For example, NIRS analysis of intact rapeseed includes, besides fatty acid composition, chlorophyll, oil, protein, glucosinolate and phytic acid contents, glucosinolate composition, sinapic acid esters content, seed color, test, weight, and seed size (Velasco, 1999). In a further step, NIRS has been adapted for the analysis of the fatty acid composition of the oil in single seeds, which is of paramount importance in plant breeding. Single-seed methods have been developed in rapeseed (Sato et al., 1998; Velasco, Möllers, and Becker, 1999) and sunflower (Sato et al., 1995; Velasco, Pérez-Vich, and Fernández-Martínez, 1999b).

Molecular Methods

The identification of superior genotypes beyond measurable phenotypic values is an indispensable requisite for efficient selection. The utilization of molecular markers allows the breeder to select on the basis of genotypic instead of phenotypic characteristics, thus eliminating environmental effects. The so-called marker-assisted selection encompasses a large number of techniques based on the presence or absence of molecular markers that are closely linked to the gene(s) controlling the target trait.

Marker-assisted selection for seed oil quality traits has so far been restricted to selection for levels of individual fatty acids. Most of the work has been conducted in rapeseed and turnip rape. In these species, markers for palmitic (Tanhuanpää, Vilkki, and Vilkki, 1995), oleic (e.g., Schierholt and Ecke, 1998; Tanhuanpää and Vilkki, 1999), linolenic (e.g., Somers, Friesen, and Rakow, 1998; Hu, Struss, and Quiros, 1999;

Rajcan et al., 1999) and erucic acid (e.g., Ecke, Uzunova, and Weissleder, 1995; Jourdren et al., 1996; Rajcan et al., 1999) have been developed. In sunflower, markers for stearic (Pérez-Vich et al., 2000a,b) and oleic acid (Dehmer and Friedt, 1998; Pérez-Vich et al., 2000b; Lacombe et al., 2000) have been identified.

BREEDING STRATEGIES

Plant breeding deals with the manipulation of existing or newly generated genetic variation. Naturally existing genetic variation is available to the plant breeder through germplasm collections, usually located in germplasm banks. The evaluation of germplasm is needed, for example, at the beginning of a breeding program focusing on a novel trait. In this case, the most logical searching sequence starts with cultivars of the crop, followed by wild accessions of the species, related species of the same genus, and finally in other genera, starting in those most phylogenetically related. The way in which the novel variation is incorporated into the target genotypes will depend upon the phylogenetic proximity between the cultivated and the donor species as well as upon the genetic nature of the trait. In the absence of naturally existing variation or when its transfer to the cultivated species is impracticable, an alternative approach is the generation of new variation. In this section we will give some examples of different breeding strategies used to improve seed oil quality traits.

The Use of Existing Genetic Variation

Intraspecific Variation

Cultivated species have been subject to selection for millennia. Farmers in ancient times practised selection on a local scale following nonstandardized criteria, which led to a huge genetic variation within each crop. Although modern agricultural practices have favored the loss of an important part of this diversity, traditional local varieties (landraces) are a major reservoir of genetic variability.

Safflower is probably the best example of the utilization of intraspecific variation for seed oil quality improvement. Most of the currently existing variation for the seed oil fatty acid profile in this crop has been developed through evaluation and utilization of genetic resources. Thus, the evaluation of germplasm led to the identification of genetic sources for high palmitic acid content (Jianguo et al., 1993), high stearic acid content (Knowles, 1965), reduced levels of total saturated fatty acids (Velasco and Fernández-Martínez, 1999), intermediate (Knowles and Hill, 1964), and high oleic acid content (Knowles and Mutwakil, 1963), as well as very high linoleic acid content (Futehally and Knowles, 1981). The great intraspecific variation for fatty acid composition has been used to develop safflower cultivars with contrasting fatty acid profiles (Fernández-Martínez, 1997).

Interspecific Hybridization

Hybridization between species of the same genus or even between species from closely related genera is an useful approach for incorporating traits that are not present in the crop species. The main problem of interspecific hybridization is the existence of pre- and post-fertilization barriers, which in many cases can be overcome by means of appropriate manipulation techniques. For example, embryo rescue in culture medium is an effective technique to overcome post-fertilization barriers.

Interspecific hybridization has been used to improve seed oil fatty acid composition in oilseed brassicas. In Ethiopian mustard, zero erucic acid lines have been obtained through interspecific crossing with rapeseed and Indian mustard (Getinet et al., 1994; Fernández-Martínez et al., 2000). Similarly, Raney, Rakow, and Olson (1995) developed low linolenic Indian mustard from crosses with a low linolenic rapeseed line. A specific application of interspecific hybridization within the genus *Brassica* is the resynthesis of amphidiploid species from their diploid parents (Engqvist and Becker, 1994). Lühs and Friedt (1995) developed high-erucic acid rapeseed by means of resynthesis from high erucic entries of *B. rapa* and *B. oleracea*.

Interspecific hybridization has also been suggested as an interesting tool for improving seed oil quality in other crops. In sunflower, wild *Helianthus* species have been identified as potential sources of genes for low saturated fatty acids (Seiler, 1996) and high linoleic acid content (De Haro and Fernández-Martínez, 1991). Similarly, the use of wild *Linum* species has been suggested to modify both the fatty acid composition (Yermanos, 1966; Friedt, Nichterlein, and Nickel, 1989) and the tocopherol composition (Velasco and Goffman, 2000) of cultivated linseed.

Somatic Hybridization

Successful use of sexual hybridization is limited by fertilization barriers. Alternative techniques are therefore needed when these barriers cannot be overcome. One of such techniques, somatic hybridization, is described in this section while a second one, gene transfer will be described in the next section.

Somatic hybrids are produced by mass fusion of protoplasts isolated from different somatic tissues, making it possible to develop hybrids regardless of sexual compatibility (Glimelius, 1999). This technique has been widely used for the alteration of the fatty acid profile of rapeseed, an oilseed crop that is particularly amenable to biotechnological manipulations. Somatic hybridization of rapeseed with *Thlaspi perfoliatum* (Fahleson et al., 1994) and *Lesquerella fendleri* (Schröder-Pontopiddan et al., 1999) led to hybrid plants exhibiting a certain amount of nervonic and ricinoleic acid, respectively in the seed oil. Neither nervonic nor ricinolenic acid are present in standard rapeseed. In both cases, the amount of the foreign fatty acid in rape seeds was very low in comparison with the donor species, suggesting that some genes involved in their synthesis were not present or not fully expressed in the hybrids (Glimelius, 1999). Somatic hybridization has also been used to facilitate the resynthesis of *Brassica* amphidiploids. With this approach, Heath and Earle (1997) developed resynthesized rapeseed exhibiting a reduced linolenic acid content.

Gene Transfer

Gene transfer technology permits the identification of individual genes as well as their isolation and transfer to a different host. In this process there are no sexual or taxonomic barriers and genes can be transferred from one species to another one phylogenetically distant. As mentioned in the previous section, rapeseed is the oilseed species on which most biotechnological work has been conducted. First studies on gene transfer in rapeseed started in the late 80s (e.g., Pua et al., 1987), and the modification of its seed oil fatty acid profile was one of the first objectives. Voelker et al. (1996) introduced in rapeseed a lauric acid (12:0) acyl carrier protein thioesterase from seeds of the California bay laurel (*Umbelluraria californica*), resulting in an accumulation of up to 50% of lauric acid in the seed oil. Lauric acid was found almost exclusively at the *sn*-1 and *sn*-3 positions of the triacylglycerols. In a further step, a lysophosphatidic acid acyltransferase from coconut was transferred to high lauric rapeseed, promoting an efficient accumulation of lauric acid also at the *sn*-2 position (Knutzon et al., 1999). This led to a further increase in lauric acid content in the seed oil (above 50%) and to the production of triacylglycerols with this fatty acid at the three *sn* positions (trilaurin). The first commercial acreage of high-laurate rape-seed was planted in 1994, and the seed oil is commercialized under the brand name Laurical™ (Friedt and Lühs, 1998). Many other types of transgenic rapeseed have been developed or are being developed nowadays, for example exhibiting high caprylic (8:0) and capric acid (10:0) contents by transfer of a thioesterase from *Cuphea hookeriana* (Dehesh et al., 1996), high stearic acid content (Knutzon et al., 1992) and high oleic acid content (DeBonte and Hitz, 1998) by using antisense expression technology, high palmitic acid by transfer of a thioesterase from *C. hookeriana* (Jones, Davies, and Voelker, 1995), and high trierucin content (Lassner et al., 1995) by transfer of an acyltransferase from meadow-foam (*Limnanthes* sp.). Much work on genetic engineering of rapeseed oil is currently ongoing and, in the medium term, many other fatty acid profiles are expected in transgenic rapeseed plants. Current research focuses on rapeseed oil with very high erucic acid content (> 65%) or expressing unusual fatty acids such as ricinoleic, petroselinic, γ-linolenic, or epoxy fatty acids (Friedt and Lühs, 1998).

Soybean is another oilseed that is very amenable to genetic engineering. Transgenic soybean plants expressing very high oleic acid (Heppard et al., 1996) or high stearic acid content (Kinney, 1998) have been developed, the former already being in the early stages of commercial production. Similarly, transgenic soybean plants with a very low palmitic acid content were produced by reducing the expression of the *FatB* gene. These lines are expected to be in commercial production in few years (Kinney, 1999). As in rapeseed, genetic engineering of soybean is also being directed to the production of unusual fatty acids with industrial applications. For example, soybean plants producing fatty acids with conjugated double bonds, through enzymes transferred from *Momordica charantia* and *Impatiens balsamica*, have been developed (Cahoon et al., 1999). Oils possessing high contents of such fatty acids are very valuable drying agents. Also, efforts to develop soybean seeds with a high vernolic acid content, an epoxy fatty acid from *Vernonia galamensis*, are under way (Kinney, 1999).

Although most of the work to improve seed oil quality through genetic engineering has been concentrated on fatty acid biosynthetic pathways, some work has also been directed to modify tocopherol content and composition. Shintani and Dellapenna (1998) succeeded in drastically increasing the levels of α-tocopherol in seeds of the experimental plant *Arabidopsis thaliana*, which naturally produces mainly γ-tocopherol. This was achieved by overexpression of the γ-tocopherol methyltransferase, which catalyses the final step in α-tocopherol synthesis from the precursor γ-tocopherol (Schultz et al., 1985).

Search for Alternative Crops or Alternative Uses of Existing Crops

Fewer than 20 plant species provide the main vegetable oils in international commerce. Other interesting sources of vegetable oil, however, remain unexploited. Some of them possess unique fatty acids of great interest for pharmaceutical or industrial applications, but in most cases they occur in the seeds of wild, undomesticated species (Earle et al., 1959). Although some authors consider that the domestication of new oilseed crops is useless, because they would be superseded by the production of similar oils in genetically engineered present-day crops (Friedt and Lühs, 1998), there is a general agreement on the benefits of new crops for agricultural sustainability (Janick et al., 1996).

There have been many attempts to domesticate wild plants to produce new oil crops. One example is the species of the genus *Cuphea,* which are excellent sources of medium-chain fatty acids for temperate regions (Hirsinger, 1985). Domestication of these species, focused on non-shattering, non-dormant, self-fertile types is under way (Crane, Tagliani, and Knapp, 1995). Other potential oil crops with unusual fatty acids being domesticated are *Stokesia laevis* (Campbell, 1981), *Euphorbia lagascae* (Pascual-Villalobos et al., 1992) and *Vernonia galamensis* (Thompson et al, 1994), which are sources of epoxy fatty acids, *Lesquerella fendleri* (Dierig et al., 1996) and *Dimorphotheca pluvialis* (Hof, 1996), containing hydroxy fatty acids, and *Limnanthes* species (Knapp, 1990), with a high concentration of very long chain fatty acids in the seed oil.

In some cases, oil-bearing plant species traditionally cultivated for other purposes (vegetables, root crops, spices, ornamental plants) are transformed into oilseed crops. For example Ethiopian mustard (*Brassica carinata* A.Braun) is a traditional African vegetable with seeds containing a high content of oil rich in erucic acid, similar to that of rapeseed, for which zero-erucic acid types have already been developed (Alonso, Fernández-Serrano, and Fernández-Escobar, 1991). Borage (*Borago officinalis* L.) is another traditional vegetable that is being bred for seed oil production, because of its high γ-linolenic acid content (Del Río, Fernández-Martínez, and De Haro, 1993). The yam beans (*Pachyrhyzus* spp.) are root crops for which breeding work is being conducted as sources of high-palmitic, high-γ-tocopherol seed oil (Grüneberg,

Goffman, and Velasco, 1999). Several species of the family Apiaceae cultivated as spices, particularly coriander (*Coriandrum sativum* L.), are being considered as potential oilseed crops, because of their oil rich in petroselinic acid (Knapp, 1990). There are also several ornamental plants with a high potential value as oilseeds, for example evening primrose (*Oenothera* spp.), an important source of γ-linolenic acid (Hudson, 1984), marigold (*Calendula officinalis*), with seed oil rich in calendulic acid (Cromack and Smith, 1998), and honesty (*Lunaria annua* L.), with a high content of very long chain fatty acids (Cromack, 1998).

The Creation of Novel Variation

Somaclonal Variation

Cell and tissue culture are routine tools for rapid propagation of genotypes, especially in vegetatively-propagated crops like potato. But *in vitro* culture not only offers a system for genotype multiplication, it is also a way to obtaining spontaneous new variability during callus differentiation. The new variation observed in such cultures is called somaclonal variation. This technique has been applied to the modification of the fatty acid composition of the seed oil in linseed (Rowland et al., 1995) and rapeseed (Craig and Millam, 1995), although no remarkable results have been obtained in this field. Nevertheless, somaclonal variation has been successfully used for the improvement of other traits in oilseed crops. For example, Australian breeders used it as one of the approaches leading to the development of low glucosinolate Indian mustard (Oram and Kirk, 1993). The amount of new variability observed in somaclonal variation can be increased by adding mutagens to the culture medium, which will be discussed in the next section.

Mutagenesis

The utilization of chemical or physical mutagens has undoubtedly been the most successful procedure used for modifying seed oil quality parameters in crops. A detailed review on the role of mutagenesis in the modification of the fatty acid profile of oilseed crops has recently been published by the authors elsewhere (Velasco, Pérez-Vich, and Fernández-Martínez, 1999a), thus this section will concentrate on the most relevant mutants developed within each crop and on some methodological aspects of mutagenesis. Mutagenesis results have been particularly remarkable in soybean, in which countless mutants with altered fatty acid profile have been developed, and also in sunflower, flax, rapeseed and Ethiopian mustard. The most relevant mutants obtained in these crops, including recombination between individual mutants, are given in Table 2.

Mutants with altered seed oil fatty acid profile are usually the result of a mutation at a single locus. Exceptions to this general rule are scarce. The most relevant one was the low linolenic acid linseed mutant developed by Rowland (1991), with 2% linolenic acid compared with about 55% in the parent, which was described as a result of two simultaneous single-gene mutations occurring at different unlinked loci.

One of the most remarkable aspects of mutagenesis is that, in many cases, mutants with a similar phenotypic expression have mutations at different genes. Therefore genetic recombination may lead to a complementary improvement of the trait. The most spectacular recombination was obtained by Green (1996a), who obtained a linseed recombinant with about 2% of linolenic acid after crossing two mutants with about

TABLE 2. Fatty acid profile (major fatty acids expressed as percent of the total fatty acids) of standard types and principal mutants developed in oilseed crops.

Crop/mutant type	Fatty acid composition (%)						Reference
	16:0[a]	18:0	18:1	18:2	18:3	22:1	
Soybean standard	11	4	31	47	7		
High 16:0	27	np[b]	np	np	np		Fehr et al. (1991b)
High 18:0	9	29	16	42	5		Bubeck et al. (1989)
Low 18:3	10	5	34	50	1		Fehr and Hammond (1998)
Rapeseed standard	5	1	62	20	9		
High oleic	4	np	80	6	6		Rücker and Röbbelen (1997)
Low linolenic	4	np	71	18	3		Rücker and Röbbelen (1997)
Ethiopian mustard standard	4	1	9	18	13	44	
Low erucic	6	2	32	30	12	7	Velasco et al. (1995)
High erucic	3	1	8	15	12	55	Velasco et al. (1998)
Sunflower standard	6	5	26	62			
High palmitic	25	4	11	55			Osorio et al. (1995)
High stearic	5	26	14	55			Osorio et al. (1995)
High oleic	np	np	79	15			Soldatov (1976)
Linseed standard	8	3	23	16	51		
High palmitic	28	4	14	6	43		Rowland and Bhatty (1990)
Low linolenic	9	5	34	51	2		Green (1986a)

[a]16:0 = palmitic-, 18:0 = stearic-, 18-1 = oleic-, 18:2 = linoleic-, 18:3 = α-linolenic-, 22:1 = erucic acid
[b]not published values

22% of linolenic acid each. Similarly, the development of soybean types producing seed oil with about 1% of linolenic acid, in comparison with levels of about 7% in standard types, has been possible after recombining mutations for reduced linolenic acid content at three different loci (Fehr and Hammond, 1998; Ross et al., 2000).

The mutagenic treatment is usually applied to seeds, which after treatment are named M_1 seeds. Since fatty acids are mainly under embryogenic control, mutants can be detected in the M_2 generation by analyzing M_2 half-seeds. Therefore only one year of plant cultivation is needed before mutants can be identified. A further improvement of this scheme consists in applying the mutagenic treatment to microspores in a culture medium (Polsoni, Kott, and Beversdorf, 1988). Since there is a good correlation between the fatty acid composition of microspore-derived embryos and that of the seeds formed by the regenerated plants (Möllers et al., 2000), identification of putative mutants can be done *in vitro* analyzing one cotyledon and leaving the rest of the embryo for plant regeneration. After analyses, only the most interesting embryos will be regenerated to plants, representing important savings for mutagenesis programs. Furthermore, this approach allows the detection of mutants shortly after the mutagenic treatment, with the main advantage that mutated haploid embryos express both dominant and recessive traits. Mutagenesis of haploid cultures has already resulted in several mutants with altered fatty acid profile in rapeseed (Turner and Facciotti, 1990; Huang et al., 1991; Wong and Swanson, 1991).

Development of Cultivars with Improved Oil Quality

The individual breeding strategies described above enable the identification or generation of oil quality characteristics demanded by the market. The commercial exploitation of such oil types, however, is not immediate but requires the adequate integration of the novel variation into a global breeding strategy including quality as well as agronomic objectives. Genotypes with improved oil quality characteristics obtained through strategies such as germplasm evaluation, interspecific crossing, somatic hybridization, gene transfer or induced mutagenesis are usually not suitable for direct commercial exploitation because of the simultaneous occurrence of undesirable agronomic or seed quality traits. Several cycles of selection are, therefore, required before a cultivar with the required yield and seed quality is developed.

The most usual scheme for the development of commercial cultivars incorporating the novel oil quality trait involves backcrossing, in which the genetic stock that possess the new trait is used as donor and a high performance cultivar, or its parental lines in the case of hybrid cultivars, are used as the recurrent parent. The complexity of the genetic transference of the novel variation depends on three main factors: the number of genes involved, their dominant or recessive character, and the presence or absence of maternal effects. Additionally other traits, related to both seed quality and agronomic performance, have to be simultaneously considered in the backcross program.

Most of the oil quality traits currently considered in breeding programmes are controlled by a reduced number of genes, usually from one to three, so that only a few backcross generations are needed. In most cases, the target traits are recessive which, unlike dominant traits, do not show up in the F_1 backcross generations. Therefore selection has to be done on the F_2 backcross generation, which increases the duration of the backcross program. Such a limitation can be overcome by using marked-assisted selection, which allows the recessive gene(s) to be identified in the F_1 backcross generation. Similarly, traits under embryogenic control, i.e., the phenotype is determined by the genotype of the developing embryo, are easier to transfer than those subject to maternal influences, since selection can be done by analyzing half seeds. If there is a substantial maternal effect, selection of single plants instead of single seeds is required, which retards the selection process in one generation per selection cycle. Most of the oil quality traits that have been genetically characterized lack maternal effects. Some interesting examples of backcross programs for transferring oil quality traits to agronomically efficient parents, usually followed by pedigreed selection, are reported by Fernández-Martínez, Dominguez-Giménez, and Jiménez-Ramírez (1988), Scarth et al. (1988), Dribnenki and Green (1996), Rücker and Röbbelen (1996), Ross et al. (2000), and Jocic et al. (2000).

Other breeding methods apart from backcrossing are also used for the development of cultivars with improved oil quality, for example, single seed descent (Fernández-Martínez et al., 1986), pedigree line development (Dribnenki and Green, 1995; Mündel and Braun, 1999), recurrent selection (Burton, Wilson, and Brim, 1983; Burton et al., 1998), or haploid breeding (Earle and Knauf, 1999).

CONCLUDING REMARKS AND FUTURE PROSPECTS

During the past forty years breeding for improved seed oil quality has concentrated on the modification of the fatty acid composition, mainly through conventional selection from naturally occurring variation and through mutagenesis. Nowadays other aspects are starting to be considered as important components of oil quality. These are the triacylglycerol profile, the concentration of antioxidant substances and the synergistic effects among different types of compounds. An increased importance of these traits in breeding programs for improved oil quality is foreseeable in the short term. Furthermore, there are many other substances in the oil, for example sterols, whose nutritional or technological implications have still to be characterized. They will be generalized breeding objectives as soon as their positive or negative value is well understood. With regard to breeding methodology, molecular approaches are gaining importance in breeding programs conducted in industrialized countries. This trend will be maintained in the coming years, despite the practical results obtained so far have been scarce in comparison with the great expectations created one decade ago. At the same time, a foreseeable renewed interest in agricultural sustainability will demand a greater biodiversity in agricultural systems, which will encourage research on recovery of neglected crops and domestication of new species.

REFERENCES

Alonso, L.C., O. Fernández-Serrano, and J. Fernández-Escobar. (1991). The outset of a new oilseed crop: *Brassica carinata* with low erucic acid content. In *Proceedings of the 8th International Rapeseed Conference.* Saskatoon, Canada: GCIRC, pp. 170-176.

Álvarez-Ortega R., S. Cantisán, E. Martínez-Force, and R. Garcés. (1997). Characterization of polar and nonpolar seed lipid classes from highly saturated fatty acid sunflower mutants. *Lipids* 32:833-837.

Anand, I.J. and R.K. Downey. (1981). A study of erucic acid alleles in digenomic rapeseed (*Brassica napus* L.). *Canadian Journal of Plant Science* 61:199-203.

Åppelqvist, L.A. (1989). The chemical nature of vegetable oils. In *Oil Crops of the World*, eds. R.K. Downey. G. Röbbelen and A. Ashri. New York: McGraw-Hill, pp. 22-37.

Auld, D.L., M.K. Heikkinen, D.A. Erickson, J.L. Sernyk, and J.E. Romero. (1992). Rapeseed mutants with reduced levels of polyunsaturated fatty acids and increased levels of oleic acid. *Crop Science* 32:657-662.

Beare-Rogers, J.L. and E.A. Nera. (1972). Cardiac fatty acids and histopathology of rats, pigs, monkeys and gerbils fed rapeseed oil. *Comparative Biochemistry and Physiology* 41:793-800.

Bickert, C., W. Lühs, and W. Friedt. (1994). Variation for fatty acid content and triacylglycerol composition in different *Linum* species. *Industrial Crops and Products* 2:229-237.

Blekas, G., M. Tsimidou, and D. Boskou. (1995). Contribution of α-tocopherol to olive oil stability. *Food Chemistry* 52:289-294.

Bouchereau, A., N. Clossais-Besnard, A. Bensaoud, L. Leport, and M. Renard. (1996). Water stress effects on rapeseed quality. *European Journal of Agronomy.* 5:19-30.

Brockman, J.A., H.A. Norman, and D.F. Hildebrand. (1990). Effects of temperature, light and a chemical modulator on linolenate biosynthesis in mutant and wild type *Arabidopsis* calli. *Phytochemistry* 29:1447-1453.

Bubeck, D.M., W.R. Fehr, and E.G. Hammond. (1989). Inheritance of palmitic and stearic acid mutants of soybean. *Crop Science* 29:652-656.

Burton, J.W., R.F. Wilson, and C.A. Brim. (1983). Recurrent selection in soybeans. IV. Selection for increased oleic acid percentage in seed oil. *Crop Science* 23:744-747.

Burton, J.W., J.R. Wilcox, R.F. Wilson, W.P. Novitzky, and G.J. Rebetzke. (1998). Registration of low palmitic acid soybean germplasm lines N94-2575 and C1943. *Crop Science* 38:1407.

Cahoon, E.B., T.J. Carlson, K.G. Ripp, B.J. Schweiger, G.A. Cook, S.E. Hall, and A.J. Kinney. (1999). Biosynthetic origin of conjugated double bonds. Production of fatty acid components of high valye drying oils in transgenic soybean embryos. *Proceedings of the National Academy of Science of USA* 96:12935-12940.

Campbell, I.A. (1981). Agronomic potential of Stokes aster. In *Monography 9: New Sources of Fats and Oils.* Ed. E.H. Pryde, L.H. Princen and K.D. Mukherjee. Champaign, IL, USA: AOCS Press, pp. 287-295.

Canvin, D.T. (1965). The effect of temperature on the oil content and fatty acid composition of the oils from several oil seed crops. *Canadian Journal of Botany* 43:63-69.

Craig, A. and S. Millam. (1995). Modification of oilseed rape to produce oils for industrial use by means of applied tissue culture methodology. *Euphytica* 85:323-327.

Craig, B.M. and N.L. Murty. (1958). The separation of saturated and unsaturated fatty acid esters by gas-liquid chromatography. *Candian Journal of Chemistry.* 36:1297-1301.

Crane, J.M., L.A. Tagliani, and S.J. Knapp. (1995). Registration of five self-fertile, partially nondormant cuphea germplasm lines: VL-90 to VL-95. *Crop Science* 35:1516-1517.

Cromack, H.T.H. (1998). The effect of sowing date on the growth and production of *Lunaria annua* in southern England. *Industrial Crops and Products* 7:217-221.

Cromack, H.T.H. and J.M. Smith. (1998). *Calendula officinalis*: production potential and crop agronomy in southern England. *Industrial Crops and Products* 7:223-229.

DeBonte, L.R. and W.D. Hitz. (1998). Canola oil having increased oleic acid and decreased linolenic acid content. *US Patent 5850026.* Date issued: December 15.

De Haro, A. and J. Fernández-Martínez. (1991). Evaluation of wild sunflower (*Helianthus*) species for high content and stability of linoleic acid in the seed oil. *Journal of Agricultural Science* 116:359-367.

Dehesh, K., A. Jones, D.S. Knutzon, and T.A. Voelker. (1996). Production of high levels of 8:0 and 10:0 fatty acids in transgenic canola by overexpression of Ch FatB2, a thioesterase cDNA from Cuphea kookeriana. *Plant Journal* 9:167-172.

Dehmer, K.J. and W. Friedt. (1998). Development of molecular markers for high oleic acid content in sunflower (*Helianthus annuus* L.). *Industrial Crops and Products* 7:311-315.

Del Río, M., J. Fernández-Martínez, and A. De Haro. (1993). Wild and cultivated *Borago officinalis* L.: sources of gamma-linolenic acid. *Grasas y Aceites* 44:125-126.

Demurin, Y. (1993). Genetic variability of tocopherol composition in sunflower seeds. *Helia* 16:59-62.

Demurin, Y., D. Škorić, and D. Karlovic. (1996). Genetic variability of tocopherol composition in sunflower seeds as a basis of breeding for improved oil quality. *Plant Breeding* 115:33-36.

Dierig, D.A., T.A. Coffelt, F.S. Nakayama, and A.E. Thompson. (1996). Lesquerella and vernonia: oilseeds for arid lands. In *Progress in New Crops*, ed. J. Janick. Alexandria, VA, USA: ASHS Press, pp. 347-354.

Downey, R.K. (1964). A selection of *Brassica campestris* L. containing no erucic acid in its seed oil. *Canadian Journal of Plant Science* 44:295.

Downey, R.K. and B.L. Harvey. (1963). Methods of breeding for oil quality in rape. *Canadian Journal of Plant Science* 43:271-275.

Dribnenki, J.C.P. and A.G. Green. (1995). Linola™ '947' low linolenic acid flax. *Canadian Journal of Plant Science* 75:201-202.

Dribnenki, J.C.P. and A.G. Green. (1996). Linola™ 989 low linolenic acid flax. *Canadian Journal of Plant Science* 76:329-331.

Earle, E.D. and V.C. Knauf. (1999). Genetic engineering. In *Biology of Brassica coenospecies*, ed. C. Gómez-Campo. Elsevier, pp. 287-313.

Earle, F.R., E.H. Melvin, L.H. Mason, C.H. Van Etten, and I.A. Wolff. (1959). Search for new industrial oils. I. Selected oils from 24 plant families. *Journal of the American Oil Chemists' Society* 36:304-307.

Ecke, W., M. Uzunova, and K. Weissleder. (1995). Mapping the genome of rapeseed (*Brassica napus* L.). 2. Localization of genes controlling erucic acid synthesis and seed oil content. *Theoretical and Applied Genetics* 91:972-977.

Engqvist, G. and H. Becker. (1994). What can resynthesized *Brassica napus* offer to plant breeding? *Sveriges Utsädesförenings Tidskrift* 104:87-92.

Fahleson, J., I. Eriksson, M. Landgren, S. Stymne, and K. Glimelius. (1994). Intertribal somatic hybrids between *Brassica napus* and *Thlaspi perfoliatum* with high content of the *T. perfoliatum*-specific nervonic acid. *Theoretical and Applied Genetics* 87:795-804.

Fehr, W.R. and E.G. Hammond. (1998). Reduced linolenic acid production in soybeans. *U.S. Patent 5850030*. Date issued: 15 December.

Fehr, W.R., G.A. Welke, E.G. Hammond, D.N. Duvick, and S.R. Cianzio. (1991). Inheritance of reduced palmitic acid in seed oil of soybean. *Crop Science* 31:88-89.

Fehr, W.R., G.A. Welke, E.G. Hammond, D.N. Duvick, and S.R. Cianzio (1991). Inheritance of elevated palmitic acid content in soybean seed oil. *Crop Science* 31:1522-1524.

Fehr, W.R., G.A. Welke, E.G. Hammond, D.N. Duvick, and S.R. Cianzio. (1992). Inheritance of reduced linolenic acid content in soybean genotypes A16 and A17. *Crop Science* 32:903-906.

Fernández-Martínez, J. (1997). Update on safflower genetic improvement and germplasm resources. In *Proceedings of the Fourth International Safflower Conference*. Bari, Italy: Arti Grafiche Savaresse, pp. 187-195.

Fernández-Martínez, J.M., M. Del Río, L. Velasco, J. Domínguez, and A. De Haro. (2000). Registration of zero erucic acid Ethiopian mustard genetic stock 25X-1. *Crop Science* (in press).

Fernández-Martínez, J., J. Domínguez-Giménez, A. Jiménez-Ramírez, and L. Hernández. (1986). Use of the single seed descent method in breeding safflower (*Carthamus tinctorius* L.). *Plant Breeding* 97:364-367.

Fernández-Martínez, J., J. Domínguez-Giménez, and A. Jiménez-Ramírez. (1988). Breeding for high content of oleic acid in sunflower (*Helianthus annuus* L.) oil. *Helia* 11:11-15.

Fernández-Martínez, J.M., M. Mancha, J. Osorio, and R. Garcés. (1997). Sunflower mutant containing high levels of palmitic acid in high oleic acid background. *Euphytica* 97:113-116.

Fernández-Moya, V., E. Martínez-Force, and R. Garcés. (2000). Identification of triacylglycerol species from high-saturated sunflower (*Helianthus annuus*) mutants. *Journal of Agricultural and Food Chemistry* 48:764-769.

Font, R., M. Del Río, J.M. Fernández-Martínez, and A. De Haro. (1998). Determining quality components in Indian mustard by NIRS. *Eucarpia Cruciferae Newsletter* 20:67-68.

Friedt, W., C. Bickert, and H. Schaub. (1995). *In vitro* breeding of high-linolenic, doubled-haploid lines of linseed (*Linum usitatissimum* L.) via androgenesis. *Plant Breeding* 114:322-326.

Friedt, W. and W. Lühs. (1998). Recent developments and perspectives of industrial rapeseed breeding. *Fett/Lipid* 100:219-226.

Friedt, W., K. Nichterlein, and M. Nickel. (1989). Biotechnology in breeding of flax (*Linum usitatissimum*). The present status and future prospects. In *Flax: Breeding and Utilization*, ed. G. Marshall. Dordrecht, The Netherlands: Kluwer Academic Publishers, pp. 5-13.

Futehally, S. and P.F. Knowles. (1981). Inheritance of very high levels of linoleic acid in an introduction of safflower (*Carthamus tinctorius* L.) from Portugal. In *Proceedings of the First International Safflower Conference*, ed. P.F. Knowles. Davis, CA: University of California, pp. 56-61.

Garcés, R. and M. Mancha. (1989). Oleate desaturation in seeds of two genotypes of sunflower. *Phytochemistry* 28:2593-2595.

Getinet, A., G. Rakow, J.P. Raney, and R.K. Downey. (1994). Development of zero erucic Ethiopian mustard through an interspecific cross with zero erucic acid Oriental mustard. *Canadian Journal of Plant Science* 74:793-795.

Glimelius, K. (1999). Somatic hybridization. In *Biology of Brassica Coenospecies*, ed. C. Gómez-Campo. Elsevier, pp. 107-148.

Goffman, F.D., L. Velasco and H.C. Becker. (1998). Tocopherols contents and other seed quality traits in several *Brassica* species. In *Advances in Plant Lipid Research*, eds. J. Sánchez, E. Cerdá-Olmedo and E. Martínez-Force. Sevilla, Spain: Secretariado de Publicaciones Universidad de Sevilla, pp. 436-438.

Goffman, F.D., L. Velasco, and W. Thies. (1999). Quantitative determination of tocopherols in single seeds of rapeseed (*Brassica napus* L.). *Fett/Lipid* 101:142-145.

Graef, G.L., W.R. Fehr, L.A. Miller, E.G. Hammond, and S.R. Cianzio. (1988). Inheritance of fatty acid composition in a soybean mutant with low linolenic acid. *Crop Science* 28:55-58.

Green, A.G. (1986a). A mutant genotype of flax (*Linum usitatissimum* L.) containing very low levels of linolenic acid in its seed oil. *Canadian Journal of Plant Science* 66:499-503.

Green, A.G. (1986b). Genetic control of polyunsaturated fatty acid biosynthesis in flax (*Linum usitatissimum* L.) seed oil. *Theoretical and Applied Genetics* 72:654-661.

Green, A.G. and D.R. Marshall. (1981). Variation for oil quantity and quality in linseed (*Linum usitatissimum*). *Australian Journal of Agricultural Research* 32:599-607.

Green, A.G. and D.R. Marshall. (1984). Isolation of induced mutants in linseed (*Linum usitatissimum*) having reduced linolenic acid content. *Euphytica* 33:321-328.

Grüneberg, W.J., F.D. Goffman, and L. Velasco. (1999). Characterization of yam bean (*Pachyrhyzus* spp.) seeds as potential sources of high palmitic acid oil. *Journal of the American Oil Chemists' Society* 76:1309-1312.

Harwood, J.L. (1996). Recent advances in the biosynthesis of plant fatty acids. *Biochimica et Biophysica Acta* 1301:7-56.

Heath, D.W. and E.D. Earle. (1997). Synthesis of low linolenic acid rapeseed (*Brassica napus* L.) through protoplast fusion. *Euphytica* 93:339-344

Heinonen, M., K. Haila, A.M. Lampi, and V. Piironen. (1997). Inhibition of oxidation in 10% oil-in-water emulsions by β-carotene with α- and γ-tocopherols. *Journal of the American Oil Chemists' Society* 74:1047-1052.

Henry, L.K., G.L. Catignani, and S.J. Schwartz. (1998). The influence of carotenoids and tocopherols on the stability of safflower seed oil during heat-catalyzed oxidation. *Journal of the American Oil Chemists' Society* 75:1399-1402.

Heppard, E.P., A.J. Kinney, K.L. Stecca, and G.H. Miao. (1996). Developmental and growth temperature regulation of two different microsomal 2-6 desaturase genes in soybean. *Plant Physiology* 110:311-319.

Hirsinger, F. (1985). Agronomic potential and seed composition of *Cuphea*, an annual crop for lauric and capric seed oils. *Journal of the American Oil Chemists' Society* 62:76-80.

Hof, L. (1996). *Dimorphoteca pluvialis*: a new source of hydroxy fatty acids. In *Progress in New Crops*, ed. J. Janick. Alexandria, VA, USA: ASHS Press, pp. 372-377.

Horrobin, D.F. (1992). Nutritional and medical importance of gamma-linolenic acid. *Progress in Lipid Research* 31:163-194.

Hu, J., G. Li, D. Struss, and C.F. Quiros. (1999). SCAR and RAPD markers associated with 18-carbon fatty acids in rapeseed, *Brassica napus*. *Plant Breeding* 118:145-150.

Huang, B., E.B. Swanson, C.L. Baszczynski, W.D. Macrae, E. Bardour, V. Armavil, L. Wohe, M. Arnoldo, S. Rozakis, M. Westecott, R.F. Keats, and R. Kimble. (1991). Application of microspore culture to canola improvement. In *Proceedings of the 8th International Rapeseed Congress*. Saskatoon, Canada: GCIRC, pp. 298-302.

Hudson, B.J.F. (1984). Evening primrose (*Oenothera* spp.) oil and seed. *Journal of the American Oil Chemists' Society* 61:540-543.

Ivanov P., D. Petakov, V. Nikolova and E. Pentchev. (1988). Sunflower breeding for high palmitic acid content in the oil. In *Proceedings of the 12th International Sun-*

flower Conference. Novi Sad, Yugoslavia: International Sunflower Association, pp. 463-465.

Janick, J., M.G. Blase, D.L. Johnson, G.D. Joliff, and R.I. Myers. (1996). Diversifying U.S. crop production. In *Progress in New Crops*, ed. J. Janick. Alexandria, VA, USA: ASHS Press, pp. 98-109.

Jianguo, Y., J. Yuzhong, L. Xuyun, and Z. Yongkang. (1993). The research on the germplasm resources of safflower with differents contents of fatty acids. In *Proceedings of the Third International Safflower Conference*. Beijing: Chinese Academy of Sciences, pp. 358-365.

Jocic, S., D. Škorić, N. Lecic, and I. Molnar. (2000). Development of inbred lines of sunflower with various oil qualities. In *Proceedings of the 15th International Sunflower Conference*. Toulouse, France: International Sunflower Association, Vol. 1, pp. A43-A48.

Jones, A., H.M. Davies, and T.A. Voelker. (1995). Palmitoyl-ACP thioesterase and the evolutionary origin of plant acyl-ACP thioesterases. *Plant Cell* 7:359-371.

Jönsson, R. (1977). Erucic-acid heredity in rapeseed (*Brassica napus* L. and *Brassica campestris* L.). *Hereditas* 86:159-170.

Jourdren, C., P. Barret, R. Horvais, N. Foisset, R. Delourne, and M. Renard. (1996). Identification of RAPD markers linked to the loci controlling erucic acid level in rapeseed. *Molecular Breeding* 2:61-71.

Kamal-Eldin, A. and L.Å. Appelqvist. (1996). The chemistry and antioxidant properties of tocopherols and tocotrienols. *Lipids* 31:671-701.

Kinney, A.J. (1998). Plants as industrial-chemical factories. New oils from genetically-engineered soybeans. *Fett/Lipid* 100:173-176.

Kinney, A.J. (1999). New and improved oils from genetically-modified oilseed plants. *Lipid Technology* 11:36-39.

Kirk, J.T.O. and R.N. Oram. (1981). Isolation of erucic acid-free lines of *Brassica juncea:* Indian mustard now a potential crop in Australia. *Journal of the Australian Institute of Agricultural Sciences.* 47:51-52.

Knapp, S.J. (1990). New temperate oilseed crops. In *Advances in New Crops*, eds. J. Janick and J.E. Simon. Portland, OR, USA: Timber Press, pp. 203-210.

Knowles, P.F. (1965). Variability in oleic and linoleic acid contents of safflower oil. *Economic Botany* 19:53-62.

Knowles, P.F. (1989). Safflower. In *Oil Crops of the World*, eds. R.K. Downey, G. Röbbelen and A. Ashri. New York: McGraw-Hill, pp. 363-374.

Knowles, P.F. and A.B. Hill. (1964). Inheritance of fatty acid content in the seed oil of a safflower introduction from Iran. *Crop Science* 4:406-409.

Knowles, P.F. and A. Mutwakil. (1963). Inheritance of low iodine value of safflower selections from India. *Economic Botany* 17:139-145.

Knutzon, D.S., T.R. Hayes, A. Wyrick, H. Xiang, H.M. Davies, and T.A. Voelker. (1999). Lysophosphatidic acid acyltransferase from coconut endosperm mediates the insertion of laurate at the *sn*-2 position of triacylglycerols in lauric rapeseed oil and can increase total laurate levels. *Plant Physiology* 120:739-746.

Knutzon, D.S., G.A. Thompson, S.E. Radke, W.B. Johnson, V.C. Knauf, and J.C. Kridl. (1992). Modification of *Brassica* seed oil by antisense expression of a stearoyl-acyl carrier protein desaturase gene. *Proceedings of the National Academy of Science of USA* 89:2624-2628.

Koester, S. and C. Paul. (1989). Application of near infrared reflectance spectroscopy (NIRS) in breeding rapeseed for improved quality. In *Book of Poster Abstracts of the XII Eucarpia Congress*. Berlin: Paul Parey Publishers, vol II, Poster 31-4.

Lacombe, S., H. Guillot, F. Kaan, C. Millet, and A. Bervillé. (2000). Genetic and molecular characterization of the high oleic content of sunflower oil in Pervenets. In *Proceedings of the 15th International Sunflower Conference*. Toulouse, France: International Sunflower Association, Vol. 1, pp. A13-A18.

Ladd, S.L. and P.F. Knowles. (1971). Interactions of alleles at two loci regulating fatty acid composition of the seed oil of safflower (*Carthamus tinctorius* L.). *Crop Science* 11:681-684.

Lambelet, P., F. Saucy, and J. Löliger. (1984). Radical exchange reactions between vitamin E, vitamin C and phosphatides in autoxidizing polyunsaturated lipids. *Free Radicals Research* 20:1-10.

Lassner, M.W., C.K. Levering, H.M. Davies, and D.S. Knutzon. (1995). Lysophosphatidic acid acyltransferase from meadowfoam meiates insertion of erucic acid at the *sn*-2 position of triacylglycerol in transgenic rapeseed oil. *Plant Physiology* 109:1389-1394.

Lühs, W. and W. Friedt. (1994). Noon-food uses of vegetable oils and fatty acids. In *Designer Oil Crops*, ed. D.J. Murphy. Weinheim, Germany: VCH Verlag gmbH, pp. 73-130.

Lühs, W. and W. Friedt. (1995). Breeding of high-erucic acid rapeseed by means of *Brassica napus* resynthesis. In *Proceedings of the 9th International Rapeseed Conference*. Cambridge, UK: GCIRC, pp. 449-451.

McGregor, D.I. (1974). A rapid and sensitive spot test for linolenic acid levels in rapeseed. *Canadian Journal of Plant Science* 54:211-213.

McKillican, M.E. (1966). Lipid changes in maturing oil-bearing plants. IV. Changes in lipid classes in rape and crambe oils. *Journal of the American Oil Chemists' Society* 43:461-464.

Mensink, R.P. and M.B. Katan. (1989). Effect of a diet enriched with monounsaturated or polyunsaturated fatty acids on levels of low-density and high-density lipoprotein cholesterol in healthy women and men. *New England Journal of Medicine* 321: 436-441.

Mensink R.P., E.H.M. Temme, and G. Hornstra. (1994). Dietary saturated and trans fatty acids and lipoprotein metabolism. *Annals of Medicine* 56:461-464.

Möllers, C., W. Lühs, E. Schaffert, and W. Thies. (1997). High-temperature gas chromatography for the detection of triercoylglycerol in the seed oil of transgenic rapeseed (*Brassica napus* L.). *Fett/Lipid* 99:352-356.

Möllers, C., B. Rücker, D. Stelling, and A. Schierholt. (2000). *In vitro* selection for oleic and linoleic acid content in segregating populations of microspore derived embryos of *Brassica napus. Euphytica* 112:195-201.

Muggli, R. (1994). Vitamin E-Bedarf bei Zufuhr von Polyenfettsäuren. *Fat Science and Technology* 96:17-19.

Mündel, H.H. and J.P. Braun. (1999). Registration of two early-maturing safflower germplasm lines with high oleic acid and high oil content. *Crop Science* 39:299.

Naczk, M., R. Amarowicz, A. Sullivan, and F. Shahidi. (1998). Current research developments on polyphenolics of rapeseed/canola: A review. *Food Chemistry* 62: 489-502.

Nickell, A.D., J.R. Wilcox, and J.F. Cavins. (1991). Genetic relationships between loci controlling palmitic and linolenic acids in soybean. *Crop Science* 31:1169-1171.

Norris, K.H. and J.R. Hart. (1965). Direct spectrophotometric determination of moisture content in grain and seeds. In *Principles and Methods of Measuring Liquids in Solids*, ed. A. Wexler. New York: Reinhold, pp. 19-25.

Ntiamoah, C., G.G. Rowland, and D.C. Taylor. (1995). Inheritance of elevated palmitic acid in flax and its relationship to the low linolenic acid. *Crop Science* 35:148-152.

Olejnik D., M. Gogolewski, and M. Nogala-Kalucka. (1997). Isolation and some properties of plastochromanol-8. *Nahrung* 41:101-104.

Ong, A.S.H. and Y.M. Choo. (1996). Carotenoids and tocols from palm oil. In *Natural Antioxidants. Chemistry, Health Effects, and Applications*, ed. F. Shahidi. Champaign, IL, USA: AOCS Press, pp. 133-149.

Oram, R.N. and J.T.O. Kirk. (1993). Induction of mutations for higher seed quality in Indian mustard. In *Proceedings of the 19th Australian Plant Breeding Conference*. Ed B.C. Imrie and J.B. Hacher. Canberra, Australia, pp. 187-191.

Osorio J., J. Fernández-Martínez, M. Mancha, and R. Garcés. (1995). Mutant sunflowers with high concentration of saturated fatty acids in the oil. *Crop Science* 35:739-742.

Padley, F.B., F.D. Gunstone, and J.L. Harwood. (1994). Occurrence and characteristics of oils and fats. In *The Lipid Handbook*, eds. F.D. Gunstone, J.L. Harwood and F.B. Padley. London: Chapman & Hall, pp. 47-223.

Pascual-Villalobos, M.J., G. Röbbelen, E. Correal, and S.E. Ehbrecht-von-Witzke. (1992). Performance test of *Euphorbia lagascae* Spreng., and oilseed species rich in vernolic acid, in southeast Spain. *Industrial Crops and Products* 1:185-190.

Pazdernik, D.L., A.S. Killam, and J.H. Orf. (1997). Analysis of amino acid composition in soybean seed using near infrared reflectance spectroscopy. *Agronomy Journal* 89:679-685.

Pérez-Vich, B., L. Velasco, and J.M. Fernández-Martinez. (1998). Determination of seed oil content and fatty acid composition in sunflower through the analysis of intact seeds, husked seeds, meal and oil by near-infrared reflectance spectroscopy. *Journal of the American Oil Chemists' Society* 75:547-555.

Pérez-Vich, B., J. Fernández, R. Garcés, and J.M. Fernández-Martínez. (1999). Inheritance of high palmitic acid content in the seed oil of sunflower mutant CAS-5. *Theoretical and Applied Genetics* 98:496-501.

Pérez-Vich, B., R. Garcés, and J.M. Fernández-Martínez. (1999). Genetic control of high stearic acid content in the seed oil of sunflower mutant CAS-3. *Theoretical and Applied Genetics* 99:663-669.

Pérez-Vich, B., R. Garcés, and J.M. Fernández-Martínez. (2000a). Epistatic interaction among loci controlling the palmitic and the stearic acid levels in the seed oil of sunflower. *Theoretical and Applied Genetics* 100:105-111.

Pérez-Vich, B., R. Garcés, and J.M. Fernández-Martínez. (2000b). Genetic relationships between loci controlling the high stearic and the high oleic acid traits in sunflower. *Crop Science* (in press).

Pérez-Vich, B., J.M. Fernández-Martínez, J. Muñoz-Ruz, S.J. Knapp, and S.T. Berry. (2000a). Progress in the development of DNA-based markers for high stearic acid content in sunflower. In *Proceedings of the 15th International Sunflower Conference.* Toulouse, France: International Sunflower Association, Vol. 1, pp. A49-A54.

Pérez-Vich, B., J.M. Fernández-Martínez, S.J. Knapp, and S.T. Berry. (2000b). Molecular markers associated with sunflower oleic and stearic acid concentrations in a high stearic acid × high oleic cross. In *Proceedings of the 15th International Sunflower Conference.* Toulouse, France: International Sunflower Association, Vol. 1, pp. A55-A60.

Polsoni, L., L.S. Kott, and W.D. Beversdorf. (1988). Large-scale microspore culture technique for mutation-selection studies in *Brassica napus. Canadian Journal of Botany* 66:1681-1685.

Pongracz, G., H. Weiser, and D. Matzinger. (1995). Tocopherole. Antioxidanten der Natur. *Fat Science and Technology* 97:90-104.

Pua, E.C., A. Mehra-Palta, F. Nagy, and N.H. Chua. (1987). Transgenic plants of *Brassica napus* L. *Bio/Technology* 5:815-817.

Rahman, S.M., Y. Takagi, and T. Kinoshita. (1996). Genetic control of high oleic acid content in the seed oil of two soybean mutants. *Crop Science* 36:1125-1128.

Rahman, S.M., T. Kinoshita, T. Anai, S. Arima, and Y. Takagi. (1998). Genetic relationships of soybean mutants for different linolenic acid contents. *Crop Science* 38: 702-706.

Rahman, S.M., Y. Takagi, and T. Kinoshita. (1997). Genetic control of high stearic acid content in seed oil of two soybean mutants. *Theoretical and Applied Genetics* 95:772-776.

Rajcan, I., K.J. Kasha, L.S. Kott, and W.D. Beversdorf. (1999). Detection of molecular markers associated with linolenic and erucic acid levels in spring rapeseed (*Brassica napus* L.). *Euphytica* 105:173-181.

Raney, P., G. Rakow, and T. Olson. (1995a). Development of low erucic, low glucosinolate *Sinapis alba.* In *Proceedings of the 9th International Rapeseed Conference.* Cambridge, UK: GCIRC, pp. 416-418.

Raney, P., G. Rakow, and T. Olson. (1995b). Development of zero erucic, low linolenic *Brassica juncea* utilising interspecific crossing. In *Proceedings of the 9th International Rapeseed Conference.* Cambridge, UK: GCIRC, pp. 413-415.

Renaud, S.C., J.C. Ruf, and D. Petithory. (1995). The positional distribution of fatty acids in palm oil and lard influences their biological effect in rats. *Journal of Nutrition 125*:229-237.

Rennie, B.D. and J.W. Tanner. (1989). Fatty acid composition of oil from soybean seeds grown at extreme temperatures. *Journal of the American Oil Chemists' Society* 66:1622-1624.

Rennie, B.D. and J.W. Tanner. (1991). New allele at the *Fan* locus in the soybean line A5. *Crop Science* 31:297-301.

Ross, A.J., W.R. Fehr, G.A. Welke, and S.R. Cianzio. (2000). Agronomic and seed traits of 1%-linolenate soybean genotypes. *Crop Science* 40:383-386.

Rowland, G.G. (1991). An EMS-induced low-linolenic-acid mutant in McGregor flax (*Linum usitatissimum* L.). *Canadian Journal of Plant Science* 71:393-396.

Rowland, G.G. and R.S. Bhatty. (1990). Ethyl methanesulphonate induced fatty acid mutations in flax. *Journal of the American Oil Chemists' Society* 67:213-214.

Rowland, G.G., A. McHughen, L.V. Gusta, R.S. Bhatty, S.L. MacKenzie, and D.C. Taylor. (1995). The application of chemical mutagenesis and biotechnology to the modification of linseed (*Linum usitatissimum* L.). *Euphytica* 85:317-321.

Rücker, B. and G. Röbbelen. (1996). Impact of low linolenic acid content on seed yield of winter oilseed rape (*Brassica napus* L.). *Plant Breeding* 115:226-230.

Rücker B. and G. Röbbelen. (1997). Mutants of *Brassica napus* with altered seed lipid fatty acid composition. In *Proceedings of the 12th International Symposium on Plant Lipids*. Dordrecht, The Netherlands: Kluwer Academic Publishers, pp. 316-318.

Sato, T., Y. Takahata, T. Noda, T. Yanagisawa, T. Morishita, and S. Sakai. (1995). Nondestructive determination of fatty acid composition of husked sunflower (*Helianthus annuus* L.) seeds by near-infrared spectroscopy. *Journal of the American Oil Chemists' Society* 72:1177-1183.

Sato, T., I. Uezono, T. Morishita, and T. Tetsuka. (1998). Nondestructive estimation of fatty acid composition in seeds of *Brassica napus* L. by near-infrared spectroscopy. *Journal of the American Oil Chemists' Society* 75:1877-1881.

Scarth, R., P.B.E. Mcvetty, S.R. Rimmer, and B.R. Stefansson. (1988). Stellar low linolenic-high linoleic acid summer rape. *Canadian Journal of Plant Science* 68:509-511.

Schierholt, A. and W. Ecke. (1998). Entwicklung von Selektionsmarken für eine Hoch-Ölsäure-Mutation bei Raps. *Vorträge für Pflanzenzüchtung* 42:110-112.

Schnebly, S.R., W.R. Fehr, G.A. Welke, E.G. Hammond, and D.N. Duvick. (1994). Inheritance of reduced and elevated palmitate in mutant lines of soybean. *Crop Science* 34:829-833.

Schröder-Pontoppidan, M., M. Skarzhinskaya, C. Dixelius, S. Stymne, and K. Glimelius. (1999). Very long chain and hydroxylated fatty acids in offspring of somatic hybrids between *Brassica napus* and *Lesquerella fendleri*. *Theoretical and Applied Genetics* 99:108-114.

Schultz, G., J. Soll, E. Fiedler, and D. Schulze-Siebert. (1985). Synthesis of prenylquinones in chloroplasts. *Physiologia Plantarum* 64:123-129.

Seiler, G.J. (1996). Search for low saturated fatty acids in wild sunflowers. In *Proceedings of the 18th Sunflower Research Workshop*. Fargo, ND: National Sunflower Association, pp. 1-3.

Shahidi, F. (1996). Natural antioxidants: an overview. In *Natural Antioxidants. Chemistry, Health Effects, and Applications*, ed. F. Shahidi. Champaign, IL, USA: AOCS Press, pp. 1-11.

Shintani, D. and D. Dellapenna. (1998). Elevating the vitamin-E content of plants through metabolic engineering. *Science* 282:2098-2100.

Shukla, V.K.S., P.K.J.P.D. Wanasundara, and F. Shahidi. (1996). Natural antioxidants from oilseeds. In *Natural Antioxidants. Chemistry, Health Effects, and Applications*, ed. F. Shahidi. Champaign, IL, USA: AOCS Press, pp. 97-132.

Soldatov, K.I. (1976). Chemical mutagenesis in sunflower breeding. In *Proceedings of the 7th International Sunflower Conference*. Krasnodar, USSR: International Sunflower Association, pp. 352-357.

Somers, D.J., K.R.D. Friesen, and G. Rakow. (1998). Identification of molecular markers associated with linoleic acid desaturation in *Brassica napus*. *Theoretical and Applied Genetics* 96:897-903.

Stefansson, B.R., F.W. Hougen, and R.K. Downey. (1961). Note on the isolation of rape plants with seed oil free from erucic acid. *Canadian Journal of Plant Science* 41:218-219.

Stymne, S. and A.K. Stobart. (1987). Triacylglycerol biosynthesis. In *The Biochemistry of Plants*, *vol. 9*, ed. P.K. Stumpf. Orlando, USA: Academic Press, pp. 175-214.

Tanhuanpää, P.K. and J. Vilkki. (1999). Marker-assisted selection for oleic acid content in spring turnip rape. *Plant Breeding* 118:568-570.

Tanhuanpää, P.K., J.P. Vilkki, and H.J. Vilkki. (1995). Identification of a RAPD marker for palmitic-acid concentration in the seed oil of spring turnip rape (*Brassica rapa* ssp. *oleifera*). *Theoretical and Applied Genetics* 91:477-480.

Tatum, V. and C.K. Chow. (1992). Effects of processing and storage on fatty acids in edible oils. In *Fatty Acids in Foods and Their Health Implications*, ed. C.K. Chow. New York: Marcel Dekker, Inc., pp. 337-351.

Taylor, D.C., S.L. MacKenzie, A.R. McCurdy, P.B.E. McVetty, E.M. Giblin, E.W. Pass, S.J. Stone, R. Scarth, S.R. Rimmer, and M.D. Pickard. (1994). Stereospecific analyses of seed triacylglycerols from high-erucic Brassicaceae: detection of erucic acid at the *sn*-2 position in *Brassica oleracea* L. genotypes. *Journal of the American Oil Chemists' Society* 71:163-167.

Terao, J., R. Yamauchi, H. Murkami, and S. Matsushita. (1980). Inhibitory effects of tocopherols and β-carotene on singlet oxygen-initiated photoxidation of methyl linoleate and soybean oil. *Journal of Food Process and Preservation* 4:79-93.

Thompson, A.E., D.A. Dierig, E.R. Johnson, G.H. Dahlquist, and R. Kleiman. (1994). Germplasm development of *Vernonia galamensis* as a new industrial oilseed crop. *Industrial Crops and Products* 3:185-200.

Turner, J. and D. Facciotti. (1990). High oleic acid *Brassica napus* from mutagenized microspores. In *Proceedings of the 6th Crucifer Genetics Workshop*. Geneva, NY, USA, p. 24.

Uppström, B. (1995). Seed chemistry. In *Brassica Oilseeds. Production and Utilization*, eds. D.S. Kimber and D.I. McGregor. Wallingford, UK: CAB International, pp. 217-242.

Velasco, L. (1999). Analysis of rapeseed quality traits by near-infrared reflectance spectroscopy (NIRS). *Lipid Technology* 11:90-93.

Velasco, L. and H.C. Becker. (1998). Estimating the fatty acid composition of the oil in intact-seed rapeseed by near-infrared reflectance spectroscopy. *Euphytica* 101: 221-230.

Velasco, L. and J.M. Fernández-Martínez. (1999). Screening for low saturated fatty acids in safflower. *Sesame and Safflower Newsletter* 14:92-96.

Velasco, L. and F.D. Goffman. (2000). Tocopherol, plastochromanol and fatty acid patterns in the genus *Linum*. *Plant Systematics and Evolution* 221:77-88.

Velasco, L., F.D. Goffman, and H.C. Becker. (1998). Variability for the fatty acid composition of the seed oil in a germplasm collection of the genus *Brassica*. *Genetic Resources and Crop Evolution* 45:371-382.

Velasco, L., J.M. Fernández-Martínez, and A. De Haro. (1995). Isolation of induced mutants in Ethiopian mustard (*Brassica carinata* Braun) with low levels of erucic acid. *Plant Breeding* 114: 454-456.

Velasco, L., J.M. Fernández-Martínez, and A. De Haro. (1997). Determination of the fatty acid composition of the oil in intact-seed mustard by near-infrared reflectance spectroscopy. *Journal of the American Oil Chemists' Society* 74:1595-1602.

Velasco, L., J.M. Fernández-Martínez, and A. De Haro. (1998). Increasing erucic acid content in Ethiopian mustard through mutation breeding. *Plant Breeding* 117:85-87.

Velasco, L., C. Möllers, and H. C. Becker. (1999). Estimation of seed weight, oil content and fatty acid composition in intact single seeds of rapeseed (*Brassica napus* L.) by near-infrared reflectance spectroscopy. *Euphytica* 106:79-85.

Velasco, L., B. Pérez-Vich, and J.M. Fernández-Martínez. (1999a). The role of mutagenesis in the modification of the fatty acid profile of oilseed crops. *Journal of Applied Genetics* 40:185-209.

Velasco, L., B. Pérez-Vich, and J.M. Fernández-Martínez. (1999b). Nondestructive screening for oleic and linoleic acid in single sunflower achenes by near-infrared reflectance spectroscopy (NIRS). *Crop Science* 39:219-222.

Velasco, L., B. Pérez-Vich, and J.M. Fernández-Martínez. (2000). Inheritance of oleic acid content under controlled environment. In *Proceedings of the 15th International Sunflower Conference*. Toulouse, France: International Sunflower Association, Vol. 1, pp. A31-A36.

Vick, B.A. and J.F. Miller. (1996). Utilization of mutagens for fatty acid alteration in sunflower. In *Proceedings of the 18th Sunflower Research Workshop*. Fargo, ND, USA: National Sunflower Association, pp. 11-17.

Voelker, T.A., T.R. Hayes, A.C. Cranmer, J.C. Turner, and H.M. Davies. (1996). Genetic engineering of a quantitative trait: metabolic and genetic parameters influencing the accumulation of laurate in rapeseed. *Plant Journal* 9:229-241.

Wardlaw, G.M. and J.T. Snook. (1990). Effects of diets high in butter, corn oil, or high-oleic acid sunflower oil on serum lipids and apolipoproteins in men. *American Journal of Clinical Nutrition.* 51:815-821.

Warner, K. and E.N. Frankel. (1987). Effects of β-carotene on light stability of soybean oil. *Journal of the American Oil Chemists' Society* 64:213-218.

Willett, W.C. and A. Ascherio. (1994). Trans fatty acids: are the effects only marginal? *American Journal of Public Health.* 84:722-724.

Wong, R.S.C. and E. Swanson. (1991). Genetic modification of canola oil: high oleic acid canola. In *Fat and Cholesterol Reduced Foods: Technologies and Strategies*, eds. C. Haberstrohn and C.F. Morris. The Woodlands, TX, USA: Portfolio Publ. Co., pp. 153-164.

Yermanos, D.M. (1966). Variability in seed oil composition of 43 *Linum* species. *Journal of the American Oil Chemists' Society* 43:546-549.

Yodice, R. (1990). Nutritional and stability characteristics of high oleic sunflower seed oil. *Fat Science and Technology* 92:121-126.

Breeding for Improved Oil Quality
in *Brassica* Oilseed Species

Peter B. E. McVetty
Rachael Scarth

SUMMARY. Oil quality in vegetable oils is determined by both nutritional and functional aspects, which are, in turn, primarily determined by the fatty acid profile (i.e., fatty acid composition) of the oil. The naturally occurring fatty acid composition of *Brassica* oils has been extensively modified using conventional plant breeding and biotechnology-based techniques to create unique and improved quality vegetable oils. New *Brassica* cultivars displaying a wide range of edible oil qualities have been developed, and commercialized in recent years. Improved *Brassica* cultivars which produce an industrial oil have also been developed and commercialized in the last two decades. World vegetable oil markets are highly competitive, so the steady improvement in oil quality of the *Brassica* oilseeds is essential to maintain or increase market share, and/or to create new niche markets. The main challenges facing *Brassica* oilseed breeders are (1) to determine the desirable fatty acid profiles of the oil for each end-use market application to create improved oils and (2) to

Peter B. E. McVetty is Professor, and Rachael Scarth is Professor and Associate Head, Department of Plant Science, University of Manitoba, Winnipeg, Manitoba, Canada, R3T 2N2.

The authors wish to thank the following individuals for their contributions to this review: Allison Ferrie, Jo Bowman, Jack Brown, Dave Charne, William Hitz, Phillippe Guerche, Laima Kott, Zenon Lisieczko, William Loh, Denis Murphy, Morten Poulsen, Gerhard Rakow, Habibur Rahman, Michael Renard, Antje Schierholt, and Keith White.

[Haworth co-indexing entry note]: "Breeding for Improved Oil Quality in *Brassica* Oilseed Species." McVetty, Peter B. E., and Rachael Scarth. Co-published simultaneously in *Journal of Crop Production* (Food Products Press, an imprint of The Haworth Press, Inc.) Vol. 5, No. 1/2 (#9/10), 2002, pp. 345-369; and: *Quality Improvement in Field Crops* (ed: A. S. Basra, and L. S. Randhawa) Food Products Press, an imprint of The Haworth Press, Inc., 2002, pp. 345-369. Single or multiple copies of this article are available for a fee from The Haworth Document Delivery Service [1-800-HAWORTH, 9:00 a.m. - 5:00 p.m. (EST). E-mail address: getinfo@haworthpressinc.com].

quickly develop improved *Brassica* cultivars which can produce the new fatty acid profile oils at competitive prices. *[Article copies available for a fee from The Haworth Document Delivery Service: 1-800-HAWORTH. E-mail address: <getinfo@haworthpressinc.com> Website: <http://www.HaworthPress. com> © 2002 by The Haworth Press, Inc. All rights reserved.]*

KEYWORDS. Fatty acid composition, end-use market suitability

INTRODUCTION

Both the nutritional and functional components of oil quality in all oilseeds are determined primarily by the fatty acid profile (i.e., fatty acid composition) of the oil. Naturally occurring fatty acids in vegetable oils range from C12 to C24 in chain length and from saturated to poly-unsaturated in degree of saturation. The major vegetable oils have distinct oil profiles in their native states (Table 1).

The objective of modifying vegetable oil quality is to develop oils with enhanced nutritional and functional properties and which will require little, if any, processing for specific end-use markets. The processes required to enhance the stability of the oil or enhance the quality of the oil are achieved using plant breeding, as an alternative to post-harvest refining or processing. The modifications can be made using conventional breeding methods, or the new biotechnologies including cell culture and gene transfer, or a combination of both techniques. The development of the new technologies for trait modification has opened the possibility of engineering a wide range of new fatty acid profiles in *Brassica*.

Nutritional and functional considerations influence the desirable levels of the monounsaturated and polyunsaturated fatty acids. New edible oils with modified levels of the monounsaturated (primarily oleic acid) and polyunsaturated fatty acids (primarily linolenic acid) are being developed by *Brassica* breeding institutions worldwide. *Brassica* oils also are being developed with modified levels of naturally occurring saturated fatty acids (primarily palmitic acid), and novel saturated fatty acids (primarily lauric acid and myristic acid).

Other modified *Brassica* oilseeds are competing for a share in the industrial vegetable oil market. Current estimates place industrial oleochemicals at less than 10% of total vegetable oil production (Murphy, 2000). However, depletion of petroleum stocks will stimulate interest in alternative renewable energy sources and renewable industrial feed

TABLE 1. Fatty acid composition of selected naturally occurring vegetable oils

Species	Crop	Ref. #	C12:0	C14:0	C16:0	C16:1	C18:0	C18:1	C18:2	C18:3	C20:0	C20:1	C22:0	C22:1	C24:0	C24:1
									Fatty acid[+] composition in percent							
B. napus	Rapeseed	11	0.0	0.0	3.0	0.3	0.8	0.0	13.5	9.8	0.6	6.8	0.7	53.6	0.0	1.0
B. rapa	Rapeseed	11	0.0	0.0	1.8	0.2	0.9	0.0	12.0	8.2	0.9	6.2	0.0	55.5	0.0	1.2
B. juncea	Mustard	11	0.0	0.0	2.5	0.3	1.2	0.0	17.7	11.4	1.2	6.4	1.2	46.6	0.7	1.9
G. max	Soybean	11	0.0	0.0	15.3	0.0	4.2	0.0	48.2	8.7	0.0	0.0	0.0	0.0	0.0	0.0
H. annuus	Sunflower	11	0.0	0.1	5.8	0.1	5.2	0.0	71.5	0.2	0.2	0.1	0.7	0.0	0.1	0.0
C. tinctorius	Safflower	11	0.0	0.0	7.6	0.0	2.0	0.0	79.6	0.0	0.0	0.0	0.0	0.0	0.0	0.0
Z. mays	Corn	11	0.0	0.0	11.5	0.0	2.2	0.0	58.7	0.8	0.2	0.0	0.0	0.0	0.0	0.0
A. hypogea	Peanut	11	0.0	9.2	0.0	0.0	3.1	0.0	23.4	0.0	1.4	1.4	2.6	0.0	1.8	0.0
E. guineensis	Palm	74	0.2	1.5	44.5	0.3	5.0	0.2	10.5	0.4	0.4	0.0	0.0	0.0	0.0	0.0
O. europaea	Olive	66	0.0	0.1	12.0	1.0	4.5	71.1	8.5	1.0	0.5	0.4	0.9	0.0	0.0	0.0

11: Downey, R. K. (1983)
74: Wuidart, W. (1996)
66: Uzzan, A. (1996)

[+] C12:0 (lauric), C14:0 (myristic), C16:0 (palmitic), C18:0 (stearic), C18:1 (oleic), C18:2 (linoleic), C18:3 (linolenic), C20:0 (arachidic), C20:1 (eicosenoic), C22:0 (behenic), C22:1 (erucic), C24:0 (lignoceric) stock sources provided by vegetable oils. Economic production of industrial vegetable oil depends on targeting specific applications with a high value product or specific markets in which the vegetable oil source is preferable for environmental or quality reasons.

The main challenges facing *Brassica* breeders are (1) to determine the desirable fatty acid profiles of the oil for each end-use market application and (2) to quickly develop new *Brassica* cultivars which can produce the new fatty acid profile oils at competitive prices.

This review will describe the fatty acid profiles of the naturally occurring and modified *Brassica* oils, the nutritional or functional background driving the development of the modified fatty acid profiles and the current status of germplasm and cultivar development efforts. The choice of categories to be reviewed is as follows: oils with reduced levels of saturated fatty acids (< 7% total saturates as the sum of palmitic acid (C16:0) + stearic acid (C18:0) + behenic acid (C20:0) + ecosenoic acid (C22:0); oils low (< 3.5%) in linolenic acid (C18:3), oils with mid (between 67 to 75%) oleic acid (C18:1) levels, oils with high (over 75%) oleic acid levels, oils with elevated naturally occurring and/or novel saturated fatty acids including lauric acid (C12:0), myristic acid (C14:0), palmitic acid (C16:0), and stearic acid (C18:0); new low erucic acid (C22:1) fatty acid profiles in the *Brassicas* and related species and high (> 50%) erucic acid rapeseed oil. These categories are not mutually exclusive but represent separate germplasm and cultivar development streams, which can then be combined to produce additional fatty acid oil profiles in the *Brassicas*.

The term "canola" is used in this review to describe the fatty acid profile of the current Canadian grown *B. napus* and *B. rapa* cultivars which produce oils with less than 2% erucic acid, as described in the trademarked name held by the Canola Council of Canada (Adolphe, 2000). Other terms used internationally to describe the canola oil fatty acid profile are low erucic acid rape or rapeseed (LEAR) and colza. A typical canola oil profile is: 3.9% palmitic acid (C16:0), 1.1% stearic acid (C18:0), 0.8% arachidic acid (C20:0), and 0.2% eicosenoic acid (C20:0), (the saturated fatty acids), 0.3% palmitoleic acid (C16:1), 59.7% oleic acid (C18:1), 23.3% linoleic acid (C18:2), 8.6% linolenic acid (C18:3), 1.8% behenic acid (C22:0), and 0.3% erucic acid (C22:1) (Table 2). The different types of fatty acids can be grouped together so that the fatty acid profiles of vegetable oils can be represented in graphical form more simply as percent saturated fat (sum of palmitic acid, stearic acid, arachidic acid and behenic acid), percent oleic acid, percent linoleic acid and percent linolenic acid. Figure 1 provides a comparison of se-

TABLE 2. Fatty acid composition of selected modified *Brassica* oils

Species or crop		Ref. #	Fatty acid[+] composition in percent													
			C12:0	C14:0	C16:0	C16:1	C18:0	C18:1	C18:2	C18:3	C20:0	C20:1	C22:0	C22:1	C24:0	C24:1
B. napus	Canola	11	0.0	0.0	3.9	0.3	1.1	59.7	23.3	8.6	0.8	1.8	0.2	0.3	0.0	0.0
B. rapa	Canola	11	0.0	0.0	1.8	0.1	1.2	58.6	24.0	10.3	0.6	1.0	0.1	0.3	0.0	0.0
B. juncea	LEA mustard	25	0.0	0.0	3.9	0.3	1.1	45.0	32.0	14.0	0.8	1.8	1.2	0.3	0.0	0.0
B. napus	Low Linolenic	53	0.0	0.0	4.1	0.1	1.4	59.1	28.9	3.3	0.5	1.4	0.4	0.1	0.2	0.0
B. napus	Mid oleic	69	0.0	0.0	3.9	0.3	1.1	67-75	15-22	<3.0	0.2	0.1	0.2	0.3	0.0	0.0
B. napus	High oleic	72	0.0	0.0	3.9	0.3	1.1	80+	5-12	<5	0.0	0.0	0.2	0.3	0.0	0.0
B. napus	High lauric	68	39.0	4.0	3.0	---	---	33.0	11.0	6.0	0.0	0.0	0.2	---	0.0	0.0
B. napus	HEAR	36	0.0	0.0	2.0	0.0	1.0	12.0	11.0	9.0	1.0	8.0	0.5	55.0+	0.0	0.0

11: Downey R.K. (1983); 25: Kirk and Oram 1981; 53: Scarth et al. 1988; 69: Warner and Mounts 1993; 72: Wong et al. 1991; 68: Voelker et al. 1996; 36: McVetty et al. 1999

lected modified *Brassica* oils in these simplified terms. As can be seen from Figure 1, canola oil is the edible vegetable oil lowest in saturated fatty acids. (Please note that high erucic acid rapeseed oil is an industrial oil.)

The methodology of development of the modified oil profiles in *Brassica* species is presented within each category of modified edible and industrial oils. These methods include development of mutant lines (Poehlman and Sleeper, 1995), crossing and selection within the segregating generations for the desired trait (Poehlman and Sleeper, 1995), microspore culture to produce doubled haploid lines (Lichter, 1982; Polsoni et al., 1988) and transgenic gene transfer or gene manipulation (Gasser and Frawley,1989). Often, several different breeding methods have been used to produce *Brassica* germplasm and cultivars with a similar oil quality. This review will provide a general description of the methodology. More detailed information is available in the original references cited for each development.

MODIFIED EDIBLE BRASSICA OILS

Reduced Saturated Fatty Acid Brassica Oils

Nutritional/Functional Background: Dietary recommendations in a number of countries focus attention on limiting total fat intake to 30% of energy and saturated fat intake to 10% of energy. These recommendations are based on the adverse effects of saturated fat on blood cholesterol and its implications for cardiovascular disease (Eskin et al., 1996). Distinctions have been made between the different saturated fatty acids in their association with elevated levels of blood cholesterol and cardiovascular disease risk. Lauric acid (C12:0), myristic acid (C14:0) and palmitic acid (C16:0) are cholesterol-raising saturated fatty acids. Stearic acid (C18:0) is apparently neutral in its effect on cholesterol (Eskin et al., 1996).

The labeling regulations in the United States and Canada allow oils with less than one gram of saturated fat per 14 grams of total fat (or less than 7.1% saturated fatty acid content in the oil) to be identified as low in saturated fatty acids. Canola oil contains a very low level (> 7.1%) of saturated fatty acids, half the level of corn oil, olive oil, and soybean, and approximately one-quarter of the level in cottonseed oil (Figure 1). Of the saturated fatty acids associated with elevated cholesterol levels (i.e., palmitic acid), canola oil contains only 4%. Aggressive marketing

FIGURE 1. Comparison of Industrial (Rapeseed) and Edible Oils

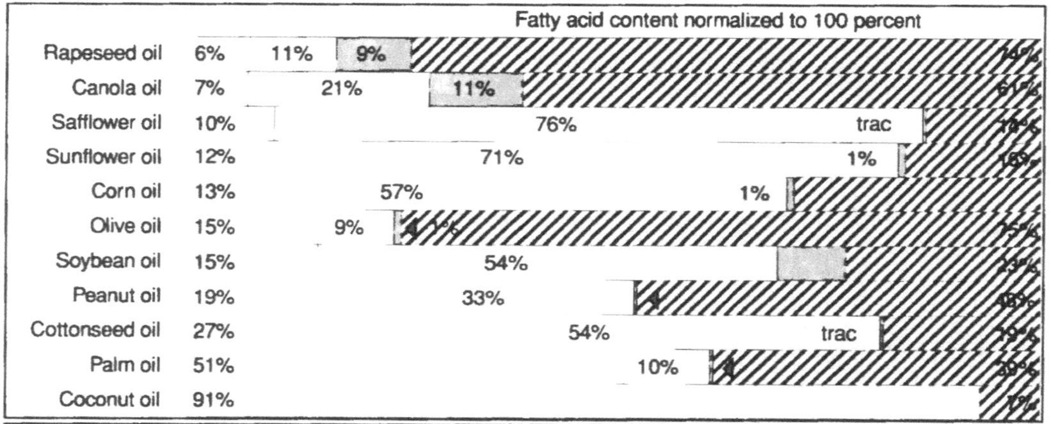

	Fatty acid content normalized to 100 percent
Rapeseed oil	6% 11% 9%
Canola oil	7% 21% 11%
Safflower oil	10% 76% trac
Sunflower oil	12% 71% 1%
Corn oil	13% 57% 1%
Olive oil	15% 9%
Soybean oil	15% 54% 54%
Peanut oil	19% 33%
Cottonseed oil	27% 54% trac
Palm oil	51% 10%
Coconut oil	91%

References: Rapeseed (Adapted from McVetty et al. 1999); POS Pilot Plant Corporation, Saskatoon, Saskatchewan, Canada
June 1994 for Canola Clouncil of Canada

Saturated Fat

Monounsaturated Fat

Polyunsaturated Fat

Alpha-Linolenic Acid
(An Omega-3 Fatty Acid)

351

of the low saturated fatty acid attribute has established canola oil as a premium quality vegetable oil in the North American market. Canola oil has captured 80% of the salad oil, 60% of the shortening and 45% of the margarine market in Canada and a steadily growing share of the United States market, according to the Canola Council of Canada (Adolphe, 2000). Defense of this market share for canola oil requires at a minimum the maintenance of the 7% upper limit on saturated fatty acids.

Canola oil is currently the lowest saturated fat vegetable oil. However, a challenger for this position has emerged in the form of soybean germplasm with reduced saturated fatty acid levels. The low saturated fatty acid trait in soybean was initially developed using chemical mutagenesis, followed by crossing and selection (Wilson, 1993). There are now soybean oils with half the normal level of palmitic acid (C16:0). Commercial production of these new low saturate (< 5% palmitic acid) soybean cultivars has begun in the USA.

An additional contributing factor to the priority of lowering the level of saturated fatty acids in canola is a significant elevation of total saturated fatty acid levels in commodity canola oil to levels of 7% (DeClerq et al., 1997). There has been a major conversion in the western Canadian canola acreage from an approximately equal division between *B. rapa* and *B. napus* species canola cultivars to a predominance of *B. napus* species canola cultivars. The *B. napus* canola cultivars currently recommended in Canada have levels of saturated fat in the seed oil approximately 1.0 to 1.5% higher than the saturated fat levels of *B. rapa* canola cultivars. The environment during seed development also influences saturated fat levels in the seed oil, with higher temperatures resulting in elevated total saturated fat levels (Deng and Scarth, 1998). Since *B. napus* canola cultivars are grown in the warmer growing areas of the Canadian Prairies, the environment effects are more likely to raise the saturated fat levels in this canola species compared to cooler growing season adapted *B. rapa* canola cultivars.

Methods of Modification: Genetic variation for significantly reduced saturated fatty acid levels in *B. napus* is not available in current *B. napus* canola or rapeseed germ plasm. The breeding approaches used to reduce saturated fatty acid levels in this species include the resynthesis of *B. napus* from its parental species *B. rapa* and *B. oleracea* using strains with reduced saturated fatty acid levels, mutagenesis in both *B. rapa* and *B. napus* and transgenic modification of *B. napus* strains.

A near term reduction in saturated fatty acid levels in *B. napus* to levels comparable to *B. rapa* may be available within *B. napus* germplasm with parentage from interspecific crosses between yellow seeded *B.*

napus, yellow seeded *B. rapa* and light brown seeded *B. alboglabra*. Doubled haploid lines screened in field trials over two years showed total saturated fatty acid contents as low as 5.4%, similar to the *B. rapa* canola cultivar check and 1.0 to 1.5% lower than the *B. napus* checks (Raney et al., 1999b).

Doubled haploid *B. rapa* lines with reduced saturated fatty acid levels (< 4%) have been developed from microspore mutagenesis conducted at the Plant Biotechnology Institute in Saskatoon. This low saturate variation is being introduced into *B. napus*, by interspecific crosses of *B. rapa* to *B. napus* and resynthesis of *B. napus* in an attempt to produce low saturate germplasm for further cultivar development at the University of Manitoba.

Current Breeding Status: Breeding institutions in Canada have focused on the reduction of saturated fatty acids in *B. napus*, initially to levels below 7%, with the long-term objective of achieving reductions in palmitic acid (C16:0) and stearic acid (C18:0) to levels below 4%. Systematic reduction in saturated fatty acid levels in *B. napus* candidate cultivars proposed for registration in Canada will be required in future years by the Western Canada Canola/Rapeseed Recommending Committee, Inc. (Gadoua, 2000).

Low Linolenic Fatty Acid Oils

Nutritional/Functional Background: Canola oil's polyunsaturated fatty acid content total of approximately 32% linolenic acid (C18:3) and linoleic acid (C18:2) is intermediate among the vegetable oils, lower than corn oil, soybean oil and sunflower oil, but higher than olive or palm oil (Figure 1). Linolenic acid (C18:3) is recognized as an essential fatty acid and has a role in reducing plasma cholesterol levels (Eskin et al., 1996). The ratio between linolenic acid (C18:3) and linoleic acid (C18:2) in canola oil (1:2) is also regarded as nutritionally favorable (Figure 1). However, in edible oil applications that require stability, vegetable oils high in polyunsaturated fatty acids such as canola are stabilized using partial hydrogenation, with the resulting formation of trans fatty acids. Nutritionists are concerned with the trans isomers of cis fatty acids raising the serum low-density lipoprotein cholesterol levels and reducing the serum high-density lipoprotein cholesterol levels. Elevated levels of low-density lipoprotein cholesterol and reduced levels of high-density lipoprotein cholesterol are associated with enhanced risk of cardio-vascular disease. The current recommendation from nutritionists is that the current levels of trans fatty acid in the diet should not be increased (Fitzpatrick and Scarth, 1998).

Low linolenic (18:3) oils were developed to increase the stability of canola oil, reducing or eliminating the requirement for hydrogenation. Low linolenic (C18:3) canola oil demonstrated improved stability under conditions of accelerated storage with no changes in overall odor intensity or pleasantness. There were also significantly lower levels of free fatty acids during frying with low linolenic canola oil with better flavor quality of the French fry product (Eskin et al., 1996). The reduction in linolenic acid is typically accompanied by an increase in linoleic acid (C18:2) and/or oleic acid (C18:1) (Table 2). Modification of the fatty acid composition in the *Brassicas*, may be accompanied by inadvertent changes in minor components such as total polar compounds, sterols and total tocopherols, with the result that the frying performance and storage stability of low linolenic oils from different *B. napus* canola lines has been variable in some studies (Przybylski et al., 1999; Przybylski and Zambiazi, 1999).

Methods of Modification: Both mutagenesis and transgenic modification are being used in *Brassica* breeding programs to reduce linolenic acid content in the oil to below 3.5%. The first low linolenic acid trait was produced by seed mutagenesis of the *B. napus* cultivar Oro, which led to the isolation of a mutation line, M11 with an altered linoleic acid (C18:2)/linolenic acid (C18:3) ratio (Rakow, 1993). A program of back crossing to the adapted Canadian canola cultivar Regent, combined with selection, led to the release of the world's first low linolenic acid canola cultivar, Stellar, with approximately 3% linolenic acid (C18:3) in the seed oil (Scarth et al. 1988). The low linolenic trait of the cultivar Stellar was found to be relatively stable over environments (Deng and Scarth, 1998). Tanhuanpaa et al. (1995) first identified a RAPD marker sequence associated with linolenic acid content in the F2 progeny of a cross of Topas × R4 (a low linolenic mutation line). Then Somers et al., (1998) used molecular DNA markers to identify three unlinked loci in the low linolenic cultivar Apollo which, when used together in a marker-based selection program, explained 51% of the variation for linolenic acid. Lines selected as carrying all three loci averaged 3.7% linolenic acid. Marker assisted selection for the low linolenic trait could therefore be used to improve the efficiency of new low linolenic cultivar development in *B. napus*.

Genetic studies conducted with Stellar as the low linolenic parent cultivar demonstrated that the L1 locus, which produced variation in linolenic acid content was associated with the fad3 gene encoding microsomal delta15 desaturase (Jourdren et al., 1996) and a second gene has been shown to be linked to a second locus L2 (Barret et al., 1998; Barret et al., 1999). A linkage study using amplified fragment length polymorphism (AFLP) analysis has located the fad3 gene homologues on two linkage groups (Somers et al., 1999).

Current Breeding Status: The following breeding organizations reported having low linolenic acid canola (*B. napus*) cultivars in production: Ag Seeds (Australia), Cargill (France and Canada), Danisco Seeds (Denmark), Limagrain (Canada) and the University of Manitoba (Canada). There is one low linolenic acid variety under Plant Variety Protection from Norddutsche Pflanzenzucht (NPZ, Germany). Plant breeders at the Agricultural and Agri-Food Research Centre in Saskatoon (Canada) have developed *B. napus* lines with a combination of the low linolenic acid trait, yellow seed coat color, and seed meal containing zero aliphatic glucosinolates (Raney et al., 1999a). Researchers at the University of Guelph (Canada) are developing low linolenic cultivars using the doubled haploid line development technique. The low linolenic profile has been developed in zero-erucic acid spring turnip rape *B. rapa* at the University of Helsinki and Boreal Plant Breeding in Finland. Current levels of linolenic acid in this *B. rapa* germplasm are 7% with a four to one ratio of linoleic acid to linolenic acid, achieved through selection (Laakso et al., 1999).

Mid and High Oleic Fatty Acid Oils

Nutritional/Functional Background: Diets rich in olive oil have been associated with a reduction in low-density lipoprotein cholesterol, the form of cholesterol which is responsible for the formation of atherosclerotic plaques. Epidemiological studies suggest that the high life expectancy with low rates of both cardiovascular disease and breast cancer in Mediterranean countries is related to the high consumption of monounsaturated olive oil (Kiritsakis, 1999).

Oils with increased levels of oleic acid in combination with reduced linoleic acid and reduced linolenic acid show a higher oxidative stability, lower oxidation products and improved stability without extensive hydrogenation. Nutritional research suggests that trans isomers of cis fatty acids may have negative nutritional effects. The Food and Drug Administration of the U.S. government will require identification of trans fatty acid content on the label of the vegetable oil products within a few years. The *Brassica* vegetable oil industry is aware of the potential for the labeling of trans fatty acids to cause loss of market share and alternatives to the use of hydrogenation for increased stability are being considered, both in processing methods and in vegetable oil composition.

The rate of oxidation of oils is influenced by the degree of unsaturation of fatty acids, light, temperature, and in the level and type of antioxidants and pro-oxidants present. Secondary oxidation products such as carbonyl compounds, aldehydes, ketones and alcohols produce unpleasant flavors and odors associated with oil rancidity (Eskin et al. 1996). Mid-oleic oils (67 to 75%) are marketed for those edible oil applications which require high cooking and frying temperature stability and for snack foods requiring long shelf-life. Comparative studies of genetically modified oils identified the optimum oil profile for frying performance or fry life of the oil, shelf-life, dietary benefits and sensory properties of end products. The optimum oil profile for these applications is composed of 5 to 7% saturates (C16:0 + C18:0 + C20:0), 67 to 75% oleic acid, 15 to 22% linoleic acid and $\leq$ 3% linolenic acid. The role of linoleic acid in enhanced sensory properties was noted as oleic acid levels over 75% results in a reduction in sensory properties (flavour and taste) and an increase in off-odors (Warner and Mounts, 1993). Research at the Institute of Food Science in Australia compared a pure mid-oleic canola oil to blends of this oil with 5 and 10% sesame oil and with 25 and 50% cottonseed oil blends in deep frying trials. It was found that the pure mid-oleic canola oil was the best in sensory attributes (Papalois et al., 1999).

Methods of Modification: Enhanced oleic acid levels have been produced through mutagenesis, applied both to seed (Auld et al., 1992) and to microspore derived embryos and through transgenic modification (Stoutjesdijk et al., 1999). Seed mutagenesis followed by crossing and selection resulted in *B. napus* genotypes with > 85% oleic and reduced levels of linoleic and linolenic acid (Wong et al., 1991). A patent filed by Cargill Incorporated described *B. napus* lines with an oleic acid content of 72 to 80%, linoleic acid of 5 to 12% and linolenic acid content of 1 to 5% (Kodali et al., 1999).

The University of Göttingen Institute fur Pflanzenbau und Pflanzenzuchtung has a project to establish high oleic (over 90%) quality in winter rapeseed developed from an ethylmethansulphonate mutation program using the winter oilseed rape cultivar 'Wotan'. A single gene controlled the enhanced oleic acid character of the mutation lines developed from the cultivar 'Wotan' and the mutations were allelic. Doubled haploid lines were developed from crosses between the high oleic mutation lines and the cultivar 'Samourai'. A study of the influence of the environment on the high oleic trait (> 64%) showed a high genetic heritability ($h^2 = 0.99$) and a low environment effect (Schierholt and Becker, 1999).

High oleic canola (> 86% oleic, < 7% linoleic, < 2.5% linolenic) has been produced using seed specific inhibition of microsomal oleate desaturase and microsomal linoleate desaturase gene expression, either through co-suppression or antisense technology. Co-suppression has been used in combination with mutation treatments to produce modified fatty acid profiles (Debonte and Hitz, 1996). Co-suppression using the oleate desaturase gene cloned from *B. juncea* (Singh et al., 1995) and the corresponding gene from *B. napus* have been transformed into *B. juncea* and *B. napus* (Stoutjesdijk et al., 1999). Transgenic *B. napus* plants were produced with over 80% oleic acid (control 60 to 65%) and *B. juncea* lines with over 62% oleic acid (control 40 to 45%). The highest expression in single seed analysis of the T1 generation was 89% oleic acid in *B. napus* and 73% in *B. juncea*. The sum of the polyunsaturated fatty acids, linoleic acid and linolenic acid was reduced to 4% (control 26%) in *B. napus* and 16% (control 44%) in *B. juncea*. The saturated fatty acid levels were similar in the transgenic and control lines (Stoutjesdijk et al., 1999).

The high oleic acid profile has been developed in spring turnip rape *B. rapa* from the germplasm source Jo4072 (Tanhuanpaa et al., 1996). Eight randomly amplified polymorphic DNA markers (RAPD) were found to discriminate the extreme high oleic acid genotypes from low oleic bulks of individual F2 plants. Six of the markers mapped to linkage group 6 (LG6), identifying a quantitative trait locus (QTL) affecting oleic acid concentration through the desaturation step from oleic acid to linoleic acid. The RAPD marker with the highest association with the QTL, OPH-17, was converted to a sequence characterized amplified region (SCAR) marker and used to select high oleic acid lines. The mean seed oleic acid content of the F2 individuals carrying the marker was 80% (Tanhuanpaa et al., 1996).

Current Breeding Status: The fatty acid profile of Cargill's Clear Valley 75 is 75% oleic acid, 10% linoleic acid and 5% linolenic acid. The Ag-Seed Research program in Australia has the trademark MONOLA registered for non-transgenic canola lines with a fatty acid profile of 70% oleic acid with reduced (< 3%) linolenic acid and also for non-transgenic canola lines with a fatty acid profile of > 70% oleic acid with normal levels of linolenic acid (White, 1998).

Elevated Saturated Fatty Acid Oils

Nutritional/Functional Background: Oils with high levels of saturated fatty acids offer alternatives to the use of liquid oils modified through hydrogenation in end-use applications such as shortenings and margarines where unmodified liquid oils cannot be used. Modified canola oil with elevated saturated fatty acid levels can replace the addition of animal fats or tropical oils in the production of margarines and shortenings. Support for this breeding objective includes the benefits of supply stability through domestic production with reduced dependence on imports of tropical high saturated fatty acid oils (Del Vecchio, 1996).

Methods of Modification: The lauroyl (12:0)-acyl-carrier protein (ACP) thioesterase was isolated from California Bay tree (*Umbellularia californica*) (Voelker et al., 1992). Transgenic *B. napus* plants were created using the California Bay tree thioesterase gene. These transgenic lines produced up to 55 to 60% lauric acid (C12:0), with levels up to 40% correlated with the number of thioesterase gene copies (Voelker et al., 1996). Above this level, thioesterase activity was not the limiting factor in lauric acid accumulation. Analysis of the stereo-positional distribution of lauric acid in the lipids of the high lauric acid lines showed an almost complete exclusion of lauric acid from the sn-2 position of the triacylglycerol (Voelker et al., 1996). The second position acyl-transferase reaction, lysophosphatidic acid acyl-transferase (LPAAT), has been shown to have strict substrate preference for the C18 fatty acids (Cao and Huang, 1987; Sun et al., 1988). The average acyl weight per seed (equivalent to oil content) showed a slight decline with the production of higher amounts of lauric acid (50 mol%) but less than was anticipated by the shift from a predominance of C18 fatty acids to the medium chain length C12 fatty acids. Analysis of the developing seeds showed an increased expression of the β-oxidation and glycolate cycle enzymes associated with germination in oilseeds to break down the stored fats. In the high lauric acid seeds producing levels of the medium chain fatty acids in excess of the capacity for storage, specific metabolic pathways corresponding to the chain length of the acyl substrate are triggered, presumably to reduce the interference with other enzymes or disruption of membrane integrity (Eccleston and Ohlrogge, 1998). The authors note that it may be necessary to regulate the metabolic induction of specific fatty acid oxidation in order to produce transgenic oils with a single predominant fatty acid component.

The isolated thioesterase gene from *Cuphea lanceolata*, ClFatB4, reported by Martini et al. (1994) was inserted into the low erucic spring oilseed rape cultivar 'Drakkar' (Rudloff and Wehling, 1997) with its native seed-specific promoter. The transgenic lines produced the novel fatty acid, myristic acid (C14:0) at levels of 12 to 14% and also displayed enhanced levels of palmitic acid production at 20%, combined with an accompanying reduction in oleic acid. Correlations between myristic acid (C14:0) and palmitic acid (C16:0) levels were very high (r = 0.8). There was good stability of expression for myristic acid and elevated palmitic acid levels in the field trials for the T3, T4, and T5 generations, with no loss of expression and low genotype by environment interaction. Temperature was found to have a major influence on expression of myristic acid (C14:0) with greenhouse grown lines expressing significantly higher levels of myristic acid than field grown lines. No detrimental phenotypic differences were observed for the transgenic lines compared to the check cultivars when both were grown in the field trials.

Overexpression of the oleic acid preferring acyl-ACP thioesterase gene from soybean (*Glycine max*) increased the palmitic acid and stearic acid content to approximately 20% of the total in transgenic lines of the canola cultivar 'Westar' (Hitz et al., 1995). Transgenic rapeseed plants with the FatA1 thioesterase gene from mangosteen (*Garcinia mangostana*) showed an increase in stearic acid up to 22%, accumulated primarily at the expense of oleic acid with a slight decline in linolenic acid (Hawkins and Kridl, 1998). Transgenic lines with the thioesterase gene C1FatB4 from *Cuphea lanceolata* have been characterized with 16% myristic acid (C14:0) content and 43% myristic acid (C14:0) + palmitic acid (C16:0) (Rudloff et al., 1999). No yield depression or differences in oil processing were noted in the myristic acid producing transgenic lines compared to the non-transgenic parent cultivar.

Current Breeding Status: High lauric acid (C12:0) canola was the world's first transgenic oilseed crop in commercial production. The typical analytical value of Monsanto's Laurical™ oil profile is 39% lauric acid, 4% myristic acid, 3% palmitic acid, 33% oleic acid, 11% linoleic acid, 6% linolenic acid and 5% other fatty acids. The Laurical™ product is suitable for use in the confectionery industry, simulated dairy products, icings and frostings (Del Vecchio, 1996).

A test crush and processing in a pilot scale plant was performed on the seed from two of the best transgenic lines and a mixture of the lines with elevated myristic acid content. Comparisons were made to seed of the low erucic acid spring oilseed rape cultivar 'Drakkar'. Expression of myristic acid in the two transgenic lines was low (1.9 and 4.2%) and higher in the mixture of the lines (9.1%). Tocopherol levels in the seed and oxidative stability at 100°C of the refined oil were enhanced in the transgenic lines in comparison to the oil from Drakkar and the modified oil was solid at 5°C (Rudloff et al., 1999).

Low Erucic Acid Profiles in Other Brassica and Related Species

Nutritional/Functional Background: Several other *Brassica* oilseed species and relatives are being modified to produce the canola oil profile and canola meal specifications. The objective is to produce the commodity canola oil in *Brassica* species with agronomic adaptation to a particular environment. *B. juncea,* for example, has been shown to have improved drought tolerance over *B. napus* (Woods, 1992).

Methods of Modification: The low erucic acid trait was identified in condiment mustard lines (*B. juncea*) from northern China, ZEM-1 and ZEM-2 (Kirk and Oram, 1981). The zero erucic acid character has been introduced into adapted cultivars through crosses (Oram et al., 1999; Agnihotri and Kaushik, 1999). Germplasm sources and variation created by mutation treatment have been used to reduce the linolenic acid levels and enhance the oleic acid levels in adapted (*B. juncea*) mustard lines. Oilseed breeders at the Agriculture and Agri-Food Canada Research Centre at Saskatoon (Canada) have developed lines of *B. juncea* with seed oil containing 60% (control 45%) oleic acid, 20% (control 32%) linoleic acid, and 10% (control 14%) linolenic acid (current levels in low erucic low glucosinolate *B. juncea* are in brackets). The longer term objectives of this breeding program are the reduction in both the levels of total saturated fatty acids and linolenic acid in the oil. The Saskatchewan Wheat Pool (Canada) had several low erucic acid, low glucosinolate *B. juncea* lines in pre-registration tests in 1998 and 1999. Enhanced oleic acid (> 55%) and reduced linoleic acid (< 18%) levels have been established in selected doubled haploid lines. The high oleic acid, low linoleic acid trait was found to be controlled by a single gene (Potts and Males, 1999). Selection was made for high oleic acid (> 55%), low allyl glucosinolate content in the seed meal, good agronomic performance, high seed oil and protein content (Potts et al., 1999). Multi-location trials over two years identified canola-quality mustard lines with acceptable fatty acid profile, low glucosinolate content, good yield and acceptable maturity relative to the current *B. napus* canola cultivar checks in western Canada pre-registration trials. The oil content in these lines was lower than the *B. napus* check cultivars and increasing oil content is therefore a major breeding objective.

Low erucic acid content has been established in *Brassica carinata* at the Agriculture and Agri-Food Research Centre at Saskatoon and in *Sinapis alba* at two breeding organizations, the Agriculture and Agri-Food Canada Research Centre at Saskatoon (Canada) and the University of Idaho, USA. The development of low erucic acid *S. alba* (Raney et al., 1995) has led to the introduction of the low erucic acid trait into adapted *S. alba* germplasm (Drost et al., 1999). Selection for low erucic acid levels using half-seed analysis has produced *S. alba* lines with high oleic acid (> 75%) and low linolenic acid (< 8%) (Raney et al., 1999c; Katepa-Mupondwa et al., 1999). The inheritance of the low erucic acid trait in *S. alba* is controlled by one gene, with partial dominance for high erucic acid content with a cytoplasmic effect on erucic acid content. The simple inheritance is encouraging for the development of canola quality *S. alba* cultivars in the near future.

Current Breeding Status: Commercial production of canola quality *B. juncea* in Canada will require changes in seed regulations and acceptance of mustard in domestic and international markets as a crop which produces canola quality (low erucic acid oil and low glucosinolate meal) products (Potts et al., 1999). The proposal is to have *B. juncea* accepted as a type of rapeseed as well as a type of mustard, to allow canola-quality mustard to be traded as double low rapeseed in countries that do not recognize the term "canola."

High Erucic Acid Oils

Functional Background: For industrial oil applications, the objective is to enhance the erucic acid levels of high erucic acid oil produced by rapeseed (HEAR) (*B. napus*) cultivars above the current upper limit of approximately 55%. Applications for erucic acid include high fluidity lubricants, slip agents, fuels, polymers, paints, inks, cosmetics and pharmaceuticals (Marcou, 1996). The erucic acid containing oil is used directly in some applications (such as in fuels and cosmetics), while in other applications, the erucic acid is extracted from the oil, modified to erucamide or other derivative and used in this modified form. The oleo-chemical market premium is primarily determined by the erucic acid available for extraction from the oil.

Methods of Modification: High erucic acid rapeseed has been developed using pedigree selection breeding techniques to produce *B. napus* cultivars with up to 55% erucic acid in the seed oil (McVetty et al., 1999).

Two genes, one from each of the diploid genomes *B. rapa* and the *B. oleracea*, control the erucic acid content in the tetraploid species, *B. napus* (Harvey and Downey, 1964). The four alleles act in an additive manner with highly active alleles contributing up to 16% erucic acid per allele. The two genes E1 and E2 controlling the concentration of erucic acid have been mapped in rapeseed (Thormann et al., 1996). Molecular characterization has supported the homology of E1 and E2 with the *Arabidopsis thaliana* gene, FAE1 which is thought to code for the condensing enzyme beta-ketoacyl-CoA synthase (KCS) component of seed specific fatty acid elongases (Barret et al., 1998; Fourmann et al., 1998). Polymorphisms in the FAE1 gene are associated with variation in erucic acid content in *Brassica* species (Luhs et al., 1999b).

Existing variation for the high erucic acid trait in oilseed rape (*B. napus*) does not exceed 66%, although some related *Brassica* species have been identified as expressing over 60% erucic acid in the seed oil (Luhs and Friedt, 1995a). The 66% upper limit is imposed by the inability of *B. napus* plants to esterify erucic acid in the middle position (sn-2) on the glycerol molecule (Taylor et al., 1992). Surprisingly, some *B. oleracea* genotypes are able to esterify erucic acid in the sn-2 position (Taylor et al., 1995). *B. oleracea* genotypes which esterify erucic acid in the sn-2 position had erucic acid levels of well below 66% and contained no trierucin in the oil, however (Taylor et al., 1995). A subsequent search for *B. rapa* genotypes which esterify erucic acid in the sn-2 position, done by Taylor et al. was successful (i.e., some *B. rapa* genotypes also esterify erucic acid in the sn-2 position, data unpublished). The two parental species of *B. napus*, both containing functional sn-2 position erucic acid esterification genes were crossed at the University of Manitoba to resynthesize *B. napus*. No resynthesized *B. napus* lines had erucic acid levels over 62%. Interspecific hybridization between *B. rapa* and *B. oleracea* has been used in Germany to produce resynthesized rapeseed with an erucic acid content of 60% (Luhs and Friedt, 1995b). While the erucic acid levels in *B. napus* were increased in these genotypes, none exceeded the 66% erucic acid upper limit of erucic acid esterification in the 1- and 3-positions.

Current Breeding Status: The following breeding organizations reported having HEAR cultivars in production: LimaGrain (Australia), Danisco Seeds, (Denmark), University of Idaho (USA), and University of Manitoba (Canada). All HEAR cultivars are spring habit *B. napus* and are grown under Identity Preserved (contract) production programs. All organizations are using pedigree selection techniques with or without doubled haploid line development. Breeding objectives include increased seed yield, increased erucic acid levels and improved disease resistance. All successful HEAR cultivars currently have erucic acid levels between 50 and 55%. The University of Manitoba has released several HEAR cultivars since 1991 for production under contract in western Canada, the first was Hero (Scarth et al., 1991), and the most recent was MillenniUM 01 (McVetty et al., 1999). The agronomic performance of the newer HEAR cultivars bred by the University of Manitoba is comparable to the current canola cultivars grown in Canada, while the erucic acid content is approximately 55%. The University of Manitoba breeding program has also been able to develop HEAR germplasm with erucic acid levels approaching 60%.

There are several organizations attempting to break the 66% erucic acid barrier in *B. napus* including the Plant Biotechnology Institute (Canada), University of Manitoba (Canada) Limagrain (U.K.) and a German group co-ordinated from the University of Geissen (Luhs et al., 1999a). Success in breaking the 66% erucic acid barrier in *B. napus* has not been achieved to date. It is presumed that the assembly of several traits in HEAR materials will be required to successfully develop very high erucic acid rapeseed cultivars.

CONCLUSIONS

Commercialization of Specialty Edible and Industrial Brassica Oils

Current Identity Preserved (contract) production of new modified oil *Brassica* species includes low linolenic, mid and high oleic canola, high saturate canola and high erucic rapeseed. There is general agreement in the industry that the current limitation to the production of the new edible oils is the lack of sufficient premium value to encourage economic contract production, and the reluctance to change the current canola commodity profile without an assurance of acceptance (Krawczyk, 1999). Segregation to maintain the specialty oilseeds in a harvest-to-market Identity Preserved system add to the costs. The premiums required for specialty vegetable oil production can be up to 20 cents (U.S.) per pound, which is a deterrent to the acceptance of the new oils in the edible oil manufacturing sector (Krawczyk, 1999). An alternative strategy to segregation is the wholesale conversion of the commodity crop, illustrated by the decision of the U.S. sunflower industry to switch the entire U.S. commodity sunflower crop of approximately three million acres to Nu-Sun mid-oleic varieties within the next two years. Some specialty oilseed companies are targeting the high-priced niche markets with lower volume sales but with the compensation of higher premiums. The current limitation to the production of new non-edible (industrial) oils is the lack of very high erucic acid contents in newly developed HEAR cultivars.

Once the new edible oil profile or non-edible (industrial) oil profile has been established in adapted cultivars, breeding objectives focus on improved agronomic performance, especially seed yield and disease resistance and improved quality characters such as seed oil and protein content. There will, however, always be a trade-off between nutrition, functionality, and the economics of specialty oil *Brassica* cultivar production, particularly in the introductory period for new and improved oil profiles.

REFERENCES

Adolphe, D. 2000. Canola: Origin and History. [Online]Available: <http:www// canola-council.org/pubs/Canola/canola_pdfs.htm> 5-5-2000.

Agnihotri, A. and N. Kaushik. 1999. Genetic enhancement for double low characteristics in Indian Rapeseed Mustard. Wratten N. and P.A., Eds. Salisbury. Canberra Australia, GCIRC. Proceedings of the 10th International Rapeseed Congress, Sept. 26, 1999.

Auld, D.L., M.K. Heikkinen, D.A. Erickson, J.L. Sernyk., and J.E. Romero. 1992. Rapeseed mutants with reduced levels of polyunsaturated fatty acids and increased levels of oleic acid. Crop Science 32, 657-662.

Barret, P., R. Delourme, M. Renard, F. Domergue, R. Lessire, M. Delseny, and T.J. Roscoe. 1998. A rapeseed FAE1 gene is linked to the E1 locus associated with variation in the content of erucic acid. Theor Appl Genet 96, 177-186.

Barret, P., R. Delourme, D. Brunel, C. Jourdren, R. Horvais, and M. Renard. 1999. Low linolenic acid level in rapeseed can be easily assessed through the detection of two single base substitutions in *fad3* genes. Wratten N. and Salisbury P.A., Eds. Canberra, Australia, GCIRC. Proceedings of the 10th International Rapeseed Congress, Sept. 26, 1999.

Cao, Y.-Z., A.H.C. Huang. 1987. Acyl Conenzyme A preference of diacylglycerol acyltansferase from the maturing seeds of *Cuphea*, maize, rapeseed and canola. Plant Physiol 84, 762-765.

Debonte, L.R. and W.D. Hitz. 1996. Canola oil having increased oleic acid and decreased linolenic acid content and its manufacture using transgenic plants. US 96-675650 960703 CAN 130:65607(CODEN: USXXAM US 5850026 A 981215).

DeClerq, D.R., J.K. Daun, and K.H. Tipples. 1997. Quality of western Canadian canola. 236, 1-17. Canadian Grain Commission. Crop Bulletin, Grain Research Laboratory.

Del Vecchio, A. J. 1996. High laurate canola. Inform 7, 230-243.

Deng, X. and R. Scarth. 1998. Temperature effects on fatty acid composition during development of low linolenic oilseed rape (*Brassica napus* L.). Journal of the American Oil Chemists Society 75, 759-766.

Downey, R.K. 1983. The origin and description of the *Brassica* oilseed crops. Chapter 1 in: High and low erucic acid rapeseed oils, J.K.G. Kramer, F.D. Sauer, and W.J. Pigden, Eds. Academic Press, Toronto.

Drost, W.J., G. Rakow, and P. Raney. 1999. Inheritance of erucic acid content in yellow mustard (*Sinapis alba* L.). Wratten N. and Salisbury P.A., Eds. Canberra, Australia. GCIRC. Proceedings of the 10th International Rapeseed Congress, Sept. 26, 1999.

Eccleston, V.S. and J.B. Ohlrogge. 1998. Expression of lauroyl-acl-carrier protein thioesterase in *Brassica napus* seeds induces pathways for both fatty acid oxidation and biosynthesis and implies a set point for fatty acid accumulation. The Plant Cell 10, 613-621.

Eskin, M., B.E. McDonald, R. Przylbylski, L.J. Malcomson, R. Scarth, T. Mag, K. Ward, and D. Adolphe. 1996. Canola oil. Vol. 2. Edible Oil and Fat Products: Oil and Oilseeds, 1-95. John Wiley&Sons Inc., NY.

Fitzpatrick, K. and R. Scarth. 1998. Improving the health and nutritional value of seed oils. PBI Bulletin January, 15-19. 1998. Saskatoon, NRC-CRC.

Fourmann, M., P. Barret, M. Renard, G. Pelletier, R. Delourme, and D. Brunel. 1998. The two genes homologous to *Arabidopsis* FAE1 co-segregate with the two loci governing erucic acid content in *Brassica napus*. Theor Appl Genet 96, 852-858.

Gadoua, R. 2000. Procedures document of the WCC/RRC Inc. for 2000. Available from R. Gadoua, secretary, WCC/RRC Inc., <http: www.gadoua@canola-council. org>.

Gasser, C.S. and R.T. Frawley. 1989. Genetically engineering plants for crop improvement. Science 244:1293.

Harvey, B.L. and R.K. Downey. 1964. The inheritance of erucic acid content in rapeseed (*Brassica napus* L.). Can J of Plant Sci 44, 104-111.

Hawkins, D. and J.C. Kridl. 1998. Characterization of acyl-ACP thioesterase of mangosteen (*Garcinia mangostana*) seed and high levels of stearate production in transgenic canola. The Plant Journal 13, 743-752.

Hitz, W.D., C.J. Mauvis, K.G. Ripp, R.J. Reiter, L.R. Debonte, and Z. Chen. 1995. The use of cloned rapeseed genes for the cytoplasmic fatty acid desaturases and the plastid acyl-ACP thioesterases to alter relative levels of polyunsaturated and saturated fatty acids in rapeseed. D5, 470-478. Cambridge, UK, GCIRC. Proceedings of the 9th International Rapeseed Congress, July 7, 1995.

Jourdren, C., P. Barret, D. Brunel, R. Delourme, and M. Renard. 1996. Specific molecular marker of the genes controlling linolenic acid content in rapeseed. Theor Appl Genet 93, 512-518.

Katepa-Mupondwa, F., G. Rakow, and P. Raney. 1999. Developing oilseed yellow mustard (*Sinapis alba* L.) in western Canada. Wratten N. and Salisbury P.A., Eds. Canberra, Australia, GCIRC. Proceedings of the 10th International Rapeseed Congress, Sept. 26, 1999.

Kiritsakis, A.K. 1999. Composition of olive oil and its nutritional and health effect. Wratten N. and Salisbury P.A., Eds. Canberra, Australia, GCIRC. Proceedings of the 10th International Rapeseed Congress, Sept. 26, 1999.

Kirk, J.T.O. and R.N. Oram. 1981. Isolation of erucic acid-free lines of *Brassica juncea* Indian mustard now a potential oilseed crop in Australia. Journal of Australian Institute of Agricultural Science 47, 51-52.

Kodali, D.R., L.R. Debonte, and Z. Fan. 1999. High stability canola oils. Cargill, Incorporated. 651,684(5,885,643). Wayzata, Minn., USA, US Patent Documents, March 23, 1999.

Krawczyk, T. 1999. Edible specialty oils–an unfilled promise. Inform 10, 552-561.

Laakso, I.J., S. Hovinen, T. Seppanen-Laakso, and R. Hitunen. 1999. Selection for low a linolenic acid content in spring turnip rape. Wratten N. and Salisbury P.A., Eds. Canberra, Australia, GCIRC. Abstracts of the 10th International Rapeseed Congress. 9-26-1999.

Lichter, R. 1982. Induction of haploid plants from isolated pollen of *Brassica napus*. Z Pflanzenphysiol Bd 105. S. 427-434.

Luhs, W.W., J. Han, A. Grafinzu Munster, D. Weier, A. Voss, W. Friedt, F.P. Wolter and M. Frentzen. 1999a. Genetics engineering of erucic acid content in rapeseed. Wratten N. and Salisbury P.A., Eds. Canberra, Australia, GCIRC. Abstracts of the 10th International Rapeseed Congress. 9-26-1999.

Luhs, W.W., A. Voss, F. Seyis, and W. Friedt. 1999b. Molecular genetics of erucic acid content in the genus *Brassica*. Wratten N. and Salisbury P.A. Canberra, Australia, GCIRC. Abstracts of the 10th International Rapeseed Congress. 9-26-1999.

Luhs, W.W. and W. Friedt. 1995a. Natural fatty acid variation in the genus *Brassica* and its exploitation through resynthesis. Eucarpia Cruciferae Newsletter 17, 14-15.

Luhs, W.W. and W. Friedt. 1995b. Breeding high-erucic rapeseed by means of *Brassica napus* resynthesis. 2, 449-551. U.K. Proceedings of the 9th International Rapeseed Congress (GCIRC) Cambridge, UK. July 7, 1995.

Marcou, L. 1996. Main industrial uses of fats. Oils and fats manual, A. Karleskind Ed. Lavoisier Publishing Inc., Secaucus, New Jersey 07094, USA.

Martini, N., J. Schell, and R. Tøpfer. 1995. Expression of medium-chain acyl-(ACP) thioesterases in transgenic rapeseed. *In* Rapeseed Today and Tomorrow. Proceedings of the 9th International. Rapeseed Congress, Cambridge, UK, 7-7-1995 vol. 2: pp. 461-463.

McVetty, P.B.E., R. Scarth, and S.R. Rimmer. 1999. Millennium 01 summer rape. Can J Plant Sci, 79, 251-252.

Murphy, D.J. 2000. Development of new oil crops in the 21st century. Inform 11, 112-117.

Oram, R.N., P.A. Salisbury, J.T.O. Kirk, and W.A. Burton. 1999. Development of early flowering, canola-grade *Brassica juncea* germplasm. Wratten N. and Salisbury P.A., Eds. Canberra, Australia, GCIRC. Proceedings of the 10th International Rapeseed Congress, Sept. 26, 1999.

Papalois, M.M., H. Tran, M. Palmer, and K. White. 1999. New specialty canola oil, Monola oil, with blends of sesame seed and cottonseed oils in deep frying trials. Wratten N. and Salisbury P.A., Eds. Canberra Australia, GCIRC. Abstracts of the 10th International Rapeseed Congress, Sept. 26, 1999.

Poehlman, J.M. and D.A. Slepper. 1995. Breeding Field Crops. 4th Ed. Iowa State University Press, Ames, Iowa.

Polsoni L., L.S. Kott, and W.D. Beversdorf. 1988. Large-scale microspore culture technique for mutation-selection studies in *Brassica napus*. Can J Bot 66:1681-1685.

Potts, D.A. and D.R. Males. 1999. Inheritance of fatty acid composition in *Brassica juncea*. Wratten N. and Salisbury P.A., Eds. Canberra, Australia, GCIRC. Proceedings of the 10th International Rapeseed Congress, Sept. 26, 1999.

Potts, D.A., G.W. Rakow, and D.R. Males. 1999. Canola-quality *Brassica juncea*, a new oilseed crop for the Canadian prairies. Wratten N. and Salisbury P.A., Eds. Canberra, Australia, GCIRC. Proceedings of the 10th International Rapeseed Congress, Sept. 26, 1999.

Przybylski, R., M. Eskin, and L. Norman. 1999. Frying performance of modified canola oils. Wratten N. and Salisbury P.A., Eds. Canberra, Australia, GCIRC. Proceedings of the 10th International Rapeseed Congress, Sept. 26, 1999.

Przybylski, R. and R. Zambiazi. 1999. Storage stability of genetically modified canola oils. Wratten N. and Salisbury P.A., Eds. Canberra, Australia, GCIRC. Proceedings of the 10th International Rapeseed Congress, Sept. 26, 1999.

Rakow, G. 1993. Selektion auf linol-und linolen-saureghalt in rapssamen nach mutagener behand-lung. Z. Planzen 69, 205-209.

Raney P., G. Rakow, and T. Olson. 1995. Development of low erucic, low glucosinolate *Sinapis alba*. Proceedings of the 9th International rapeseed Congress, Cambridge, UK, July 7, 1995, 416-418.

Raney, J.P., G. Rakow, R.K. Gugel, and T.V. Olson. 1999a. Low linolenic 'zero' aliphatic glucosinolate *Brassica napus*. Wratten N. and Salisbury P.A., Eds. Canberra, Australia, GCIRC. Proceedings of the 10th International Rapeseed Congress, Sept. 26, 1999.

Raney, J.P., G. Rakow, and T.V. Olson. 1999b. Identification of *Brassica napus* germ plasm with seed oil low in saturated fat. Wratten N. and Salisbury P.A., Eds. Salisbury. Canberra, Australia, GCIRC. Proceedings of the 10th International Rapeseed Congress, Sept. 26, 1999.

Raney, J.P., G. Rakow, and T.V. Olson. 1999c. Selection for high oleic acid in zero erucic acid *Sinapis alba*. Wratten N. and Salisbury P.A., Eds. Canberra, Australia, GCIRC. Proceedings of the 10th International Rapeseed Congress, Sept. 26, 1999.

Rudloff, E. and P. Wehling. 1997. Release of transgneic oilseed rape (*Brassica napus* L.) with altered fatty acids. *In* Thomas, G. and Mionteiro, A.A. ed. Proceedings of the International Symposium on Brassicas, Rennes France., Acta Hort. 459, pp. 379-385. Sept. 22, 1997.

Rudloff, E., H.U. Jurgens, B. Ruge, and P. Wehling. 1999. Selection in transgenic lines of oilseed rape (*Brassica napus* L.) with modified seed oil composition. Wratten N. and Salisbury P.A., Eds. Canberra, Australia, GCIRC. Proceedings of the 10th International Rapeseed Congress, Sept. 26, 1999.

Scarth, R., P.B.E. McVetty, S.R. Rimmer, and B.R. Stefansson. 1988. Stellar low linolenic-high linoleic acid summer rape. Can J Plant Science 68, 509-510.

Scarth, R., P.B.E. McVetty, S.R. Rimmer, and B.R. Stefansson. 1991. Hero summer rape. Can J of Plant Science 71, 865-866.

Schierholt, A. and H.C. Becker. 1999. Genetic and environmental variability of high oleic acid content in winter oilseed rape. Wratten N. and Salisbury P.A., Eds. Canberra, Australia, GCIRC. Proceedings of the 10th International Rapeseed Congress, Sept. 26, 1999.

Singh, S.P., T. van der Heide, S. McKinney, and A. Green. 1995. Nucleotide sequence of a cDNA from *Brassica juncea* encoding a microsomal omega-6 desaturase. Plant Physiology 109, 1498.

Somers, D.J., K.R.D. Friesen, and G. Rakow. 1998. Identification of molecular markers associated with linoleic acid desaturation in *Brassica napus*. Theor Appl Genet 96, 897-903.

Somers, D., G.R. Akow, P.J. Raney, V. Prabhu, G. SeguinSwartz, R. Rimmer, R.K. Gugel, D. Lydiate, and A. Sharpe. 1999. Developing marker-assisted breeding for quality and disease resistance traits in *Brassica* oilseeds. Wratten N. and Salisbury P.A., Eds. Canberra, Australia, GCIRC. Proceedings of the 10th International Rapeseed Congress, Sept. 26, 1999.

Stoutjesdijk, P.A., C. Hurlstone, S.P. Singh, and A.G. Green. 1999. Genetic manipulation for altered oil quality in *Brassicas*. Wratten N. and Salisbury P.A., Eds. Canberra, Australia, GCIRC. Proceedings of the 10th International Rapeseed Congress, Sept. 26, 1999.

Sun, C., Y.Z. Cao, and A.H.C. Huang. 1988. Acyl coenzyme A preference of the glycerol phosphate pathway in the microsomes from maturing seeds of palm, maize and rapeseed. Plant Physiol 88, 56-60.

Tanhuanpaa, P.K., J.P. Vilkki, and H.J. Vilkki. 1995. Association of a RAPD marker with linolenic acid concentration in the seed oil of rapeseed (*Brassica napus* L.). Genome 38, 414-416.

Tanhuanpaa, P.K., J.P. Vilkki, and H.J. Vilkki. 1996. Mapping of a QTL for oleic acid concentration in spring turnip rape (*Brassica rapa* ssp. *oleifera*). Theor Appl Genet 92, 952-956.

Taylor, D.C., N. Weber, L.R. Hogge, E.W. Underhill, and M.K. Pomeroy. 1992. Formation of trierucoylglycerol (Trierucin) from 1,2-dierucoylglycerol by a homogenate of microspore-derived embryos of *Brassica napus* L. Journal of the American Oil Chemists Society 69, 355-358.

Taylor, D.C., D.L. Barton, E.M.N. Goblin, S.L. MacKenzie, C.G.J. van den Berg, and P.B.E. McVetty. 1995. Microsomal lyso-phosphatidic acid acyl transferase from a *Brassica oleracea* cultivar incorporates erucic acid into the sn-2 position of seed triacylglycerols. Plant Physiology 109, 409-420.

Thormann, C.E., J. Romero, J. Mantet, and T.C. Osborn. 1996. Mapping loci controlling the concentrations of erucic acid and linolenic acid in seed oil of *Brassica napus* L. Theor Appl Genet 93, 282-286.

Uzzan, A. 1996. The olive and olive oil. pp. 225-233 in Chapter 3: Sources of the main fats and relevant monographs in the Oils and Fats Manual, Intercept Ltd., Hampshire, UK, 1996.

Voelker, T., A. Worrel, L. Anderson, J. Bleibaum, C. Fan, D. Hawkins, S. Radke, and M. Davies. 1992. Fatty acid biosynthesis redirected to medium chains in transgenic oilseed plants. Science 257, 72-74.

Voelker, T.A., T.R. Hayes, A. Cranmer, J.C. Turner, and H.M. Davies. 1996. Genetic engineering of a quantitative trait: metabolic and genetic parameters influencing the accumulation of laurate in rapeseed. Plant J 9, 229-241.

Warner, K. and T.L. Mounts. 1993. Frying stability of soybean and canola oils with modified fatty acid composition. Journal of the American Oil Chemists Society 60, 983-988.

White, K. 1998. Specialty Oilseed Information Day. 1-40. Horsham, Victoria, Australia, Ag-Seed Research, Dovuro Pty Ltd., Horsham IAMA. 10-28-1998.

Wilson, R.F. 1993. Practical biotechnology strategies for improving the quality and value of soybeans. Inform 4, No. 2:193-200.

Wong, R., J.D. Patel, I. Grant, J. Parker, D. Charne, M. Elhalwagy, and E. Sys. 1991. The development of high oleic canola. A16, 53-54. Saskatoon, Canada, GCIRC. Proceedings of the 8th International Rapeseed Congress, Nov. 7, 1991.

Woods, D. 1992. Comparative performance of mustard and canola in the Peace River region. Can J Plant Sci 72, 829-830.

Wuidart, W. 1996. Palm oil and its fractions. pp. 233-243 in Chapter 3: Sources of the main fats and relevant monographs in the Oils and Fats Manual, Intercept Ltd., Hampshire, UK.

Quality Improvement of Upland Cotton (*Gossypium hirsutum* L.)

O. Lloyd May

SUMMARY. Quality of cotton can be defined through seed or fiber properties, but is most often associated with fiber properties that influence processing into yarn and textile products. Global competition in the production and consumption of cotton fiber combined with technological evolution of yarn manufacturing has spurred renewed efforts to enhance cotton fiber quality. Cotton fiber quality can be improved through genetics, crop management, and postharvest processing. Knowledge of the effects of fiber properties on processing and their inheritance, relationships, and environmental influences is necessary to formulate improvement strategies. Breeding to improve fiber quality has traditionally focused on enhancing measures of the longest fibers or fiber strength for ring yarn manufacturing systems. With the technological evolution of yarn manufacturing from solely ring-based spinning to predominately rotor and potentially in the near future air-jet spinning, needs for fiber profiles have been revised for these spinning systems. Successful rotor spinning requires high fiber strength for all yarn counts, along with fiber fineness for fine count yarns. The even more productive air-jet spinning requires a minimum, but uniform fiber length, fiber fineness, and to a lesser extent strong fiber. In contrast, ring spinning requires a minimum fiber length, fiber strength, and to a lesser extent fiber fineness. Breeders do not conduct direct selection for yarn properties because of impracticalities, thus they select for fiber properties that influence processing, so-called *indirect* selection. The inherent environmentally induced vari-

O. Lloyd May is Assistant Professor, Department of Crop & Soil Science, University of Georgia, Tifton, GA 31793 USA.

[Haworth co-indexing entry note]: "Quality Improvement of Upland Cotton (*Gossypium hirsutum* L.)." May, O. Lloyd. Co-published simultaneously in *Journal of Crop Production* (Food Products Press, an imprint of The Haworth Press, Inc.) Vol. 5, No. 1/2 (#9/10), 2002, pp. 371-394; and: *Quality Improvement in Field Crops* (ed: A. S. Basra, and L. S. Randhawa) Food Products Press, an imprint of The Haworth Press, Inc., 2002, pp. 371-394. Single or multiple copies of this article are available for a fee from The Haworth Document Delivery Service [1-800-HAWORTH, 9:00 a.m. - 5:00 p.m. (EST). E-mail address: getinfo@haworthpressinc.com].

ability in fiber properties presents challenges to enhance them through breeding or biotechnological approaches. Because variability in fiber properties is problematic to fiber processing, future-breeding and biotechnological approaches should simultaneously focus on enhancing fiber properties and reducing variation. This paper will review strategies to enhance fiber profiles through genetic approaches while ameliorating their variation. *[Article copies available for a fee from The Haworth Document Delivery Service: 1-800-HAWORTH. E-mail address: <getinfo@haworthpressinc. com> Website: <http://www.HaworthPress.com> © 2002 by The Haworth Press, Inc. All rights reserved.]*

KEYWORDS. Cotton fiber quality, yarn manufacture, fiber strength, length uniformity, fiber variability, selection response

INTRODUCTION

Quality of cotton is associated with properties of the seed and fiber. Since over 95% of the value of the cotton crop is in the fiber, fiber quality mostly defines quality of cotton (National Cotton Council, 1999). In fact, fiber properties that define fiber quality are the basis for the marketing and sale of cotton throughout the World (Hake et al., 1990). Fiber quality is defined in broad terms by linear density, length, and tensile properties that influence yarn and textile manufacture. The significance of cotton fiber quality varies to producers, ginners, marketers, or consumers of cotton. Producers view quality in relation to the fiber properties with which the marketing system dictates premiums and discounts they receive for fiber delivered to the yarn mill, merchant, or cooperative marketing association. Ginners focus on creating the cleanest cotton bale through various ginning operations that remove non-lint content from seed-cotton or ginned lint for producers to receive maximum profit (Anthony, 1999). Merchants attempt to purchase at the lowest price the highest quality cotton bales frequently based on contracts to deliver certain minimum fiber qualities to then resell to yarn or other manufacturers of cotton fiber based products. Yarn manufacturers relate fiber properties at the bale and lay-down level to the size (e.g., count) and type of yarn they need to produce, given the yarn spinning system they possess. The purpose of this review is to synthesize needs of the processors of raw fiber with classical and new biotechnology based genetic improvement strategies to achieve these goals. Because upland cotton (*Gossypium hirsutum* L.) dominates as a textile fiber compared with extra-long-staple upland, short-staple diploid cottons, or Pima-types (*Gossypium hirsutum* L.), this paper will concentrate on upland cotton.

The cotton industry, like many others, has experienced globalization in recent years, increasing competition among fiber producers and consumers. Cotton fiber must also compete with synthetic fibers for its share of the market. Increasingly, cotton buyers and consumers are emphasizing fiber quality in purchase decisions, as yarn-manufacturing systems are updated and require ever more stringent fiber profiles to economically produce high quality yarn. Poor quality cotton bales are being left unsold in warehouses around the globe, reflecting the needs of fiber processors for better fiber quality. Yarn manufacture is the first step in the construction of cotton textile products, and it continues to undergo technological advance to produce more yarn at less unit cost to remain profitable (Deussen, 1992; Faerber, 1995). Among the many properties of cotton fiber, fiber length, strength, and fineness are several key properties that influence its conversion into yarn, and the resulting quality of the yarn. With the evolution of yarn manufacturing systems this century from entirely ring spinning, to predominately rotor (Smith and Zhu, 1999) and possibly air-jet in the near future (El-Mogahzy, 1998) has been the definition of different fiber profiles for each type of yarn manufacture. While breeders need not become fiber technologists, they should understand fiber property needs of the different spinning systems. These somewhat disparate fiber profiles for air-jet, ring, and rotor spinning guide breeding objectives that must evolve in tandem with processing methods.

EVOLUTION OF YARN MANUFACTURE AND FIBER PROPERTY NEEDS

Rotor spinning has become the dominant U.S. yarn manufacturing process because of its economic advantage over ring spinning, a result of more yarn produced per unit of time combined with less labor and fewer pre-spinning fiber preparation operations (Faerber, 1995). Worldwide, rotor spinning has increased at the expense of ring spinning as well, but to a lesser extent. Air-jet spinning may be the next technological revolution in yarn manufacture. The Murata Vortex form of air-jet spinning is particularly promising because it can spin a fine count yarn of similar quality to that of the same size ring spun yarn (El-Mogahzy, 1998), yet it has an exponentially greater productivity. Each yarn manufacturing method has somewhat disparate needs in terms of the basic fiber properties to produce strong yarn for subsequent textile construction. Studies of yarn strength and quality as affected by fiber properties at the bale level have helped to prioritize fiber property needs for each spinning system (Backe, 1996; Ramey, Lawson, and Worley, 1977). The top three fiber properties desired for rotor spinning are fiber strength, fiber fineness, and fiber length, while air-jet spinning works best with fiber of high length uniformity (low short fiber content), fineness, and fiber strength (Deussen, 1992), in descending order of importance. Yarn properties cannot be directly selected through breeding because their measurement is precluded on early generation breeding material by available lint sample size, genetic population size, and cost. Additionally, small-scale air-jet spinning systems amenable for breeding work are not developed, so only germplasm at or near release as a cultivar can be evaluated for air-jet yarn performance. Small-scale rotor (Steadman, 1997) and ring spinning (Landstreet, Ewald, and Kerr, 1959) systems do exist, but only the micro-ring spinning system is generally available to breeders, albeit at a level of expense that necessitates evaluation of only advanced generation breeding lines. Breeders must therefore, conduct *indirect* selection for yarn properties, by selection of one or more fiber properties that influence yarn manufacture (Meredith et al., 1991; May and Taylor, 1998). Therein lies the challenge to the breeder to effect selection response of yarn strength, the most important yarn property influencing textile performance.

FIBER PROPERTIES IMPORTANT TO PROCESSING

Since cotton is processed into yarn in groups of fibers rather than as individual fibers, properties such as length variability, short fiber content, fineness, maturity, and bundle strength among others influence yarn quality and strength, and resulting textile products (Perkins, Ethridge, and Bragg, 1984). Steady gains in longer fiber and higher bundle strength have been accomplished through breeding in the 20th century (Niles and Feaster, 1984; Sasser and Shane, 1996; May, 1999). Breeders have not had access until recently to fiber measurements such as fineness, maturity, and short fiber content. Thus, these traits essentially varied at random in breeding populations, precluding directed efforts at their manipulation. The Advanced Fiber Information System (AFIS; Bragg and Shofner, 1993) provides facile measurement of short fiber content, fineness, maturity, and immature fiber content that may allow these traits to be manipulated through crop management and genetics. A further complication in designing genetic studies to enhance fiber properties is that they are not independent entities, and thus cannot be manipulated individually without effects on other properties. An example is fiber length and fiber fineness. As fiber length is increased, finer fiber often results that may not be a desirable outcome (Meredith, 1984). Thus, selection for one property may result in changes in fiber properties not intentionally selected, so-called *correlated* response to selection (Falconer, 1989). It is helpful in setting goals to understand relationships among fiber properties and the idiosyncrasies with which they are measured. This paper shall review the impact of key fiber properties on cotton quality and make recommendations as to future efforts at their improvement.

Fiber Strength

The importance of fiber strength to yarn manufacture and quality of the yarn produced depends on the yarn spinning process. Rotor spun yarns need the highest strength fiber (Deussen, 1992) because less of the inherent fiber strength is preserved in the strength of a rotor spun yarn compared with that of a ring spun yarn. Deussen (1992) terms the effect of fiber strength on yarn strength as the strength-yield relationship, or degree of fiber strength preserved in the yarn strength. Heritability of fiber strength is normally moderate to high (>50%), reflecting the greater influence of genetic vs. non-genetic sources of variation (Meredith, 1984; May, 1999). Consequently, fiber strength has been slowly, but steadily improved through breeding, increasing 0.44% annually between 1980 and 1995 (Taylor et al., 1995). The environment does influence fiber strength, as evidenced by strength variation among fruiting zones exceeding the standard deviation of the measurement (Lewis, 1998). Generally though, genotype × environment interactions are not of a magnitude or type (e.g., rank changes) such that superior breeding lines cannot be identified (Meredith, Sasser, and Rayburn, 1996). Breeders have alleviated a portion of the confounding effects of environmentally induced fruiting site variation on genetic differences among breeding lines by preferential harvest of fiber for testing from certain boll positions in the main fruiting zone. Lewis (1998) showed that fiber property determination on fiber harvested from the initial eight first position bolls from cotton plants facilitates separation of genetic variation in fiber quality.

Fiber strength is measured on a bundle of fibers with HVI, Stelometer (Hertel, 1953), or Pressley (Pressley, 1942) instruments (Taylor, 1986) and on single fibers with the Mantis Single Fiber Tester (Sasser, 1992; Suh, Cui, and Sasser, 1994). The breaking strength of a bundle of fibers is predictive of yarn strength (Steadman, 1997), a key indicator of subsequent textile performance. Single fiber strength is not necessarily predictive of yarn performance, but may have value as a component of bundle fiber strength. The choice of instrument with which bundle fiber strength is measured can be a consideration in developing cotton germplasm with fiber profiles to meet current and future needs of yarn manufacturers. Most commercial breeding firms employ HVI testing to select for fiber properties (Latimer, Wallace, and Calhoun, 1996) because of speed and cost effectiveness. The HVI system of fiber testing was developed as the basis of a bale classification system with which to rapidly and objectively measure certain basic fiber properties for marketing purposes (Hake et al., 1990). Fiber testing by HVI is rapid because of mechanization that allows a relatively complete fiber profile to be measured on the same fiber sample (Taylor, 1986). In contrast, instruments such as Stelometer or Pressley, measure one or a few related fiber properties, and are known as *single instruments*. Single instrument testing is more expensive, by as much as 100%, slower due to less mechanization, and requires manual sample preparation. Single instrument testing has a reputation among breeders for greater accuracy than that of HVI testing (Cooper, Oakley, and Dobbs, 1988), despite few studies showing selection response varies. Selection response (Δ_G) of one trait after one cycle of selection is the product of:

$$\Delta_G = i h^2 \sigma_p$$

selection intensity (i), heritability (h^2), and the square root of phenotypic variability (σ_p). Thus, heritability is a key statistic indicating potential for trait modification (Falconer, 1989). Latimer, Wallace, and Calhoun (1996) reported heritability of fiber strength by HVI measurement to be greater than that measured by Pressley. By contrast, May and Jividen (1999) reported fiber strength by Stelometer measurement to be 15% greater on average than that measured by HVI, but that direct response to selection varied little in two genetic populations. This finding of equivalent strength improvement in populations selected for higher strength with either instrument despite higher heritability by Stelometer measurement implies greater phenotypic variability for HVI fiber strength (since selection intensity did not vary). Greater phenotypic variability for HVI fiber strength could reflect more variation in the underlying fiber properties that impart bundle fiber strength and/or more measurement error. Overall, selection for fiber strength by HVI would seem adequate for most breeders.

A discussion of the selection response of cotton fiber properties would be incomplete without considering correlated responses to selection. Correlated response to selection describes changes in traits not directly selected for when another trait is selected (Falconer, 1989). As an example, consider two traits designated trait 1 and trait 2. The following formula (Becker, 1985) models change in trait 2 when trait 1 is selected ($\Delta_{G2.1}$) after one cycle of selection,

$$\Delta_{G2.1} = i \ r_{g(1,2)} h_1 h_2 \ \sigma_{p2}$$

and is the product of selection intensity (i), genetic correlation between traits 1 and 2 ($r_{g(1,2)}$), square roots of heritability of traits 1 and 2 ($h_1 h_2$), and phenotypic standard deviation of trait 2 (σ_{p2}). Fiber bundle strength and linear density are examples of two fiber traits that tend to be genetically correlated–as bundle strength increases, linear density tends to decrease (e.g., greater fineness–more on this issue later in this paper). Correlated responses of underlying fiber properties contributing to bundle fiber strength have implications for developing fiber profiles to benefit rotor and air-jet yarn manufacture. Among the Stelometer, Pressley, and HVI instruments, each employs different methods of fiber sample preparation, mass measurement, and breaking of the fiber bundle (Taylor, 1982; Steadman, 1997). Fiber mass of the sample broken to determine fiber strength is measured directly with the Stelometer and Pressley instruments during sample preparation, while HVI employs light or airflow attenuation to *estimate* mass (Steadman, 1997). The consequence of these idiosyncrasies is different properties influencing the breaking strength of a bundle of cotton fibers may be measured and thus, exploited by each instrument. May and Jividen (1999) found that selection for fiber strength by Stelometer and HVI in the same genetic population can produce a correlated change in fiber fineness. In one genetic population, selection for higher fiber strength by Stelometer measurement resulted in finer fiber (e.g., smaller AFIS fineness in millitex) compared with selection for fiber strength by HVI. Though the difference in population means shown by May and Jividen (1999) were small, they become more significant if one projects this finding over additional cycles of breeding. Selection for fiber strength by Stelometer may result in higher bundle strength in part through preferential selection for finer fiber. Meredith et al. (1991) reached a similar conclusion in a study of relationships among fiber properties with a set of genetically diverse Upland cotton germplasm. Finer fiber would generally, in comparison to coarser fiber, place more fibers in the 3.25 g fiber sample broken on the Stelometer to measure bundle strength contributing to overall increased bundle fiber strength. In contrast, selection for bundle fiber strength by HVI may preferentially exploit fiber properties other than fiber fineness. The influence of indirectly measuring fiber mass on HVI bundle strength has not been explored from a breeding perspective. It has been suggested that greater progress in breeding stronger fiber bundles might occur if bundle fiber strength variation was adjusted for among sample variance in linear density.

Fiber strength and yield have historically been difficult to simultaneously improve (Scholl and Miller, 1976; Culp, Moore, and Pitner, 1985; Culp, 1992). Theories of why an antagonistic genetic relationship exists within breeding populations between fiber yield and strength include energy demands on the plant to produce high fiber strength. Alternatively, if Stelometer tends to select for fiber fineness (e.g., smaller perimeter) to enhance bundle strength, the resulting finer fiber can be associated with lower yields (Meredith et al., 1991).

Fiber Length

Enhancing fiber length is a complex issue because fiber samples from field research or cotton bales contain a range or distribution of fiber lengths (Behery, 1993), the variability of which is generally not measured. Such variation in fiber length arises because both genotype and environment impart variation in fiber length even on the same seed (Bradow et al., 1997). Cotton fiber length is most frequently measured with HVI and fibrograph (Hertel, 1940) instruments providing various mean length distributions (Steadman, 1997). Therefore, breeding to change 'fiber length' typically involves selecting for one or more measures of longest length distribution rather than a specific fiber length. Measures of mean fiber length distribution include the upper-half-mean (UHM) length from HVI, defined as the mean length of the longer one-half of the fibers in a sample; or as uniformity index, expressed as the ratio of the mean length (mean length of all fibers in the sample) to the UHM (Steadman, 1997). The fibrograph provides analogous length distributions including the 2.5% span length (length at which 2.5% of the fibers are at least this length), and the 50% span length (length at which one-half of the fibers are at least this length). Uniformity ratio (50% span length divided by 2.5% span length), is analogous to the uniformity index from HVI, but takes lower value (Steadman, 1997). The recent development of the AFIS allows more detailed information concerning fiber lengths contained in a sample of cotton, including and perhaps most importantly, direct measurement of fiber length variability. Overall, knowledge of the significance of measures of length distribution to processing and yarn properties aids in linking breeding objectives with processor needs as fiber length related traits tend to be moderately to highly heritable (Meredith, 1984; May, 1999), and are thus amenable to modification through breeding. Yarn manufacturers use 2.5% span length or UHM to set distances between draft-rolls in roller drafting of the pre-spinning drawing operation (Perkins, Ethridge, and Bragg, 1984). A minimum or specific range of UHM and uniformity indices along with fiber strength and micronaire readings are often employed to assemble cotton bales into lots for yarn manufacture with bale selection software such as the Engineered Fiber Selection System® (Chewning, 1994). In terms of its effect on yarn strength, length variability influences degree of fiber-to-fiber cooperation in yarn spun through ring, rotor, or air-jet systems (Deussen, 1992). Rotor spinning can tolerate a relatively short UHM yet produce a strong coarse count yarn. Finer count rotor spun yarns become stronger with longer UHM, while strength of air-jet yarns requires effective fiber wrapping because of its 'false twist' yarn structure (El Mogahzy, 1998). Thus, uniformity of length is a key breeding goal for rotor and air-jet spinning and overall cost control for yarn manufacturers. Breeding has resulted in steady increases in measures of the longest fibers (2.5% span length, UHM; Faerber and Deussen, 1994). Given these gains, what then should be breeding priorities for 'fiber length' related traits to meet the needs of a rotor and air-jet dominated spinning industry. Maintaining existing gains in UHM or 2.5% span lengths, while seeking more uniform fiber length within bales through less short fiber content produced by the cotton plant is one objective to positively influence processing. If the goal is to breed fiber with ever-longer UHM, then care must be taken that overall length variability does not increase proportionately.

Short fiber content, defined as the proportion by weight or number of fibers, of length 12.7 mm (about 1/2 inch) or less results in fiber waste to yarn manufacturers and weaker rotor and especially, air-jet yarns (El-Mogahzy, 1998). Reduction of short fiber content is currently one of the top priorities of the U.S. cotton industry (Rowland, 1999).

Breeding to reduce short fiber content has been hampered by lack of access to direct short fiber measurements. Among the fiber length information available to breeders, uniformity index from HVI and uniformity ratio from fibrograph do not directly measure short fiber content (Behery, 1993), despite continued attempts to refine mathematical algorithms relating these quantities (Knowlton, 1999; Rowland, 1999). Additionally, neither the length of the longest fibers (UHM or 2.5% span length) directly relate to short fiber content, but extremely short UHM or 2.5% span length can be associated with increased bale short fiber (Behery, 1993). With the development of the AFIS, short fiber content by weight and number can now be evaluated in breeding efforts if this instrument is made available to breeders on a scale allowing direct selection to reduce short fiber content. Short fiber content by AFIS has been shown to be heritable (May and Jividen, 1999), albeit mean heritability from parent-offspring regression in two populations was only 0.19. Low heritability can be caused by lack of genetic variation or high environmental variation or combination of both factors. Lewis (1998) provides a detailed assessment of the genetic contribution to short fiber content. He showed how short fiber content by AFIS measurement varies among bolls on the same plant. Overall, results of Lewis (1998) indicated that reduction of short fiber content through breeding and controlling plant density may be possible.

One difficulty in breeding for less short fiber content is the necessity of ginning breeding material with laboratory scale gins. Laboratory gins were intended to only separate lint and seed and thus, lack lint dryers and cleaners among other equipment commercial gins employ to prepare lint to enter the marketplace. Lint cleaning during the commercial ginning operation is regarded as the primary contributor to fiber breakage and thus, short fiber content in the bale (Anthony and Bragg, 1987), while the genotypic contribution to bale short fiber content as influenced by management system requires further research to quantify. Thus, breeder-scale gins that substantially replicate commercial ginning might facilitate genetic and crop management research aimed at reducing bale short fiber content.

This author advocates that breeding objectives with respect to fiber length be revised to consider not only the mean length of the longest fibers, but also the proportion of the short fiber population. Simply selecting germplasm for longer length distribution based on the longest fibers (e.g., UHM or 2.5% span length) would seem insufficient to meet current and future textile industry needs for more uniform fiber length and less short fiber. May (2000) reported that selection for longer 2.5% span length or uniformity ratio had little impact on reducing short fiber content. This finding is perhaps not surprising considering that information on the length of the longest fibers may not be related to short fiber content. Thus, new strategies to enhance fiber length properties are needed. A companion study currently underway is to compare selection response of short fiber content based on selection for longest mean fiber length from HVI measurement (mean length = UHM × uniformity index) and least short fiber by AFIS with selection only for longer UHM length. Theoretically, it should be possible to enhance overall fiber length uniformity by selecting for longest mean fiber length and lowest short fiber content by weight (Steadman, 1997).

A further motivation to reduce short fiber content is that this may be an under exploited means of increasing fiber yields by potentially increasing fiber weight per seed. Fiber weight per seed and seeds produced per unit area are recognized as the most basic yield components of cotton (Coyle and Smith, 1997; Lewis, 2000). Lewis (2000) presents data showing breeders should focus on increasing weight of fibers per seed as a strategy to enhance cotton lint yields. Indeed, an interesting experiment might assess the contribution of short fiber content to variation in the weight of fibers per seed in a population divergently selected for high and low weight of fiber per seed. Short fiber content as a quality factor and its contribution to fiber yields illustrates how fiber quality and yield improvement can be complementary rather than disparate goals.

Fiber Fineness

Fiber fineness, or coarseness as Steadman (1997) terms this property, influences yarn manufacture through the number of fibers per yarn cross section. Micronaire reading is often used to estimate fiber fineness (Hake et al., 1990), but the complexities of this issue are discussed further in the next paragraph. Cotton with high (5.0+) and low (< 3.5) micronaire is problematic for fiber processors because fiber with high micronaire reading can be coarse fiber, while fiber with low micronaire reading may be immature. Smaller fineness values (millitex) by AFIS actually denote finer fiber, but larger English count values (> 40 s) denote finer yarns, leading to some degree of confusion in use of the term fineness. As discussed earlier, cotton fiber is processed into yarn in groups, thus the number of fibers per cross section affects spinnability (number of ends down) and the spin limit, or finest count yarn that can be spun considering quality and ends down. Rotor spinning of fine count yarns would benefit from finer fiber through more fibers per yarn cross-section as long as fiber maturity is not sacrificed (Deussen, 1992; El-Mogahzy, 1998).

Breeders do not typically have access to fiber fineness measurements per se, but instead often rely on micronaire readings (Hake et al., 1990). Micronaire reading is the resistance of a plug of fibers to airflow in the micronaire instrument. Kerr (1966) indicates that resistance of a given weight of cotton to airflow reflects fiber surface area per weight. Relating micronaire reading to fineness (e.g., low micronaire reading = fine fiber) is frequently done, but micronaire reading reflects the number of fibers in the sample and their maturity, or degree of secondary wall thickening. Therefore, the two properties can be confounded in a breeding population where maturity and fiber perimeter segregate. Steadman (1997) suggests that micronaire reading can be decomposed into the effects of linear density (two-thirds) and maturity (one-third). Meredith (1991) indicates that genetic variation for micronaire reading reflects nearly equal contributions of maturity and perimeter. In contrast to the single airflow reading of the micronaire instrument, the Arealometer was designed to measure resistance of a group of fibers at two air pressures, from which maturity and perimeter can be separately resolved (Hertel and Craven, 1951). Maturity and fiber perimeter by Arealometer measurement have been shown to be heritable characteristics (Bishr, 1954; May and Taylor, 1998), but the Arealometer instrument has been little used in breeding. Breeding smaller perimeter, but mature fiber, could be accomplished by selection with Arealometer data, rather then relying on a composite property, micronaire reading. Greater fineness (e.g., lower millitex) and less immature fiber content could be achieved through AFIS measurement, but is more costly than Arealometer data. The utility of the Arealometer instrument as a breeding tool to modulate high and low micronaire readings deserves further study. As we have seen, the chief deterrent to breeding Upland germplasm with finer fiber and thus lower micronaire reading is frequently lower yield (Meredith et al., 1996) often associated with reducing fiber weight per unit length.

RELATIONSHIPS AMONG FIBER PROPERTIES

Linear density, length, and tensile properties of cotton fiber are not independent qualities (Meredith et al., 1991; Meredith et al., 1996) such that one can be manipulated without effects on another property. For example, as fiber becomes longer there is a tendency towards smaller perimeter, producing finer fiber that can coincidentally be immature (incomplete secondary fiber cell wall development). Finer fiber can result in higher bundle strength as we have seen but also more neps (small tangled fibers) that result in dye defects in textiles (Meredith et al., 1996).

Meredith (1992) reports results of statistical analysis of commonality to establish contributions of individual fiber properties to variation in bundle fiber strength. These data implicate variation in 50% fiber span length as a contributor to variation in bundle fiber strength, along with variation in single fiber strength. The contribution of longer 50% span length to bundle fiber strength was hypothesized to reflect less short fiber content as indirectly estimated by span length. Interestingly, Meredith (1992) found little contribution of fineness variation as estimated with fiber perimeter by Arealometer measurement to variation in bundle strength. This foundational study suggests that breeding enhanced bundle fiber strength might result in greater progress if the components of bundle strength were considered in the selection scheme. Breeders must have access to the appropriate fiber measurement instrumentation for indirect selection to be possible.

As we saw earlier in this paper, breeders do not normally select for yarn properties such as yarn strength until advanced stages of breeding. Thus, fiber properties genetically correlated with yarn tenacity are selected for soon after genetic population construction followed by yarn strength testing in advanced generations. Genetic studies establishing fiber properties as selection criteria to improve yarn properties have been performed by necessity with advanced generation breeding lines because of the expensive nature of yarn testing (Meredith et al., 1991; May and Taylor, 1998). These studies are about the best that can be done, as it would be prohibitive indeed, to construct populations to test hypotheses concerning individual and joint selection for fiber traits to enhance yarn properties. Meredith et al. (1991) in a study relating fiber properties to yarn strength among released germplasm lines and cultivars implicated bundle fiber strength as the most important contributor to yarn strength, but interactions between 2.5% span length, bundle strength, and fiber perimeter contributed as well. May and Taylor (1998) conducted a study of selection criteria to improve ring yarn strength with a set of F5 and F6 lines from the same cross that had not been previously selected for fiber properties. They reported the most improvement in yarn strength with a selection index simultaneously considering low micronaire reading, longer 50% span length, and highest bundle fiber strength. The influence of longer 50% fiber span length on yarn strength was hypothesized to reflect less short fiber, similar to the effect of 50% fiber span length on bundle fiber strength (Meredith, 1992).

GENETIC AND ENVIRONMENTALLY INDUCED VARIABILITY IN FIBER PROPERTIES

Variability of cotton fibers in a bale is both a blessing and a curse in that the inherent variability in the properties of length, strength, and fineness confers cotton fabric 'character' through small imperfections in the fabric. In contrast, fabric made of all synthetic fibers is very uniform, and sometimes lacks consumer appeal due to its uniformity (Deussen, 1992). The high uniformity of synthetic fibers makes them easier to process compared with cotton, sometimes giving a competitive edge of synthetic fibers over cotton. The curse of cotton fiber property variation can be seen in its effects on yarn imperfections and dye defects in textiles. The very nature of the fruiting habit of Upland cotton contributes to environmentally induced variation in fiber properties among fruiting sites on the same plant (Lewis, 1992; Lewis, 1996; Lewis, 1998). Fruit are produced at predictable intervals such that yield is attained over an extended time period. The result of an extended fruiting period is to have fibers borne in the initial fruit set develop under environmental conditions possibly quite variable compared with fruit set later. When harvested, the variability in fiber properties among bolls and plants in a cotton field is agglomerated into each bale of cotton, resulting in processing difficulties for cotton consumers. An unanswered question is whether it is possible to reduce variability in fiber properties through genetics.

Lewis (1996) and Lewis (1998) provide detailed information on cultivar variation for short fiber content as related to boll position. These data show cultivar variation for short fiber content by fruiting zones and by composite mean short fiber content derived by weighting short fiber content for fruiting zones by fraction of lint yield set in the same fruiting zone. Composite mean short fiber content varied two-fold among eight cultivars, indicating that there exists genetic differences for propensity to produce short fiber. This finding is heartening in that it may be possible to identify germplasm with genes controlling fiber development whose expression is less sensitive to environmental variations. Such genes might be manipulated more broadly through breeding with biotechnological approaches in the future. Lewis (1998) illustrates also how management and genetics can interact to ameliorate short fiber content. As plant population was varied from 1-4 plants per 0.3 m, the proportion of yield produced in first position bolls on fruiting branches 1-8 (fruiting sites that typically contribute most to yield) varied from about 41% at one plant per 0.3 m to 74% at three plants per 0.3 m. Combined with cultivar variation for propensity to produce short fiber by fruiting zone, these data suggest further study to identify germplasm with less short fiber as influenced by crop management that encourages first position boll set.

Just as short fiber content can vary by fruiting site, so can bundle fiber strength and properties that contribute to bundle strength. Lewis (1998, 1999) evaluated bundle fiber strength variation and that of several of its components, single fiber strength and linear density on high and low bundle strength cultivars. He found that as the genetic level of bundle strength increased that variability in bundle strength among fruiting zones increased proportionately. Interestingly, as the level of genetic single fiber strength (grams to single fiber break from Mantis tester) increased, its variation decreased proportionately. Lewis (1999) reports that the popular cultivar Deltapine 50 with low fiber bundle strength exhibits much boll-to-boll variation in micronaire reading compared with higher bundle strength cultivars MD51 and Pima S-7 (*Gossypium barbadense* L.). He therefore implicates variation in linear density (tex) among fruiting zones as a contributor to variation in bundle strength, as bundle strength (grams force to break bundle per tex) can be considered to be a trait itself composed of various components. It is biologically and mathematically probable that variation in bundle strength is at least partly related to variation in linear density. Perhaps these data suggest consideration of variation in fiber linear density among fruiting sites in breeding for higher and less variable bundle fiber strength. Single fiber testing with the Mantis instrument (Sasser, 1992) is slow and labor intensive, and not generally available to breeders, thus increasing single fiber strength is not yet a feasible criterion to enhance single fiber strength.

Overall, Lewis (1996, 1998) provides valuable algorithms for fiber property comparisons among genotypes and new ideas to facilitate fiber quality improvement of cotton.

BIOTECHNOLOGICAL APPROACHES

Biotechnological approaches to enhance cotton fiber quality might involve application of structural or functional genomic tools and/or transformation technology. Much research effort is currently directed at developing molecular markers and transformation systems for cotton improvement.

John (1992, 1996, 1999) reported the first efforts to enhance cotton fiber properties through transformation approaches. These great technological achievements included identification of fiber specific promoters that drive genes imparting higher fiber strength and altered fiber composition, including production of polyester in the fiber lumen (John, 1996). The fiber containing polyester or 'bioplastics' was intended to increase thermal retention properties over that of cotton only fabrics (John, 1999). May, John, and Wofford (2000) reported the first attempts at exploiting through breeding a transgene imparting enhanced fiber strength. This gene enhanced fiber strength up to 70% in descendents of the original biolistically transformed descendents, but inheritance of enhanced fiber strength was unstable. The authors concluded after examining numerous parent-offspring relationships over several field seasons that the enhanced fiber strength was inherited in a random fashion, precluding further development of the direct transformed germplasm. The expression of this gene as influenced by background genotype is the subject of current study (May, unpublished data). Jenkins et al. (1997) and Sachs et al. (1998) reported genetic background influenced expression of insect tolerant transgenes, suggesting transgene X genetic background interactions may be a source of variation to exploit in crop improvement. Additionally, these scientists recommended transgene expression in the donor parent be established prior to use of this parent for introgression efforts and that once candidate cultivars have been introgressed with a transgene, that careful expression analysis be conducted prior to commercialization. Since purchase in 1997 of Agracetus by Monsanto Company (now Pharmacia), efforts to engineer traits that enhance the value of the harvested crop, so-called output traits, have been de-emphasized to concentrate on transformation-mediated insect and herbicide tolerance input traits. Genetic engineering output traits as a means of crop improvement apparently remains for the future.

The Agracetus methodology to enhance fiber quality through transformation involved identification of native or non-native genes that impact or impart fiber properties followed by their transfer to elite germplasm and subsequent evaluation of the effect on fiber quality. Other scientists have adopted an alternative approach to immediate transformation efforts to enhance cotton fiber quality that focuses on a more thorough understanding of the regulation of fiber development.

Once genes regulating fiber development are identified, they become candidates for manipulation through biotechnological approaches (Wilkins and Jernstedt, 1999). Wilkins (1996) highlights the limitations of transformation efforts with 'fiber genes' identified through functional genomic analysis when the role of these genes in fiber development is not well understood. She also points out that the success of transformation approaches with 'fiber genes' is likely dependent on whether the genes are regulated at the transcriptional, translational, or further downstream levels.

Another approach to enhancing fiber quality may be to increase the nutritional status of developing fiber. Haigler et al. (2000a,b) report findings from transformation experiments aimed at increasing substrates for cellulose synthesis when cotton is produced under cool night temperatures on the Texas High Plains. Cool (< 15°C) night temperatures inhibit the production of cellulose for fiber growth often resulting in incomplete development of the fiber secondary wall. These scientists have succeeded in isolating a key enzyme involved in cellulose synthesis, sucrose phosphate synthase (SPS) from spinach. Spinach, being a cool-season crop, has a SPS enzyme with greater expression at low temperatures than cotton SPS. Wilkins and Jernstedt (1999) provide details of the role of SPS in cotton fiber development. The initial year of field trials on the High Plains demonstrated enhanced length, strength, and linear density properties of germplasm transformed with the spinach SPS enzyme. A current project involves production of these transgenic cottons all across the cotton belt to assess effects of upregulating SPS in cotton under higher night temperatures. Efforts to introgress the SPS gene into locally adapted germplasm have also commenced.

Transformation to effect enhanced fiber properties may be a reality in the future as fiber development is better understood, but much current research emphasizes application of structural genomic tools to cotton improvement. Application of molecular markers to cotton improvement is in its infancy as molecular biologists seek facile marker systems useable by breeders and genetic maps of sufficient density within and portability among populations to allow traits and portions of chromosomes to be associated, and thus manipulated. Molecular markers promise to increase breeding progress because they allow selection to operate on the genotype rather than solely on the phenotype. In this manner, a portion of the environmental variation and genotype × environment interaction can be removed from the selection response equation. The selection response equation we examined earlier can be expanded to include selection for molecular markers linked to exons imparting fiber quality. Briefly, the success of maker-assisted selection for a quantitatively inherited trait is predicated on the amount of the additive genetic variance explained by variation for markers at marker loci (Lande and Thompson, 1990). Fiber quality and yield traits are considered to be quantitatively inherited, and thus are genetically controlled by numerous genes each imparting small effects. Manipulation of loci (quantitative trait loci or QTLs) imparting fiber properties by selection for associated molecular markers is the source of much current research effort. Reinisch et al. (1994) reported the first dense cotton linkage map, comprising 705 loci at about 7 cM spacing. Since then, this map has become denser as additional markers have been found and new marker systems have been applied to cotton (Paterson and Smith, 1999). Existing dense linkage maps have been created in interspecific *G. hirsutum* × *G. barbadense* populations (Reinisch et al., 1994) because of insufficient marker polymorphism in intra-specific populations. Despite a relatively dense linkage map, marker assisted breeding or manipulation of QTLs is not yet a widespread reality because extant marker systems do not reveal sufficient genetic variation within *G. hirsutum* (Shappley et al., 1998; Jiang et al., 1998; Wright et al., 1999) and are not technologically feasible for most plant breeders. More facile marker systems amenable to high throughput by automation include single nucleotide polymorphism (Landegren, Nilsson, and Kwok, 1998; Lemieux, Aharoni and Schena, 1998) or SNPs. A SNP is the most detailed genetic difference between two genetic types and is envisioned as a key to understanding effects of complex gene interactions on plant phenotypes. Their abundance in the cotton genome is not yet established, but in theory SNPs may reveal more intra-specific variation than marker systems to date.

There is little doubt that genomic tools will soon identify chromatin imparting fiber qualities and allow breeders to accumulate this chromatin into genotypes through manipulation of molecular markers. These tools combined with knowledge of which fiber properties to target, their relationships, and methods to phenotype selections that minimize confounding effects of environment should be a powerful combination to effect better fiber quality.

CONCLUSIONS

Given the complexities involved in fiber quality improvement due to genetic and environmental influences, combined with their non-independence and the changing fiber needs of a textile industry adopting new technology, the question arises as to what should goals be to enhance fiber quality. It seems clear that rotor spinning and perhaps air-jet spinning will dominate global yarn manufacture in the near future, and thus the fiber needs of these systems should set priorities. Given the fiber profile needs of these yarn-spinning systems, this writer recommends that fiber length and its variability be given highest priority. Fiber length should be 'improved' by maintaining existing breeding gains in UHM while seeking more uniformity of fiber length in cotton bales. It is no longer sufficient to simply breed germplasm with longer UHM or 2.5% span length without considering variability in length uniformity. Higher fiber length uniformity in cotton bales could be achieved through germplasm that produces less short fiber, crop management systems that encourage fiber yield to be produced at first position fruiting sites, ginning systems that break fewer fibers, and possibly biotechnological approaches aimed at precisely regulating genes involved in fiber development. Fiber bundle strength, yarn strength, and perhaps fiber yields should positively respond to ameliorating short fiber content. Reducing short fiber content should coincidentally eliminate some degree of dye defects in textile products, as short fibers can be immature and consequently contribute to uneven dye uptake. Certainly, fiber quality variation should be given equal consideration to enhancing levels of fiber quality.

The pressure to enhance fiber quality has recently risen to the point that efforts are underway to revise the current cotton marketing system to encourage rather than discourage production of the highest quality fiber. The National Cotton Council recently passed a resolution to change the U.S. cotton marketing system to place more value on fiber length uniformity, fiber strength, and lower micronaire readings (less than 5.0). This resolution was prompted by cotton yarn and textile manufacturer concerns over fiber not meeting their needs and some U.S. grower's inability to sell fiber on world markets. Clearly, the World cotton industry is at a cross roads with respect to its current and future economic health if concerted efforts are not made to enhance fiber quality, including variability in fiber properties. New strategies to enhance cotton fiber quality are encouraged to maintain cotton's dominance as a textile fiber, lest this industry be left behind.

REFERENCES

Anthony, W.S. (1999). Post-harvest management of fiber quality. In *Cotton Fibers: Developmental Biology, Quality Improvement, and Textile Processing*, ed. A.S. Basra. Binghamton, NY: Food Products Press, pp. 293-333.

Anthony, W.S. and C.K. Bragg. (1987). Response of cotton fiber length distribution to production and ginning practices. *Transactions American Society of Agricultural Engineers* 30:290-296.

Backe, E.E. (1996). The importance of cotton fiber elongation on yarn quality and weaving performance. In *Proc. 9th Annual Engineered Fiber Selection System® Research Conference*, ed. C.H. Chewning. Raleigh, NC: Cotton Incorporated, pp. 1-13.

Becker, W.A. (1985). *Manual of Quantitative Genetics*. Pullman, WA: Academic Enterprises.

Behery, H.M. (1993). Short fiber content and uniformity index in cotton. *International Cotton Advisory Committee and Center for Agriculture and Biosciences Review Article 4*.

Bishr, M.A. (1954). Inheritance of perimeter and wall thickness of fiber in a cross between two varieties of upland cotton. *Ph.D. dissertation*, Baton Rouge, Lousiana State University.

Bradow, J.M., L.H. Wartelle, P.J. Bauer, and G.F. Sassenrath-Cole. (1997). Small-sample cotton fiber quality quantitation. *Journal of Cotton Science* 1:48-60.

Bragg, C.K. and F.M. Shofner. (1993). A rapid, direct measurement of short fiber content. *Textile Research Journal* 63:171-176.

Chewning, C.H., Jr. (1994). Cotton fiber management using Cotton Incorporated's Engineered Fiber Selection System and high volume instrument testing. In *Proc. 7th Annual Engineered Fiber Selection® System Conf.*, ed. C.H. Chewning. Raleigh, NC: Cotton Incorporated. pp. 16-18.

Cooper, H.B., S.R. Oakley, and J. Dobbs. (1988). Fiber strength by different test methods. In *Proceedings of the Beltwide Cotton Conference*, eds. D.J. Herber and D.A. Richter. Memphis, TN: National Cotton Council, pp.138-139.

Coyle, G.G. and C.W. Smith. (1997). Combining ability for within-boll yield components in cotton, *Gossypium hirsutum* L. *Crop Science* 37:1118-1122.

Culp, T.W., R.F. Moore, and J.B. Pitner. (1985). Simultaneous improvement of lint yield and fiber strength in cotton. *South Carolina Agricultural Experiment Station Bulletin 1090*.

Culp, T.W. (1992). Simultaneous improvement of lint yield and fiber quality in upland cotton. In *Cotton Fiber Cellulose: Structure, Function and Utilization Conference*, eds. C.R. Benedict and G.M. Jividen. Memphis, TN: National Cotton Council, pp. 247-287.

Deussen, H. (1992). Improved cotton fiber properties–The textile industry's key to success in global competition. In *Cotton Fiber Cellulose: Structure, Function and Utilization Conference*, eds. C.R. Benedict and G.M. Jividen. Memphis, TN: National Cotton Council, pp. 43-63.

El-Mogahzy, Y.E. (1998). Cotton Blending: How the EFS® system can help in producing optimum yarn quality. In *Proc. 10th Annual Engineered Fiber Selection® System Conf.*, ed. C.H. Chewning. Raleigh, NC: Cotton Incorporated. pp. 79-104.

Faerber, C. and H. Deussen. (1994). Improved cotton fiber quality and improved spinning technology–a profitable marriage. Part I. Progress in rotor spinning and progress in the quality profile of U.S. upland cotton. Part II. The contributions of improved cotton quality and those of rotor spinning developments to higher profits in cotton production and in spinning. In *Proceedings of the Beltwide Cotton Confer-*

ence, eds. D.J. Herber and D.A. Richter. Memphis, TN: National Cotton Council, pp. 1615-1621.

Faerber, C. (1995). Future demands on cotton fiber quality in the textile industry, technology-quality-cost. In *Proceedings of the Beltwide Cotton Conference*, eds. D.A. Richter and J. Armour. Memphis, TN: National Cotton Council, pp.1449-1454.

Falconer, D.S. (1989). *Introduction to Quantitative Genetics*. New York, NY: John Wiley & Sons.

Haigler, C.H., E.F. Hequet, D.R. Krieg, R.E. Strauss, B.G. Wyatt, W. Cai, T. Jaradat, N.G. Srinivas, C. Wu, A.S. Holoday, and G.M. Jividen. (2000a). Transgenic cotton with improved fiber micronaire, strength, and length, and increased fiber weight. In *Proceedings of the Beltwide Cotton Conference*, eds. P. Dugger and D.A. Richter. Memphis, TN: National Cotton Council, In press.

Haigler, C.H., W. Cai, K. Martin, J. Tummala, R. Anconetani, A.S. Holoday, G.M. Jividen, and J.R. Gannaway. (2000b). Mechanisms by which fiber quality and fiber and seed weight can be improved in transgenic cotton growing under cool night temperatures. In *Proceedings of the Beltwide Cotton Conference*, eds. P. Dugger and D.A. Richter. Memphis, TN: National Cotton Council, In press.

Hake, K., B. Mayfield, H. Ramey, and P. Sasser. (1990). Producing quality cotton. Memphis, TN: National Cotton Council.

Hertel, K.L. (1940). A method of fibre-length analysis using the fibrograph. *Textile Research Journal* 10:510-525.

Hertel, K.L. (1953). The Stelometer, it measures fiber strength and elongation. *Textile World* 103:97-260.

Hertel, K.L. and C.J. Craven. (1951). Cotton fineness and immaturity as measured by the arealometer. *Textile Research Journal* 21:765-774.

Jenkins, J.N, J.C. McCarty, Jr., R.E. Buehler, J. Kiser, C. Williams, and T. Wofford. (1997). Resistance of cotton with delta-endotoxin genes from *Bacillus thuringiensis* var. *kurstaki* on selected lepidopteran pests. *Agronomy Journal* 89:768-780.

Jiang, C., R. Wright, K. El-Zik, and A.H. Paterson. (1998). Polyploid formation created unique avenues for response to selection in *Gossypium* (cotton). Proceedings of the National Academy of Sciences 95:4419-4424.

John, M.E. (1992). Genetic engineering of cotton for fiber modification. In *Cotton Fiber Cellulose: Structure, Function and Utilization Conference*, eds. C.R. Benedict and G.M. Jividen. Memphis, TN: National Cotton Council, pp. 91-105.

John, M.E. (1996). Metabolic pathway engineering in cotton: Biosynthesis of polyester in fiber. In *Proceedings of the Beltwide Cotton Conference*, eds. P. Dugger and D.A. Richter. Memphis, TN: National Cotton Council, p. 1679.

John, M.E. (1999). Genetic engineering strategies for cotton fiber modification. In *Cotton Fibers: Developmental Biology, Quality Improvement, and Textile Processing*, ed. A.S. Basra. Binghamton, NY: Food Products Press, pp. 271-292.

Kerr, T. (1966). Yield components in cotton and their interrelations with fiber quality. *Proceedings of the 18th Cotton Improvement Conference*. Memphis, TN: National Cotton Council, pp. 276-287.

Knowlton, J.F. (1999). Short fiber measurement. In *Proc. 12th Annual Engineered Fiber Selection® System Conf.*, ed. C.H. Chewning. Raleigh, NC: Cotton Incorporated. <http://www.cottoninc.com/EFSConference/>.

Lande, R. and R. Thompson. (1990). Efficiency of marker-assisted selection in the improvement of quantitative traits. *Genetics* 124:743-756.

Landegren, U., M. Nilsson, and P.-Y. Kwok. (1998). Reading bits of genetic information: Methods for single nucleotide polymorphism analysis. *Genome Research* 8:769-776.

Landstreet, C.B., P.R. Ewald, and T. Kerr. (1959). A miniature spinning test for cotton. *Textile Research Journal* 29:701-706.

Latimer, S.L., T.P. Wallace, and D.S. Calhoun. (1996). Cotton breeding: High volume instrument versus conventional fiber quality testing. In *Proceedings of the Beltwide Cotton Conference*, eds. D.J. Herber and D.A. Richter. Memphis, TN: National Cotton Council, p.1681.

Lemieux, B., A. Aharoni, and M. Schena. (1998). Overview of DNA chip technology. *Molecular Breeding* 4:277-289.

Lewis, H.L. (1992). Future horizons in cotton research. In *Cotton Fiber Cellulose: Structure, Function and Utilization Conference*, eds. C.R. Benedict and G.M. Jividen. Memphis, TN: National Cotton Council, pp. 5-18.

Lewis, H.L. (1996). Variation in cotton fiber quality. In *Proc. 9th Annual Engineered Fiber Selection System® Research Conference*, ed. C.H. Chewning. Raleigh, NC: Cotton Incorporated, pp. 29-33.

Lewis, H.L. (1998). Genetic short fiber content of American upland cotton. In *Proc. 11th Annual Engineered Fiber Selection System® Research Conference*, ed. C.H. Chewning. Raleigh, NC: Cotton Incorporated, pp. 61-67.

Lewis, H.L. (1999). How should cotton fiber be improved ? In *Proc. 12th Annual Engineered Fiber Selection® System Conf.*, ed. C.H. Chewning. Raleigh, NC: Cotton Incorporated. <http://www.cottoninc.com/EFSConference/>.

Lewis, H.L. (2000). Cotton yield and quality–yesterday today and tomorrow. In *Proc. 13th Annual Engineered Fiber Selection® System Conf.*, ed. C.H. Chewning. Raleigh, NC: Cotton Incorporated. <http://www.cottoninc.com/EFSConference/>.

May, O.L. (1999). Genetic variation in fiber quality. In *Cotton Fibers: Developmental Biology, Quality Improvement, and Textile Processing*, ed. A.S. Basra. Binghamton, NY: Food Products Press, pp. 183-229.

May, O.L. (2000). Breeding cotton–public sector outlook. Keys to a profitable U.S. cotton industry–yield/quality issues to address through breeding. In *Proc. 13th Annual Engineered Fiber Selection® System Conf.*, ed. C.H. Chewning. Raleigh, NC: Cotton Incorporated. <http://www.cottoninc.com/EFSConference/>.

May, O.L. and G.M. Jividen. (1999). Genetic modification of cotton fiber properties with single- and high-volume instruments. *Crop Science* 39:328-333.

May, O.L., M.E. John, and T.J. Wofford. (2000). Breeding transformed cotton expressing enhanced fiber strength. *Journal of New Seeds* 2: In press.

May, O.L. and R.A. Taylor. (1998). Breeding cottons with higher yarn tenacity. *Textile Research Journal* 68:302-307.

Meredith,W.R., Jr. (1984). Quantitative genetics. In *Cotton*, eds. R.J. Kohel and C.F. Lewis. Madison, WI: Crop Science Society of America, pp.131-150.

Meredith,W.R., Jr. (1992). Improving fiber strength through genetics and breeding. In *Cotton Fiber Cellulose: Structure, Function and Utilization Conference*, eds. C.R. Benedict and G.M. Jividen. Memphis, TN: National Cotton Council, pp. 289-302.

Meredith, W.R., Jr., T.W. Culp, K.Q. Robert, G.F. Ruppenicker, W.S. Anthony, and J.R. Williford. (1991). Determining future cotton variety fiber quality objectives. *Textile Research Journal* 61:715-720.

Meredith, W.R., Jr., P.E. Sasser, and S.T. Rayburn. (1996). Regional high quality fiber properties as measured by conventional and AFIS methods. In *Proceedings of the Beltwide Cotton Conference*, eds. P. Dugger and D.A. Richter. Memphis, TN: National Cotton Council, pp.1681-1684.

National Cotton Council. (1999). Cotton: Profile of a resourceful industry. Memphis, TN: National Cotton Council <http://www.cotton.org/ncc/education/profile.pdf>.

Niles, G.A and C.V. Feaster. (1984). Breeding. In *Cotton*, eds. R.J. Kohel and C.F. Lewis. Madison, WI: Crop Science Society of America. pp. 201-231.

Paterson, A.H. and R.H. Smith. (1999). Future horizons: Biotechnology for cotton improvement. In *U.S. Cotton: Origin, History, Technology, and Production*, eds. C.W. Smith and J.T. Cothren. New York, NY: John Wiley & Sons, pp. 415-432.

Perkins, H.H., Jr., D.E. Ethridge, and C.K. Bragg. (1984). Fiber. In *Cotton*, eds. R.J. Kohel and C.F. Lewis. Madison, WI: Crop Science Society of America, pp. 437-509.

Pressley, E.H. (1942). A cotton fiber strength tester. *American Society for Testing and Materials Bulletin* 118:13-18.

Ramey, H.H., Jr., R. Lawson, and S. Worley, Jr. (1977). Relationship of cotton fiber properties to yarn tenacity. *Textile Research Journal* 47:685-691.

Reinisch, A.J., J. Dong, C.L. Brubaker, D.M. Stelly, J.F. Wendel, and A.H. Paterson. (1994). A detailed RFLP map of cotton, *Gossypium hirsutum* × *Gossypium barbadense*: Chromosome organization and evolution in a disomic polyploid genome. *Genetics* 138:829-847.

Rowland, J.D. (1999). Cotton short fiber testing. In *Proc. 12th Annual Engineered Fiber Selection® System Conf.*, ed. C.H. Chewning. Raleigh, NC: Cotton Incorporated. <http://www.cottoninc.com/EFSConference/>.

Sachs, E.S., J.H. Benedict, D.M. Stelly, J.F. Taylor, D.W. Altman, S.A. Berberich, and S.K. Davis. (1998). Expression and segregation of genes encoding CryIa insecticidal proteins in cotton. *Crop Science* 38:1-11.

Sasser, P.E. (1992). The physics of fiber strength. In *Cotton Fiber Cellulose: Structure, Function and Utilization Conference*, eds. C.R. Benedict and G.M. Jividen. Memphis, TN: National Cotton Council, pp. 19-27.

Sasser, P.E. and J.L. Shane. (1996). Crop quality–a decade of improvement. In *Proceedings of the Beltwide Cotton Conference*, eds. P. Dugger and D.A. Richter. Memphis, TN: National Cotton Council, pp. 9-12.

Scholl, R.L. and P.A. Miller. (1976). Genetic association between yield and fiber strength in upland cotton. *Crop Science* 16:780-783.

Shappley, Z.W., J.N. Jenkins, J. Zhu, and J.C. McCarty, Jr. (1998). Quantitative trait loci associated with agronomic and fiber traits of upland cotton. *Journal of Cotton Science* 4:1-28.

Smith, H. and R. Zhu. (1999). The spinning process. In *Cotton*, eds. C. Wayne Smith and J. Tom Cothren. New York, NY: John Wiley & Sons, Inc., pp. 729-749.

Steadman, R.G. (1997). Cotton Testing. *Textile Progress* 27:1-66.

Suh, M.W., X.-L. Cui, and P.E. Sasser. (1994). New understanding on HVI tensile data based on Mantis single fiber test results. In *Proceedings of the Beltwide Cotton*

Conference, eds. P. Dugger and D.A. Richter. Memphis, TN: National Cotton Council, pp.1400-1402.

Taylor, R.A. (1982). Measurement of cotton fiber tenacity on 1/8 gage HVI tapered bundles. *Journal of Engineering for Industry* 104:169-174.

Taylor, R.A. (1986). Cotton tenacity measurements with high-speed instruments. *Textile Research Journal* 56:92-101.

Taylor, R.A., L.C. Godbey, D.S. Howle, and O.L. May. (1995). Why we need a standard strength test for cotton variety selection. In *Proceedings of the Beltwide Cotton Conference*, eds. P. Dugger and D.A. Richter. Memphis, TN: National Cotton Council, pp. 1175-1178.

Wilkins, T.A. (1996). Bioengineering fiber quality: Molecular determinants of fiber length and strength. In *Proceedings of the Beltwide Cotton Conference*, eds. P. Dugger and D.A. Richter. Memphis, TN: National Cotton Council, pp. 1679-1680.

Wilkins, T.A. and J.A. Jernstedt. (1999). Molecular genetics of developing cotton fibers. In *Cotton Fibers: Developmental Biology, Quality Improvement, and Textile Processing*, ed. A.S. Basra. Binghamton, NY: Food Products Press, pp. 231-269.

Wright, R., P. Thaxton, A.H. Paterson, and K. El-Zik. (1999). Molecular mapping of genes affecting pubescence of cotton. *J. Heredity* 90:215-219.

Sugar Beet Quality Improvement

Larry G. Campbell

SUMMARY. More than one-third of the sugar (sucrose) consumed by humans is obtained from sugar beet (*Beta vulgaris* L.). Sucrose extraction begins with the production of a dark opaque juice from strips of sugar beet. This juice is purified with lime and carbon dioxide, thickened by evaporation, and crystallized under a vacuum. Soluble nonsucrose constituents of sugar beet, referred to collectively as impurities, impede sucrose crystallization in normal factory processes. Sucrose concentration and the ratio of sucrose to total soluble solids (sucrose plus impurities) determine processing quality of sugar beet. Among the more important impurity components are sodium, potassium, and amino-nitrogen. Sucrose and impurity concentrations can be altered in breeding programs. However, a negative association between root yield and sucrose concentration and interactions among impurity components and between impurity components and yield or sucrose concentration have complicated breeding efforts. Also, almost any cultural practice may affect the quality of the crop. Nitrogen fertilizer management is a challenge wherever sugar beet is grown. Producers' returns can be increased with proper nitrogen application but even moderate over-fertilization may result in a costly reduction in crop quality. Both producers and processors operate on small profit margins. In this economic environment, producing a high quality crop is a necessity. *[Article copies available for a fee from The Haworth Document Delivery Service: 1-800-HAWORTH. E-mail address: <getinfo@haworthpressinc.com> Website: <http://www.HaworthPress.com>]*

KEYWORDS. *Beta vulgaris* L., nitrogen fertilizer, sucrose, sugar, sugar beet breeding, sugar beet processing, sugar beet production

Larry G. Campbell is Research Geneticist, USDA-Agricultural Research Service, Northern Crop Science Laboratory, Fargo, ND 58105-5677 USA.

[Haworth co-indexing entry note]: "Sugar Beet Quality Improvement." Campbell, Larry G. Co-published simultaneously in *Journal of Crop Production* (Food Products Press, an imprint of The Haworth Press, Inc.) Vol. 5, No. 1/2 (#9/10), 2002, pp. 395-413; and: *Quality Improvement in Field Crops* (ed: A. S. Basra, and L. S. Randhawa) Food Products Press, an imprint of The Haworth Press, Inc., 2002, pp. 395-413. Single or multiple copies of this article are available for a fee from The Haworth Document Delivery Service [1-800-HAWORTH, 9:00 a.m. - 5:00 p.m. (EST). E-mail address: getinfo@haworthpressinc.com].

INTRODUCTION

Approximately 255 million tons of sugar beet (*Beta vulgaris* L.) are produced worldwide on 7.76 million hectares. Thirty-five to 40% of the sucrose consumed by humans is extracted from sugar beet with the remainder from sugarcane (*Saccharum* sp. Hyb.). Sugar beet is grown as a summer crop in maritime, prairie, and semi-continental climates and as a winter or summer crop in Mediterranean and some semi-arid environments (Draycott, 1972). In regions with mild climates, sugar beet is harvested, delivered to the factory, and processed within a few days. In regions with cold winters, harvest is delayed until freezing temperatures are anticipated, fields are harvested within a short period, and sugar beet is stored in large exposed piles for up to five months awaiting processing (Bugbee, 1993).

To begin processing, the beet is washed and cut into cossettes (slender strips). A dark opaque juice is extracted from the cossettes with hot water. This 'raw juice' is purified with lime and carbon dioxide to produce a clear filtrate or 'thin juice.' Through evaporation, the thin juice becomes a 'thick juice,' containing approximately 60% dissolved solids. Crystallization of sucrose from the thick juice occurs under a vacuum (Harvey and Dutton, 1993). The white sugar produced for sale is 99.9% sucrose (Bichsel, 1988). The molasses remaining after all the sucrose (that can be extracted economically) is removed is a by-product used primarily as animal feed or as a culture media for fermentation. Approximately 50% of the dry matter of molasses is sucrose that can not be extracted during normal processing (Harland, 1993). Some of the sucrose in the molasses can be extracted (Perschak, 1998); however, this requires additional equipment and time and does not diminish the value of high quality sugar beet to processors.

Payments to growers for sugar beet constitutes the major expense for processors and most of their revenue is realized from the sale of crystallized sucrose. Therefore, sugar beet quality generally refers to sucrose concentration and concentrations of naturally occurring soluble nonsucrose constituents that impede sucrose crystallization, as determined at harvest or time of delivery. Changes during sugar beet storage that reduce sucrose extraction rate are also important to processors. Growers' attempts to optimize the quality of their crop by choice of hybrids and cultural practices are beneficial; however, weather conditions have a substantial and often unpredictable impact on root yield and quality.

QUALITY COMPONENTS AND THEIR MEASUREMENT

At harvest, 73 to 77% of the weight of a typical sugar beet root is water, 4.5 to 5% insoluble solids, and 16 to 22% soluble solids. Almost all of the insoluble solids are readily removed, leaving the soluble solids as the basic raw material for processing. Sucrose makes up approximately 80% (14 to 20% of root weight) of the soluble solids (Bichsel, 1988; Bohn et al., 1998). The remaining nonsucrose soluble components are collectively referred to as impurities. Each kilogram of impurities prevents crystallization of 1.5 to 1.8 kg of sucrose that consequently is lost to molasses (Alexander, 1971). Hence, sucrose content and purity, the percent of total soluble solids which is sucrose, define sugar beet quality.

The soluble nonsucrose components of sugar beet include: (1) nitrogenous compounds such as proteins, betaine, amino acids, and amides; (2) nitrogen-free organic compounds including pectic substances, organic acids, lipids, and saponins; (3) inorganic cations (potassium, sodium, calcium, and magnesium) and anions (chlorides, sulfates, phosphates, silicates, iron, and aluminum); and (4) monosaccharides of which glucose and fructose comprise the major fraction (Anderson and Barfoed, 1988; Bohn et al., 1998). Glucose and fructose are products of metabolism and also are formed through sucrose hydrolysis. An equal molar mixture of these two hexoses is referred to as invert sugar (Hartmann, 1977).

Commercial processing operations and many research situations require determination of sucrose concentration and quality characteristics of numerous samples. The goal is to obtain a reasonable estimate of processing value with a few laboratory measurements as quickly and accurately as feasible. Typically a 10 to 15 beet sample is delivered to the laboratory. Soil adhering to the roots is washed off and any remaining leaf material removed. A brei (fine beet particles) sample is obtained by passing roots over a beet rasp or through a multi-bladed beet saw. For determining sucrose concentration, brei is mixed with aluminum chloride solution, blended, and filtered. Most laboratories use polarimetry to determine sucrose concentration of the resulting clarified aqueous root extract and calculate sucrose percent based upon fresh weight of the sugar beet (Martin et al., 1980; Hecker and Martin, 1981). Dexter et al. (1967) presented a method for estimating purity and extractable sucrose percent. With their procedure, juice is squeezed from a brei sample, heated, and adjusted to proper pH in a series of steps. Total solids in the final solution are determined using a refractometer and a polarimeter is used to measure sucrose content. By making reasonable assumptions regarding factory losses and molasses purity (Carruthers et al., 1963), one can use these measurements to calculate percent extractable sucrose and hence extractable sucrose per ton or hectare.

Carruthers et al. (1962) and Last and Draycott (1977) demonstrated that the concentration of three major impurity components, sodium, potassium, and amino-nitrogen, could be combined to estimate purity or calculate percent sucrose loss to molasses (Hilde et al., 1983). These three impurities are easily measured using a portion of the clarified root extract obtained for determining sucrose concentration (Martin et al., 1980). Sodium and potassium concentrations are frequently determined using flame photometry; amino-nitrogen with a colorimetric procedure.

Knowledge of the concentration of other impurity components may be useful for regulating some factory operations. Additional nitrogen-containing compounds (McGinnis, 1982; Bohn, 1988), especially betaine (Smith et al., 1977), are often considered because of their effects on sugar extraction and the tendency of the crop to accumulate these compounds when available soil nitrogen is excessive. Increased concentrations of invert sugar and raffinose and gum formation are associated with beet deterioration during storage or freezing (Oldfield et al., 1971; Bugbee, 1993).

The basic methods for determining sucrose and impurity component concentration or percent purity have been used by the sugar industry for some time with only slight modification. However, advances in instrumentation, automation, and computer technology (Schiweck and Steinle, 1988; van der Poel et al., 1998) have increased the efficiency of obtaining needed measurements and accuracy of results. Hobbis et al. (1982) described the design, operation, and maintenance of a quality laboratory capable of processing 1,080 samples per hour. With this capacity the company can sample 40% of the truck loads received and determine each grower's payment based upon recoverable sucrose delivered.

Of the 160 kg of sugar in a typical ton of sugar beet with a 16% sucrose concentration at harvest, 83% or 133 kg will be recovered for sale as sugar, 20 kg will be lost to molasses, and 7 kg will be lost during storage and processing and not accounted for (Bichsel, 1988). Additionally, 55 kg of pulp and 42 kg molasses will need to be stored, processed, and sold as by-products. The 42 kg of molasses from each ton of sugar beet will contain 13.4 kg of impurities and 8.6 kg of water, in addition to the 20 kg of sugar. Modern factories process thousands of tons of sugar beet each day. Therefore, relatively small increases in sucrose concentration or decreases in impurities can have significant economic impact (Oldfield et al., 1977; Hilde et al., 1983; van Geijn, 1983; Viccari et al., 1988; van den Hil and de Nie, 1989). Quality of sugar beet being processed has minimal impact upon processing cost per ton, within wide ranges.

PLANT CHARACTERISTICS RELATED TO QUALITY

An inverse relationship between root yield (tonnage) and sucrose concentration has hampered progress of sugar beet breeders and frustrated agronomists and growers since the beginning of the beet sugar industry. Campbell and Kern (1983) found that while variability in sucrose concentration had a major effect upon recoverable sucrose per ton; the major factor influencing recoverable sucrose per hectare was root yield, followed by sucrose concentration. Sodium, potassium, and amino-nitrogen concentration were positively correlated with each other and with root yield and negatively associated with sucrose concentration. The consistency of relative amino-nitrogen concentrations across environments and its relatively large influence on recoverable sucrose per ton suggested that it be given prime consideration in any attempt to lower impurity levels. Carter (1986) found high sucrose concentration was associated with low to moderate nitrogen uptake, low sodium concentration, high potassium to sodium ratio, and low water concentration.

Artschwager (1930) examined the effects of cambial ring number and structure on sucrose concentration and yield. Anatomical configurations associated with low or high sugar sucrose concentration were not consistent in all cultivars, suggesting they were of little value as selection criteria in a breeding program. Milford (1973) observed an association between smaller cell size and increased sucrose concentration. Doney et al. (1981) hypothesized that a positive correlation between sucrose concentration and ring number and density was largely due to differences in cell size, particularly size of the inter-vascular parenchyma cells. Pack (1930) examined relationships among 50 characteristics and found sucrose concentration to be positively associated with high density, high dry matter content, shoulder extension, firmness of tissue, root length, roughness of skin, and darkness of leaf color.

What is often referred to as a sugar beet root consists of a crown, hypocotyl, and true root (Wyse, 1982). The crown is stem tissue that supports the leaves. The hypocotyl is the transition zone between crown and true root. The relatively contributions of crown and root to sugar beet yield and quality have been studied extensively (Carruthers et al., 1966). Up to 20 percent of sugar beet is crown tissue (Cole, 1980). Zielke and Snyder (1974) reported that cultivar rankings for impurity concentration in the crown were different from rankings based upon concentration in the root. The concentration of impurities in the crowns averaged 70% more than in the roots. Furthermore, crowns have lower sucrose concentrations than roots (Cole and Seiler, 1976); however, complete removal of the crown during harvest would decrease recoverable sugar yields by 900 to 1120 kg ha^{-1} (Zielke, 1973). Campbell and Cole (1986) suggested that relationships among yield and quality traits were strongly influenced by crown size. Root yield and sucrose loss to molasses were positively correlated with each other and negatively associated with sucrose concentration. Much of the ability to respond to more favorable environmental conditions appeared to be related to production of larger crowns.

The skin or peel of sugar beet also has a lower sucrose concentration and higher impurity content than the main portion of the root. Edwards et al. (1989a, 1989b) found that removing 3.5 to 5.8% of the weight of the root by peeling increased purity of the extract by as much as 1.6%; however, it also resulted in a loss of 1.8 to 2.8% of the sugar. Peeling has not been economically feasible in commercial processing. Commercial production of smooth root sugar beet (sugar beet roots without the two vertical groves and associated mass of fibrous roots characteristic of current cultivars) would make peeling less difficult (Theurer, 1993; Saunders et al., 2000).

GENETIC IMPROVEMENT OF QUALITY TRAITS

The negative correlation between root yield and sucrose concentration (Wyse, 1979; Campbell and Kern, 1983; Campbell and Cole, 1986) has complicated quality improvement efforts. In general, the genetic variance for sucrose concentration is predominantly additive. However, nonadditive genetic variance and specific combining ability components are significant in determining root yield, and, therefore have a role in determining sucrose yield (MacLachlan, 1972; Smith et al., 1973; Hecker, 1991; Ahmadi and Assad, 1998). Increasing sugar yields by increasing root yield often has been more fruitful than attempting to increase sucrose concentration, perhaps because a large portion of the dry weight of current commercial sugar beet hybrids is sucrose (Bohn et al.,

1998). However, continued improvement within elite breeding populations may require increased emphasis on sucrose concentration and purity (Smith and Hecker, 1973). Selecting for increased sugar concentration usually results in a concurrent increase in purity. Some processors pay premiums for higher extractable sucrose concentrations, hence, hybrids producing the most sugar per hectare are not necessarily the most profitable for growers. These premiums are not uniform across the industry, complicating breeding efforts to maximize economic return to growers.

Among the soluble nonsucrose constituents or impurity components, sodium, potassium, and amino-nitrogen have received the most attention in cultivar development programs (Smith et al., 1977). Finkner and Bauserman (1956) found that selection for sodium, per se, had little effect upon sucrose concentration and concluded that selecting for low sodium was of little value, in most situations. In contrast, Wood et al. (1958) presented evidence that suggested sucrose content could be increased more efficiently by selecting for both high sucrose and low sodium than selecting for sucrose concentration alone. Dudley and Powers (1960) concluded that it should be possible to develop lines with low concentrations of both sodium and potassium. Sodium, potassium, and amino-nitrogen levels can be shifted dramatically with only a few cycles of selection (Powers et al., 1962; Coe, 1987; Smith and Martin, 1989), suggesting that additive genetic variance is important in determining relative levels of these traits (Smith et al., 1973). Interactions among impurity components, sucrose concentration, and root yield complicate selection for optimum levels of yield and quality traits. Selection for low sodium concentration was accompanied by an increase in extractable sugar per ton. In contrast, selection for low amino-nitrogen concentration caused a decrease in extractable sugar per ton and selection for low potassium had no effect on extractable sugar (Smith and Martin, 1989). Coe (1987) found that roots selected for the lowest concentrations of nonsucrose solubles tended to be the smallest roots and concluded that root size must be considered if one wishes to minimize reductions in root yield that would otherwise accompany selection for low impurities. Multiple character selection will be required when attempting to increase extractable sugar per ton and also maintain or increase root yield (Smith, 1988).

Most commercial sugar beet hybrids are best characterized as 3-way topcross hybrids (Bosemark, 1993; Campbell, 2000). The female parent in the final cross is an F_1 hybrid between a cytoplasmic male sterile line and an unrelated O-type line (often referred to as a B-line or maintainer line in other crops). A tetraploid pollinator and a diploid cytoplasmic male sterile F_1 are routinely crossed to produce triploid hybrids for commercial production. An advantage of triploid hybrids is that only one of three genomes of the hybrid originate from the cytoplasmic male sterile parent, the parent that often is more difficult to improve. Continuing availability of commercial diploid hybrids suggests that triploids are not clearly superior to diploids. Neither a general beneficial nor a detrimental effect on root yield or sucrose appears to be associated with the addition of a genome (Hecker et al., 1970; McFarlane et al., 1972; Smith et al., 1979). In instances where triploids enhanced sugar production, the benefit almost always was related to an increase in root yield, with little effect upon sucrose concentration. In comparisons of diploids, triploids, and tetraploids, ploidy status influenced the level of activity of several enzymes important in sugar beet metabolism (Spettoli et al., 1976).

Environmental variation (Milford and Thorne, 1973; Burley, 1990; Scott and Jaggard, 1993; Campbell, 1995) and cultivar × environment interactions (Ulrich, 1961) must be considered in any plant breeding program. Since sugar beet production does not require completion of the plant's reproductive cycle, these interactions may be less important in sugar beet than in crops requiring seed production. Cultivar × location, cultivar × year, and cultivar × location × year variance components for sucrose, sodium, and potassium concentration were relatively small compared to the cultivar variance component, indicating consistency in relative expression of these characters in an analysis including ten commercial hybrids and 20 environments (Campbell and Kern, 1982). Cultivar × planting date interactions for yield and sucrose concentration (Yonts et al., 1999) indicated that some hybrids are better than others for late planting or replanting. Lasa et al. (1989) found sucrose yields of triploids to be more stable across environments than that of diploids.

The early beet sugar industry was dependent upon a few open-pollinated lines that had been selected for high sugar concentration. With this limited base as a beginning and utilization of single, or at best a few, sources for monogerm seed, cytoplasmic male sterility, and resistance to some major diseases, the genetic base of the commercial sugar beet crop is quite narrow (Lewellen, 1992; Stander, 1993; McGrath et al., 1999). Sugar beet breeders constantly are seeking germplasm that will increase extractable sugar concentration without reducing root yield. Also, since heterosis generally is enhanced by increasing genetic diversity of the parents, desirable germplasm from previously unused sources would benefit long-term commercial hybrid breeding efforts. L19, an inbred line with a Polish cultivar in its parentage (Theurer, 1981), is noted for its high sucrose concentration and ability to increase sugar concentration in hybrids in which it is a parent. Doney and Theruer (1990) found that the osmolality of extracts of L19 root tissue was consistently higher than other lines examined and also observed a unique seasonal sucrose accumulation pattern for L19 (Theurer and Doney, 1989). Accessions from the USDA collection have not been used extensively in breeding programs and, thus, may provide unique genetic combinations for increased root yield and sucrose concentration while increasing genetic diversity in the crop. *Beta vulgaris* accessions from this collection were used as a source population in a selection program with the objective of extracting agronomically desirable germplasm from previously untapped sources (Campbell, 1989). This endeavor resulted in the release of five germplasm lines that had sucrose concentrations at least as high as current commercial hybrids (Campbell, 1990; 1992). Attempts to introduce diversity from wild sea beet (*B. vulgaris* ssp. *maritima*) have had limited success. Test crosses involving four germplasm lines selected from a cultivated × wild cross exhibited significant heterosis. Root yields were equal or greater than commercial hybrids; however, sugar concentrations were slightly lower (Doney, 1995).

EFFECTS OF CULTURAL PRACTICES ON QUALITY

Cultural practices significantly influence root yield, sucrose concentration, and quality characteristics of sugar beet. Obtaining an adequate uniform initial stand is essential (Winter, 1980). This usually involves planting as soon as soil temperatures are adequate for germination and risk of newly emerged seedlings being damaged by frost is low (Yonts et al., 1983; Lee et al., 1986; Campbell and Enz, 1991). In general, sucrose content and purity will increase as plant spacing (Eckhoff et al., 1991) and distance between rows (Yonts and Smith, 1997) decreases. Narrow rows and dense stands reduced amino-nitrogen concentration and sugar loss to molasses even when water stress reduced root yields substantially (Winter, 1989). Excessively high plant populations should be avoided because they may reduce the number of harvestable roots (Yonts and Smith, 1997).

Proper nitrogen fertilization is of utmost importance in producing quality sugar beet. Producers' returns can be increased with proper application of nitrogen and even moderate over-fertilization may be costly

(Adams et al., 1983; Eckhoff, 1999; Smith, 2000). The amount of nitrogen required to optimize recoverable sucrose yields is substantially below the level required to maximize tonnage. High levels of nitrogen reduce the sucrose portion of root dry weight (Milford and Watson, 1971). High rates of nitrogen increase root cell volumes, but have no effect upon number of cambial rings. The amount of sucrose entering a root is not affected by excess nitrogen; however, more sucrose is metabolized for root growth than when nitrogen is limited. Crown tissue yield increases and sucrose concentration decreases in both root and crown offset the benefits of increased tonnage obtained with excessive nitrogen (Halverson et al., 1978). Sugar beet cultivars differ in amount of crown tissue they produce; however, differences in crown tissue and sucrose production due to nitrogen fertilizer rate are larger than differences among cultivars, according to Halverson and Hartman (1980). The percentage of each whole beet that was crown tissue increased linearly in response to increased available nitrogen, sucrose concentration and purity decreased linearly, and root yield increased curvilinearly. Sucrose production was maximized at the same nitrogen rate for all cultivars examined. When determining the amount of nitrogen fertilizer to apply one must take into account the available residual nitrogen. The detrimental effects of excessive deep residual nitrogen are especially difficult to overcome (Winter, 1998). Deep-rooted crops, such as alfalfa (*Medicago sativa* L.), in the crop rotation with sugar beet may be required for production of quality sugar beet on fields with excessive deep nitrogen (Winter, 1984). If nitrogen in the upper 60 to 90 cm of the soil is low, fertilizer probably will be necessary for maximum sugar beet production even though deep nitrogen may be detrimental later in the season (Moraghan, 1982). Carter and Traveller (1981) recommended that nitrogen fertilizer be applied early. Excessive or late nitrogen applications increased impurities and decreased refined sucrose production. The crop preceeding sugar beet may affect the amount of nitrogen fertilizer needed. This effect cannot always be explained by differences in residual nitrogen or soil moisture (Christenson and Butt, 2000).

In production regions where irrigation is common, the amount of fertilizer needed by the crop will depend upon the amount of water to be applied. Availability of water and the cost of irrigating has prompted examination of irrigation amounts below those required to fully satisfy evapotranspiration (Winter, 1988). Drier irrigation treatments increased impurity concentrations, especially amino-nitrogen concentrations, even when less nitrogen was applied. Winter (1990) reported significant nitrogen × irrigation interactions for all yield and quality components in fields with low residual nitrogen. Serious impurity problems occurred when excess applied or residual nitrogen accompanied reduced irrigation. Production of low impurity sugar beet with reduced irrigation may not be possible on some soils without sacrificing yield of prior crops in the rotation. An economic analysis revealed that relatively large changes in irrigation costs, sugar beet prices, and fertilizer costs changed optimal irrigation and nitrogen levels only slightly (Lansford et al., 1989).

Nutrients other than nitrogen must be available in adequate amounts; however, their management generally is less complicated than that of nitrogen. Potassium and sodium can partially replace each other. However, optimum performance requires an adequate supply of both elements (Draycott, 1993). Potassium and sodium fertilization increases sugar beet yields in much of Britain (Farley and Draycott, 1974). Optimum amounts of potassium and sodium fertilizer increased sucrose concentration and root weight to top weight ratio. Neither element affected processing quality, although each element increased its own concentration it proportionally decreased amino-nitrogen concentration. Ludwick et al. (1980) reported that applying potassium chloride fertilizer to fields with existing high levels of potassium and chloride had no effect on root yield, sucrose concentration, or purity. Other research (Giroux and Tran, 1989) has indicated that adding potassium to soils high in potassium will decrease sucrose extractability. Carter (1986) and Bravo et al. (1989) observed a positive correlation between potassium and sodium concentrations in the roots and nitrogen uptake. Differences among cultivars were noted also (Carter, 1986). The response of sugar beet to phosphorus fertilization is usually small in cropping systems were phosphorus has been applied for a number of years (Draycott, 1993). Although an adequate supply of phosphorus is required for normal growth and development its effects upon sugarbeet processing quality per se are minimal.

Factories in areas were sugar beet is stockpiled for processing during the winter often begin processing a month or so before the end of the growing season. During this time growers deliver only enough sugar beet to operate the factories on a day-to-day basis. This allows for efficient use of the factory but limits productivity on a portion of the acreage. A harvest date $\times$ nitrogen interaction for recoverable sucrose was observed by Lauer (1995), suggesting that fertilizer rates should be decreased in fields slated for early harvest. Held et al. (1994) found that net returns increased if nitrogen fertilizer rates were reduced by 13 to 16 kg ha^{-1} for each week prior to normal harvest date. Inconsistent or nonsignificant harvest date $\times$ cultivar interactions indicated that cultivar choice should not be affected by anticipated harvest date (Lauer, 1994; Smith et al., 2000).

CONCLUSIONS

Production costs have increased at a faster rate than commodity prices for many agricultural products, including sugar beet. To remain viable, producers must increase per hectare yields and/or increase the value of their products. Similarly, the wholesale price of sugar has remained relatively stable while processing costs have increased. Therefore, production of quality sugar beet will benefit both producers and processors and allow the beet sugar industry to remain competitive in a sweetener market that also includes cane sugar and high-fructose corn sweeteners.

Enhancing sugar beet quality through breeding is slow and laborious. The negative correlation between root yield and sucrose concentration, interactions among impurity components and between impurity components and yield or sucrose concentration, and the requirement that commercial hybrids have resistance to prevalent diseases (Smith and Campbell, 1996) make the task difficult. Root yield and quality traits are not simply inherited and, in many cases, the needed disease resistance is controlled by a number of genes.

Almost any cultural practice can influence sugar beet quality, but nitrogen fertility management poses a significant challenge wherever sugar beet is grown. Applying excess nitrogen will not only increase input costs unnecessarily, but frequently will decrease recoverable sugar concentration and income if payment is based upon quality. Variable rate fertilizer applicators, grid soil sampling, and yield monitors mounted on harvesters are tools that facilitate production of a uniform high quality crop. Proper water management is crucial when irrigation is required or available. Diseases, insects, and weeds will reduce root yields and quality if not controlled. Uncontrollable environmental conditions that can alter sugar beet quality and effect relationships among individual quality factors (Afanasiev, 1964; Snyder, 1971) consistently frustrate attempts to produce a quality crop. Processors face the additional task of maintaining quality during storage (Bugbee, 1993). Growers can assist in this effort by delivering clean roots with no attached leaf material and taking care to minimize mechanical injury to roots (de Vletter and van Gils, 1976; Cole, 1977).

No attempt has been made to recommend specific cultivars or cultural practices. The intent was to introduce readers to relevant considerations in developing or producing quality sugar beet. References cited may provide information applicable to a specific situation, but in most cases local extension personnel or sugar company agronomists should be relied upon for information and recommendations regarding specific locations or problems.

REFERENCES

Adams, R. M., P. J. Farris, and A. D. Halverson. (1983). Sugar beet N fertilization and economic optima: recoverable sucrose vs. root yield. *Agron. J.* 75: 173-176.

Afanasiev, M. M. (1964). The effect of simulated hail injury on yield and sugar content of beets. *J. Am. Soc. Sugar Beet Technol.* 13(3): 225-237.

Ahmadi, M. and M. T. Assad. (1998). Estimating genetic parameters of agronomic and quality traits in a diallel cross of sugarbeet. *Iran Agric. Res.* 17: 19-34.

Alexander, J. T. (1971). Factors affecting quality. In *Advances in Sugarbeet Production: Principles and Practices*, eds. R. T. Johnson, J. T. Alexander, G. E. Rush, and G. R. Hawkes, Ames, Iowa: Iowa State Univ. Press, pp. 371-381.

Anderson, V. and S. Barfoed. (1988). An integrated juice purification system. In *Chemistry and Processing of Sugarbeet and Sugarcane*, eds. M. A. Clarke and M. A. Godshall, Amsterdam: Elsevier Science Publishers B. V., pp. 35-45.

Artschwager, E. (1930). A study of the structure of sugar beets in relation to sugar content and type. *J. Agric. Res.* 40: 867-915.

Bichsel, S. E. (1988). An overview of the U. S. beet sugar industry. In *Chemistry and Processing of Sugarbeet and Sugarcane*, eds. M. A. Clarke and M. A. Godshall, Amsterdam: Elsevier Science Publishers B. V., pp. 1-8.

Bohn, K., M. A. Clark, K. Buchholz, K.-M. Bliesener, R. Buczys, and K. Thielecke. (1998). Composition of sugarbeet and sugarcane and chemical behavior of constituents in processing. In *Sugar Technology, Beet and Cane Sugar Manufacture*, eds. P. W. van der Poel, H. Schiweck, and T. Schwartz, Berlin: Verlag Dr. Albert Bartens KG, pp. 115-208.

Bosemark, N. O. (1993). Genetics and breeding. In *The Sugar Beet Crop*, eds D. A. Cooke and R. K. Scott, London: Chapman & Hall, pp. 67-119.

Bravo, S., G. S. Lee, and W. R. Schmehl. (1989). The effect of planting date, nitrogen fertilizer rate, and harvest date on seasonal concentration and total content of six micronutrients in sugarbeet. *J. Sugar Beet Res.* 26(1): 34-49.

Bugbee, W. M. (1993). Storage. In *The Sugar Beet Crop*, ed. D. A. Cooke and R. K. Scott, London: Chapman & Hall, pp. 551-570.

Burley, J. B. (1990). Sugarbeet productivity model for Clay County Minnesota. *J. Sugar Beet Res.* 27(3-4): 50-57.

Campbell, L. G. (1989). *Beta vulgaris* NC-7 collection as a source of high sucrose germplasm. *J. Sugar Beet Res.* 26(1): 1-9.

Campbell, L. G. (1990). Registration of F1010 sugarbeet germplasm. *Crop Sci.* 30: 429-430.

Campbell, L. G. (1992). Registration of four sugarbeet germplasms from the NC-7 *Beta* collection. *Crop Sci.* 32: 1079.

Campbell, L. G. (1995). Long-term yield patterns of sugarbeet in Minnesota and eastern North Dakota. *J. Sugar Beet Res.* 32(1): 9-22.

Campbell, L. G. (2000). Sugarbeet breeding and improvement. In *Crop Improvement: Challenges in the 21st Century*, ed. M. S. Kang, Binghamton, New York: Food Products Press, pp. (in press).

Campbell, L. G. and D. F. Cole. (1986). Relationships between taproot and crown characteristics and yield and quality traits in sugarbeets. *Agron. J.* 78: 971-973.

Campbell, L. G. and J. W. Enz. (1991). Temperature effects on sugarbeet seedling emergence. *J. Sugar Beet Res.* 28(3-4): 129-140.

Campbell, L. G. and J. J. Kern. (1982). Cultivar × environment interactions in sugarbeet yield trials. *Crop Sci.* 22: 932-935.

Campbell, L. G. and J. J. Kern. (1983). Relationships among components of yield and quality of sugarbeets. *J. Am. Soc. Sugar Beet Technol.* 22: 135-145.

Carruthers, A., J. V. Dutton, J. F. T. Oldfield, C. W. Elliot, R. K. Heaney, and H. J. Teague. (1963). Estimation of sugars in beet molasses. *Int. Sugar J.* 65: 234-237, 266-270.

Carruthers, A., J. F. T. Oldfield, and H. J. Teague. (1962). Assessment of beet quality. *15th Tech. Conf, British Sugar Corp.* 28 p.

Carruthers, A., J. T. F. Oldfield, and H. J. Teague. (1966). The influence of crown removal on beet quality. *Int. Sugar J.* 68: 297-302.

Carter, J. N. (1986). Potassium and sodium uptake effects on sucrose concentration and quality of sugarbeet roots. *J. Am. Soc. Sugar Beet Technol.* 23(3-4): 183-202.

Carter, J. N. and D. J. Traveller. (1981). Effect of time and amount of nitrogen uptake on sugarbeet growth and yield. *Agron. J.* 73: 665-671.

Christenson, D. R. and M. B. Butt. (2000). Response of sugarbeet to applied nitrogen following field bean (*Phaseolus vulgaris* L.) and corn (*Zea mays* L.). *J. Sugar Beet Res.* 37(1): 1-16.

Coe, G. E. (1987). Selecting sugarbeets for low content of nonsucrose solubles. *J. Am. Soc. Sugar Beet Technol.* 24: 41-48.

Cole, D. F. (1977). Effect of cultivar and mechanical damage on respiration and storability of sugarbeet roots. *J. Am. Soc. Sugar Beet Technol.* 19(2): 240-245.

Cole, D. F. (1980). Effect of complete crown removal on quality of sugarbeets. *J. Am. Soc. Sugar Beet Technol.* 20(5): 449-454.

Cole, D. F. and G. J. Seiler. (1976). Effect of crown material on yield and quality of sugar beet roots: A grower survey. *J. Am. Soc. Sugar Beet Technol.* 19: 131-137.

de Vletter, R. and W. van Gils. (1976). Influence of mechanical handling of sugarbeets on sugar yield in the factory. *Sugar J.* 38(8): 8-13.

Dexter, S. T., M. G. Frakes, and F. W. Snyder. (1967). A rapid and practical method of determining extractable white sugar as may be applied to the evaluation of agronomic practices and grower deliveries. *J. Am. Soc. Sugar Beet Technol.* 14(5): 433-454.

Doney, D. L. (1995). Registration of four sugarbeet germplasms: y317, y318, y322, and y387. *Crop Sci*. 35: 947.

Doney, D. L. and J. C. Theurer. (1990). Osmolality of L19 type sugarbeet germplasm. *J. Sugar Beet Res*. 27: 81-89.

Doney, D. L., R. E. Wyse, and J. C. Theurer. (1981). The relationship between cell size, yield, and sucrose concentration of the sugarbeet root. *Can. J. Plant Sci*. 61: 447-453.

Draycott, A. P. (1972). *Sugar-Beet Nutrition*. New York: John Wiley and Sons, Inc.

Draycott, A. P. (1993). Nutrition. In *The Sugar Beet Crop*, ed. D. A. Cooke and R. K. Scott, London: Chapman & Hall, pp. 239-278.

Dudley, J. W. and L. Powers. (1960). Population genetic studies on sodium and potassium in sugar beets (*Beta vulgaris* L.). *J. Am. Soc. Sugar Beet Technol*. 9(2): 97-127.

Eckhoff, J. L. A. (1999). Sugarbeet response to nitrogen at four harvest dates. *J. Sugar Beet Res*. 36(4): 33-45.

Eckhoff, J. L. A., A. D. Halverson, M. J. Weiss, and J. W. Bergman. (1991). Seed spacing for nonthinned sugarbeet production. *Agron. J*. 83: 929-932.

Edwards, R. H., J. M. Randall, W. M. Camirand, and D. W. Wong. (1989a). Pilot plant scale high pressure steam peeling of sugarbeets. *J. Sugar Beet Res*. 26(2): 40-54.

Edwards, R. H., J. M. Randall, and L. W. Rodel. (1989b). Peeling of sugarbeets by use of high pressure steam. *J. Sugar Beet Res*. 26(1): 63-76.

Farley, R. F. and A. P. Draycott. (1974). Growth and yield of sugar beet in relation to potassium and sodium supply. *J. Sci. Food Agric*. 26: 385-392.

Finkner, R. E. and H. M. Bauserman. (1956). Breeding sugar beets with reference to sodium, sucrose, and raffinose content. *J. Am. Soc. Sugar Beet Technol*. 9:170-177.

Giroux, M. and T. S. Tran. (1989). Effect of potassium fertilization and N-K interaction on sugarbeet quality and yield. *J. Sugar Beet Res*. 26(2): 11-23.

Halverson, A. D. and G. P. Hartman. (1980). Response of several sugarbeet cultivars to N fertilization: yield and crown tissue production. *Agron. J*. 72: 665-669.

Halverson, A. D., G. P. Hartman, D. F. Cole, V. A. Haby, and D. E. Baldridge. (1978). Effect of N fertilization on sugarbeet crown tissue production and processing quality. *Agron. J*. 70: 876-880.

Harland, J. I. (1993). By-products. In *The Sugar Beet Crop*, ed. D. A. Cooke and R. K. Scott, London: Chapman & Hall, pp. 619-647.

Hartmann, E. M. (1977). Technical vocabulary for American beet-sugar processors. *J. Am. Soc. Sugar Beet Technol*. 19: 345-364.

Harvey, C. W. and J. V. Dutton. (1993). Root quality and processing. In *The Sugar Beet Crop*, ed. D. A. Cooke and R. K. Scott, London: Chapman & Hall, pp. 571-617.

Hecker, R. J. (1991). Effect of sugarbeet root size on combining ability of sucrose yield components. *J. Sugar Beet Res*. 28: 41-48.

Hecker, R. J. and S. S. Martin. (1981). Effects of sugarbeet sample preparation and handling on sucrose, nonsucroses, and purity analyses. *J. Am. Soc. Sugar Beet Technol*. 21(2): 184-197.

Hecker, R. J., R. E. Stafford, R. H. Helmerick, and G. W. Magg. (1970). Comparison of the same sugarbeet F_1 hybrids as diploids, triploids, and tetraploids. *J. Am. Soc. Sugar Beet Technol*. 16: 106-116.

Held, L. J., P. A. Burgener, J. G. Lauer, and D. J. Menkhaus. (1994). A economic analysis of reducing nitrogen an early harvest sugarbeets. *J. Prod. Agric.* 7: 422-428.

Hilde, D. J., S. Bass, R. W. Levos, and R. L. Ellingson. (1983). Grower practices system promotes beet quality improvement in the Red River Valley. *J. Am. Soc. Sugar Beet Technol.* 22(1): 73-88.

Hobbis, J., J. Kysilka, and M. Holle. (1982). Design and operating characteristics of a new beet quality measuring system. *La Sucrerie Belge* 101: 49-59.

Lansford, V. D., S. R. Winter, and W. L. Harman. (1989). Irrigated sugarbeet root yield response in the Texas High Plains. *J. Sugar Beet Res.* 26(1): 50-61.

Lasa, J. M., I. Romagosa, R. J. Hecker, and J. M. Sanz. (1989). Combining ability in diploid and triploid sugarbeet hybrids from diverse parents. *J. Sugar Beet Res.* 26(1): 10-18.

Last, P. J. and A. P. Draycott. (1977). Relationships between clarified beet juice purity and easily-measured impurities. *Int. Sugar J.* 79: 183-185.

Lauer, J. G. (1994). Early harvest of sugarbeet: yield and quality response to irrigation, cultivar and nitrogen. *J. Sugar Beet Res.* 31(3-4): 117-133.

Lauer, J. G. (1995). Plant density and nitrogen rate effects on sugar beet yield and quality early in harvest. *Agron. J.* 87: 586-591.

Lee, G. S., G. Dunn, and W. R. Schmehl. (1986). Effect of date of planting and nitorgen fertilization on growth components of sugarbeet. *J. Am. Soc. Sugar Beet Technol.* 24(1): 81-100.

Lewellen, R. T. (1992). Use of plant introductions to improve populations and hybrids of sugarbeet. In *Use of Plant in Cultivar Development. Part 2*. CSSA Spec. Publ. 20. Madison, Wisconsin: Crop Science Society of America, pp. 117-136.

Ludwick, A. E., W. E. Gilbert, and D. G. Westfall. (1980). Sugarbeet quality as related to KCL fertilization. *Agron. J.* 72: 453-456.

MacLachlan, J. B. (1972). Estimation of genetic parameters in a population of monogerm sugar beet (*Beta vulgaris*). 1. Sib-analysis of mother-line progeny. 2. Offspring/parent regression analysis of mother-line progenies. 3. Analysis of a diallel set of crosses among heterozygous population. *Irish J. Agric. Res.* 11: 237-246, 319-325, 327-338.

Martin, S. S., R. J. Hecker, and G. A. Smith. (1980). Aluminum clarification of sugarbeet extracts. *J. Am. Soc. Sugar Beet Technol.* 20(6): 597-609.

McFarlane, J. S., I. O. Skoyen, and R. T. Lewellen. (1972). Performance of sugarbeet hybrids as diploids and triploids. *Crop Sci.* 12: 118-119.

McGinnis, R. A. (1982). Chemistry of the beet and processing materials. In *Beet Sugar Technol. 3rd edition*, ed. R. A. McGinnis, Fort Collins, Colorado: Beet Sugar Development Foundation, pp. 25-63.

McGrath, J. M., C. A. Derrico, and Y. Yu. (1999). Genetic Diversity in selected, historic US sugarbeet germplasm and *Beta maritima* ssp. *maritima*. *Theor. Appl. Genet.* 98: 968-976.

Milford, G. F. J. (1973). The growth and development of the storage root of sugar beet. *Ann. Appl. Biol.* 75: 427-438.

Milford, G. F. J. and G. N. Thorne. (1973). The effect of light and temperature late in the season on the growth of sugar beet. *Ann. Appl. Biol.* 75: 419-425.

Milford, G. F. and D. J. Watson. (1971). The effect of nitrogen on the growth and sugar content of sugar-beet. *Ann. Bot.* 35: 287-300.

Moraghan, J. T. (1982). The influence of residual deep soil nitrate on sugarbeet production. *J. Am. Soc. Sugar Beet Technol.* 21(4): 362-373.

Oldfield, J. F. T., J. V. Dutton, and H. J. Teague. (1971). Significance of invert and gum formation in deteriorate beet. *Int. Sugar J.* 73:3-8, 35-40, 66-68.

Oldfield, J. F. T., M. Shore, J. V. Dutton, B. J. Houghton, and H. J. Teague. (1977). Sugar beet quality–factors of importance to the U. K. industry. *Int. Sugar J.* 79: 37-43, 67-71.

Pack, D. A. (1930). Selection characters as correlated with percent of sucrose, weight, and sucrose content of sugar beet. *J. Agric. Res.* 40: 523-546.

Perschak, F. (1998). Molasses desugarizing. In *Sugar Technology, Beet and Cane Sugar Manufacture*, eds. P. W. van der Poel, H. Schiweck, and T. Schwartz, Berlin: Verlag Dr. Albert Bartens KG, pp. 939-950.

Powers, L., W. R. Schmehl, W. T. Federer, and M. G. Payne. (1962). Chemical genetic and soils studies involving thirteen characters in sugar beets. *J. Am. Soc. Sugar Beet Technol.* 12(5): 393-448.

Saunders, J. W., J. M. McGrath, J. M. Halloin, and J. C. Theurer. (2000). Registration of SR95 sugarbeet germplasm with smooth root. *Crop. Sci.* 40: 1205-1206.

Scott, R. K. and K. W. Jaggard. (1993). An analysis of the efficiency of the sugar beet crop in exploiting the environment. *J. Sugar Beet Res.* 30(1-2): 37-56.

Schiweck, H. and G. Steinle. (1988). Analytical methods of sugar factories–new developments. In *Chemistry and Processing of Sugarbeet and Sugarcane*, eds. M. A. Clarke and M. A. Godshall, Amsterdam: Elsevier Science Publishers B. V., pp. 146-161.

Smith, G. A. (1988). Effect of plant breeding on sugarbeet composition. In *Chemistry and Processing of Sugarbeet and Sugarcane*, eds. M. A. Clarke and M. A. Godshall, Amsterdam: Elsevier Science Publishers B. V., pp. 9-19.

Smith, G. A. and L. G. Campbell. (1996). Association between *Cercospora* resistance and yield in commercial sugarbeet hybrids. *Plant Breed.* 115: 28-32.

Smith, G. A. and R. J. Hecker. (1973). Components of yield and recoverable sugar in random and improved sugarbeet populations. *Can. J. Plant Sci.* 53: 665-670.

Smith, G. A., R. J. Hecker, G. W. Maag, and D. M. Rasmuson. (1973). Combining ability and gene action estimates in an eight parent diallel cross of sugarbeet. *Crop Sci.* 13: 312-316.

Smith, G. A., R. J. Hecker, and S. S. Martin. (1979). Effects of polyploidy level on the components of sucrose yield and quality in sugarbeet. *Crop Sci.* 19: 319-323.

Smith, G. A. and S. S. Martin. (1989). Effect of selection for sugarbeet purity components on quality and extractions. *Crop Sci.* 29: 294-298.

Smith, G. A., S. S. Martin, and K. A. Ash. (1977). Path coefficient analysis of sugarbeet purity components. *Crop Sci.* 17: 249-253.

Smith, L. J. (2000). Nitrogen management–the key to profitability. In *Sugarbeet Research and Extension Reports*, Cooperative Extension Service, North Dakota State Univ. 30: 105-110.

Smith, L. J., T. Cymbaluk, and J. Nielsen. (2000). Sugarbeet profitability as affected by nitrogen rate, variety, and harvest date. In *Sugarbeet Research and Extension Reports*, Cooperative Extension Service, North Dakota State Univ. 30: 111-114.

Snyder, F. W. (1971). Some agronomic factors affecting processing quality of sugarbeet. *J. Am. Soc. Sugar Beet Technol.* 16(6): 496-507.

Spettoli, P., G. Cacco, and G. Ferrari. (1976). Comparative evaluation of the enzyme multiplicity in a diploid, a triploid and a tetraploid sugar beet variety. *J. Sci. Food Agric.* 27: 341-344.

Stander, J. R. (1993). Pre-breeding from the perspective of the private plant breeder. *J. Sugar Beet Res.* 30(4): 197-207.

Theurer, J. C. (1993). Pre-breeding to change sugarbeet root architecture. *J. Sugar Beet Res.* 30(4): 221-239.

Theurer, J. C. (1981). Registration of eight germplasm lines of sugarbeet. *Crop Sci.* 18: 1101.

Theurer, J. C. and D. L. Doney. (1989). Sugar accumulation in L19 type sugarbeet germplasm. *J. Sugar Beet Res.* 26(2): 55-64.

Ulrich, A. (1961). Variety climate interactions of sugar beet varieties in simulated climates. *J. Am. Soc. Sugar Beet Tech.* 11(5): 376-387.

van den Hil, J. and L. H. de Nie. (1989). Beet quality: Technological and economic values and a payment system. *Zuckerindustrie* 114: 645-650.

van der Poel, P. W., H. Schiweck, and T. Schwartz, ed. (1998). *Sugar Technology, Beet and Cane Sugar Manufacture.* Berlin: Verlag Dr. Albert Bartens KG.

Van Geijn, N. L., L. C. Giljam, and L. H. deNie. (1983). α-Amino-nitrogen in sugar processing. In *Proceedings of Nitrogen and Sugar Beet Symposium,* Brussels, Inst. Beet Res., pp. 13-25.

Vaccari, G., M. G. Marzola, and G. Mantovani. (1988). Chemical and enzymatic changes in strongly damaged beets. *Food Chemistry* 27: 203-211.

Winter, S. R. (1980). Planting sugarbeets to stand when establishment is erratic. *Agron. J.* 72: 654-656.

Winter, S. R. (1984). Cropping systems to remove excess soil nitrate in advance of sugarbeet production. *J. Am. Soc. Sugar Beet Technol.* 22(3-4): 285-290.

Winter, S. R. (1988). Influence of seasonal irrigation amount on sugarbeet yield and quality. *J. Sugar Beet Res.* 25(1): 1-9.

Winter, S. R. (1989). Sugarbeet yield and quality response to irrigation, row width, and stand density. *J. Sugar Beet Res.* 26(1): 26-33.

Winter, S. R. (1990). Sugarbeet response to nitrogen as affected by seasonal irrigation. *Agron. J.* 82: 984-988.

Winter, S. R. (1998). Sugarbeet response to residual and applied nitrogen in Texas. *J. Sugar Beet Res.* 35(1-2): 43-62.

Wood, R. R., H. L. Bush, and R. K. Oldemeyer. (1958). The sucrose-sodium relationship in selecting sugar beets. *J. Am. Soc. Sugar Beet Technol.* 10: 133-137.

Wyse, R. E. (1979). Parameters controlling sucrose content and yield of sugarbeet roots. *J. Am. Soc. Sugar Beet Technol.* 20: 368-385.

Wyse, R. E. (1982). The sugarbeet and chemistry. In *Beet Sugar Technology, 3rd edition,* ed. R. A. McGinnis, Fort Collins, Colorado, Beet Sugar Development Foundation, pp. 17-24.

Yonts, C. D., K. J. Fornstorm, and R. J. Edling. (1983). Sugarbeet emergence affected by soil moisture and temperature. *J. Am. Soc. Sugar Beet Technol.* 22(2): 119-134.

Yonts, C. D. and J. A. Smith. (1997). Effects of pant population and row width on yield of sugarbeet. *J. Sugar Beet Res.* 34(1-2): 21-30.

Yonts, C. D., R. G. Wilson, and J. A. Smith. (1999). Influence of planting date and stand on yield and quality of sugarbeet. *J. Sugar Beet Res.* 36(3): 1-13.

Zielke, R. C. (1973). Yield, quality, and sucrose recovery from sugarbeet root and crown. *J. Am. Soc. Sugar Beet Technol.* 17: 332-344.

Zielke, R. C. and F. W. Snyder. (1974). Impurities in sugarbeet crown and root. *J. Am. Soc. Sugar Beet Technol.* 18: 60-75.

Index